L. W. Hart

ACOUSTIC WAVEGUIDES

ACOUSTIC WAVEGUIDES

Applications to Oceanic Science

C. ALLAN BOYLES
Applied Physics Laboratory
Johns Hopkins University

A Wiley-Interscience Publication
JOHN WILEY & SONS
New York   Chichester   Brisbane   Toronto   Singapore

Copyright © 1984 by John Wiley & Sons, Inc.

All rights reserved. Published simultaneously in Canada.

Reproduction or translation of any part of this work beyond that permitted by Section 107 or 108 of the 1976 United States Copyright Act without the permission of the copyright owner is unlawful. Requests for permission or further information should be addressed to the Permissions Department, John Wiley & Sons, Inc.

Library of Congress Cataloging in Publication Data:

Boyles, C. Allan (Charles Allan), 1936–
  Acoustic waveguides.

  "A Wiley-Interscience publication."
  Includes bibliographical references and index.
  1. Underwater acoustics. 2. Wave guides. 3. Wave equation. 4. Oceanography. I. Title.
  QC233.B75 1984   534′.23   83-17000
  ISBN 0-471-88771-4

Printed in the United States of America

10 9 8 7 6 5 4 3 2 1

PREFACE

A systematic and detailed exposition of the theory of acoustic propagation in oceanic waveguides is given in this book. In particular, this book presents a very detailed, rigorous derivation of the acoustic wave equation and its exact and approximate solutions for various models of oceanic waveguides, starting with the simple model of a homogeneous layer and ranging up to an ocean in which the speed of sound varies horizontally as well as vertically and the sea surface is randomly rough.

This book originated as a set of lecture notes for a course that I taught several years ago at The Johns Hopkins University Applied Physics Laboratory. The reason for writing these lecture notes was because no book existed that covered the derivation of the wave equation and up-to-date methods for solving it in the kind of detail that a novice to the field requires. Consequently, this book is intended for two types of individuals. It is meant to be used at the first-year graduate level as an introductory text to the theory of sound propagation in the ocean. I feel that this book covers the gap between undergraduate physics and mathematics courses and state-of-the-art research. For this same reason, it should also fill the need of scientists who are starting to work in this field, but who have received no formal college training in it.

The book is organized so that chapters should be read in sequence. The first chapter starts by introducing the reader to the hydrodynamics and thermodynamics of perfect fluids. No previous knowledge of this is required. These results are then used to derive in extreme detail the acoustic wave equation, the conservation equation for acoustic energy, and the boundary conditions associated with the wave equation. Such a detailed derivation lays bare all the physical and mathematical assumptions that are inherent in the wave equation.

The second chapter gives a comprehensive review of those aspects of the theory of differential equations that are needed to solve the boundary value problem associated with acoustic propagation in an oceanic waveguide. Since there was no single source for this material that the reader could be referred to, the material was collected from multiple sources and put into a form that suited my needs. The topics include separation of variables technique, theory of eigenfunctions, a thorough discussion on completeness, Green's function, and Bessel and Airy functions.

The third and fourth chapters apply these mathematical techniques to obtain an exact solution of the wave equation for a series of waveguide models that start from the simplest, which is a single, homogeneous layer, to a more complicated ocean in which the sound speed is allowed to vary with depth. In all these models the sea surface is assumed to be flat.

The fifth chapter deals with approximate solutions to the wave equation. The approach in this chapter is to start with the exact solution developed in Chapters 3 and 4 and make a series of approximations of increasing severity. The approximate solutions presented are the WKB solution, the ray solution, and the ray solution corrected for smooth caustics.

In Chapter 6 I show application of the mathematical solutions developed in Chapters 3, 4, and 5 to three realistic ocean environments. The first is a North Pacific summer environment, the second is a North Atlantic summer environment, and the third is a North Atlantic winter environment exhibiting deep surface ducts.

The material described in Chapters 3 through 6 pertains to an ocean with a perfectly flat sea surface and a sound speed which varies with depth below the ocean's surface only. Now a real ocean certainly does not have a flat-sea surface and over ranges of hundreds of miles there can be extreme horizontal variations in the oceanographic conditions. The mathematical solutions developed in Chapters 3 to 6, which were based on the separation of variables technique, cannot handle these realistic problems.

In Chapter 7, I present some original work which is the solution of the wave equation in an oceanic wave guide with a randomly rough sea surface and a sound speed that can vary horizontally as well as vertically. Some of this material has appeared in journals and some appears here for the first time. A numerical algorithm and computer code for the theory appearing in Chapter 7 was written by Dr. L. B. Dozier of Science Applications, Inc., McLean, Va. The work is so new that at the time of writing this manuscript there were no numerical examples available.

I would like to acknowledge the many people who helped prepare the manuscript. At Vitro Laboratories they are Jean Blitz, editor, Millie Waller, page makeup, Barbara Albin, Carolyn Roher, Phyllis Hipp, Katherine Thompson, and Barbara Reutemann, typists. At the Applied Physics Laboratory they are Lorna Weaver, Judy Rockey, and Denise Phillips. I would also like to express my appreciation to Donald Yeager who did many of the drawings.

PREFACE

Finally, I would like to thank Geraldine Joice and Albert Biondo for their many helpful discussions on the material.

C. ALLAN BOYLES

Laurel, Maryland
October 1983

CONTENTS

1. THE FUNDAMENTAL EQUATIONS   1

- 1.0. Introduction, 1
- 1.1. Fluid Motion, 1
- 1.2. The Jacobian, 5
- 1.3. Reynolds Transport Theorem, 13
- 1.4. Conservation of Mass, 13
- 1.5. Conservation of Momentum for a Perfect Fluid, 14
- 1.6. Conservation of Energy for a Perfect Fluid, 16
- 1.7. The Acoustic Wave Equation, 22
- 1.8. Conservation of Energy in the Acoustic Approximation, 30
- 1.9. Boundary Conditions, 35
- References, 37

2. MATHEMATICAL REVIEW   39

- 2.0. Introduction, 39
- 2.1. Separation of Variables, 39
- 2.2. Sturm-Liouville Theory, 44
- 2.3. Calculus of Variations, 49
- 2.4. Eigenfunctions and the Variational Principle, 57
- 2.5. Completeness of a Set of Eigenfunctions, 59

- 2.6. The Continuous Spectrum, 71
- 2.7. Green's Functions, 77
- 2.8. Method of Steepest Descent, 82
- 2.9. Bessel's Differential Equation of Arbitrary Order, 90
- 2.10. Airy's Differential Equation, 107
  References, 110

3. PROPAGATION IN THE OCEAN AS A BOUNDARY VALUE PROBLEM: THE HOMOGENEOUS LAYERED MODEL 111

- 3.0. Introduction, 111
- 3.1. The Complex λ-Plane Representation of the Green's Function for a Homogeneous Layer Bounded by a Pressure-Release Surface and a Rigid Bottom, 111
- 3.2. Eigenfunction Expansion of the Green's Function for a Homogeneous Layer Bounded by a Pressure-Release Surface and a Rigid Bottom, 127
- 3.3. The Complex λ-Plane Representation of the Green's Function for an Ocean Consisting of Two Homogeneous Layers Bounded by a Pressure-Release Surface and a Rigid Bottom, 132
- 3.4. Eigenfunction Expansion of the Green's Function for an Ocean Consisting of Two Homogeneous Layers Bounded by a Pressure-Release Surface and a Rigid Bottom, 139
- 3.5. The Complex λ-Plane Representation of the Green's Function for an Ocean Consisting of a Homogeneous Layer with a Pressure-Release Surface Overlying a Homogeneous Halfspace—The Discrete Plus Continuous Spectrum, 144
- Appendix 3.A. Diffraction Theory and the Sommerfeld Radiation Condition, 155
- Appendix 3.B. Evaluation of the Complex Integral Solution of the Wave Equation Using the Contour C_r^+, 160
  References, 162

4. PROPAGATION IN THE OCEAN AS A BOUNDARY VALUE PROBLEM: THE INHOMOGENEOUS LAYERED MODEL 163

- 4.0. Introduction, 163
- 4.1. A Normal Model for N-Inhomogeneous Layers with Discontinuous Properties, 163
  Reference, 177

CONTENTS xi

5. APPROXIMATE SOLUTIONS OF THE WAVE EQUATION 178

5.0. Introduction, 178
5.1. Ray Equations as a Quasi-Plane Wave Approximation, 178
5.2. WKB Approximation, 197
5.3. Correction for Turning Points in the WKB Method, 204
5.4. WKB Solution in a Waveguide, 211
5.5. WKB Green's Function for an Unbounded Media, 214
5.6. Ray Theory Approximation by the Method of Steepest Descent, 217
5.7. Physical Description of the Formation of a Smooth Caustic, 220
5.8. Correction to Ray Theory for Smooth Caustics, 227
Appendix 5.A. Derivation of the -90-Degree Phase Shift as a Wave Front Passes Through a Focal Point, 235
References, 238

6. APPLICATION TO CONVERGENCE ZONE AND SURFACE DUCT PROPAGATION 240

6.0. Introduction, 240
6.1. Speed of Sound in the Ocean, 240
6.2. Convergence Zone Propagation—A Single Channel North Atlantic Profile, 241
6.3. Convergence Zone Propagation—A Double Channel North Atlantic Profile, 263
6.4. Surface Duct Propagation—A North Atlantic Profile, 273
References, 293

7. AN OCEANIC WAVEGUIDE WITH A RANGE- AND DEPTH-DEPENDENT REFRACTIVE INDEX AND A TIME VARYING, RANDOMLY ROUGH SEA SURFACE 294

7.0. Introduction, 294
7.1. The Coupled Second-Order System of Differential Equations in Cylindrical Coordinates, 294
7.2. The Coupled First-Order System of Differential Equations in Cylindrical Coordinates, 302
7.3. A Model for a Time Varying, Randomly Rough Sea Surface, 305

7.4. The Coupled Second-Order System of Differential Equations in Cartesian Coordinates, 307
7.5. The Coupled First-Order System of Differential Equations in Cartesian Coordinates, 312
Appendix 7.A. The Narrow-Band Approximation to the Wave Equation, 315
References, 316

INDEX 317

1 THE FUNDAMENTAL EQUATIONS

1.0. INTRODUCTION

The goal of this chapter is to derive the acoustic wave equation and conservation of energy in the acoustic approximation for a perfect fluid that is vertically stratified by the influence of a gravitational field. To accomplish this end, we first derive the hydrodynamic equations for the motion of a perfect fluid with no heat conductivity. The derivation of the hydrodynamic equations of motion is presented in Sections 1.1 through 1.6. The thermodynamic background material presented in Section 1.6, used to derive the equation for conservation of energy for a perfect fluid, is taken from the very excellent book by H. Callen [1.1].

It will be assumed that the reader has a background in differential and integral calculus, elementary differential equations, complex variable theory, vector algebra, and vector calculus. Suitable books on these subjects are those of Courant and John [1.2], Boyce and Di Prima [1.3], Churchhill, Brown, and Verhey [1.4], Coburn [1.5], and Hay [1.6]. Finally, it will be assumed that the reader has been introduced to basic concepts in acoustics. The clearest presentation of basic acoustic concepts that this author knows is the book by Kinsler and Frey [1.7].

1.1. FLUID MOTION

Let us consider a fluid in motion. Instead of supposing a fluid to be composed of individual molecules, we imagine the fluid to be a continuum whose velocity at any point is the average velocity of the molecules in a neighborhood of that point. At some instant we note that a certain particle is at position ξ and at a

2 THE FUNDAMENTAL EQUATIONS

later time that same particle is at position x. We can take the first instant to be time $t = 0$ and the later time to be time t. We say that x is a function of time t and the initial position $\boldsymbol{\xi}$, and we write this relationship in vector notation as

$$\mathbf{x} = \mathbf{x}(\boldsymbol{\xi}, t) \tag{1.1.1}$$

or, in component notation,

$$x_i = x_i(\xi_i, t), \quad i = 1, 2, 3 \tag{1.1.2}$$

where x_1, x_2, and x_3 are the Cartesian components of the position vector x, and ξ_1, ξ_2, and ξ_3 are the Cartesian components of the vector $\boldsymbol{\xi}$.

The initial coordinate $\boldsymbol{\xi}$ of a particle is called the "material coordinate" of the particle. The material coordinate is also called the "convected coordinate" or the "Lagrangian coordinate." The material coordinate system is convected with the fluid. The spatial coordinate x of the particle is called its "position." It will be assumed that the motion of the fluid is continuous and single valued. Further, it will be assumed that Eq. (1.1.1) can be inverted to yield

$$\boldsymbol{\xi} = \boldsymbol{\xi}(\mathbf{x}, t) \tag{1.1.3}$$

or, in component notation,

$$\xi_i = \xi_i(x_i, t), \quad i = 1, 2, 3 \tag{1.1.4}$$

The following theorem, which is proven in Courant and John [1.8], tells us the exact conditions under which the transformation given by Eq. (1.1.1) can be inverted to yield Eq. (1.1.3).

Theorem 1.1.1. Let $x_i(\xi_i, t)$ be continuous functions of ξ_i with continuous first partial derivatives with respect to ξ_i in a neighborhood of a point $(\xi_1^0, \xi_2^0, \xi_3^0)$ for which the Jacobian J given by

$$J = \frac{\partial (x_1, x_2, x_3)}{\partial (\xi_1, \xi_2, \xi_3)} = \begin{vmatrix} \frac{\partial x_1}{\partial \xi_1} & \frac{\partial x_1}{\partial \xi_2} & \frac{\partial x_1}{\partial \xi_3} \\ \frac{\partial x_2}{\partial \xi_1} & \frac{\partial x_2}{\partial \xi_2} & \frac{\partial x_2}{\partial \xi_3} \\ \frac{\partial x_3}{\partial \xi_1} & \frac{\partial x_3}{\partial \xi_2} & \frac{\partial x_3}{\partial \xi_3} \end{vmatrix} \tag{1.1.5}$$

is not zero at $(\xi_1^0, \xi_2^0, \xi_3^0)$. Put $x_i^0 = x_i^0(\xi_i^0, t)$. Then there exists a neighborhood N of ξ_i^0 and N' of x_i^0 such that the mapping

$$x_i = x_i(\xi_i, t)$$

has a unique inverse

$$\xi_i = \xi_i(x_i, t)$$

mapping N' into N. The functions $\xi_i(x_i, t)$ satisfy the identities

$$x_i = x_i[\xi_i(x_i, t), t]$$

for x_i in N' and the equations

$$\xi_i^0 = \xi_i^0(x_i^0, t)$$

The inverse functions $\xi_i(x_i, t)$ also have continuous derivatives $\partial \xi_i / \partial x_i$ in the neighborhood N'.

Physically, the mathematical requirement of continuity means that a continuous arc of particles does not break up during the motion or that the particles in the neighborhood of a given particle continue in that particle's neighborhood during the motion. The equations are single valued in that a particle cannot split up and occupy two places and two distinct particles cannot occupy the same place.

The transformation given by Eq. (1.1.1) may be considered as parametric equations in the parameter t. These curves are called the particle paths. Any property of the fluid may be followed along the particle paths. This material description of the change of some physical property of the fluid, say $\mathscr{G}(\xi, t)$, can be changed into a spatial description $\mathscr{G}(\mathbf{x}, t)$ by Eq. (1.1.3):

$$\mathscr{G}(\mathbf{x}, t) = \mathscr{G}[\xi(\mathbf{x}, t), t] \qquad (1.1.6)$$

The physical interpretation of Eq. (1.1.6) says that the value of the property at position $\mathbf{x}$ and time t is the value associated with the particle when it is at position $\mathbf{x}$ at time t.

We now need to ask how the property $\mathscr{G}(\mathbf{x}, t)$ varies with time. First, let us define the particle velocity $\mathbf{v}'$ by

$$\mathbf{v}' = \frac{d\mathbf{x}}{dt} \qquad (1.1.7)$$

or, in component notation,

$$v_i' = \frac{dx_i}{dt} \qquad (1.1.8)$$

To find the time rate of change of the property $\mathscr{G}(\mathbf{x}, t)$ of the fluid in the neighborhood of a specified fluid particle whose position is given by $\mathbf{x}$ at time t, we cannot simply compute $\partial \mathscr{G}/\partial t$, the partial derivative of $\mathscr{G}$ with respect to time at point **x**, because the particle does not usually stay at one point. The change we wish to determine is the difference between the value of $\mathcal{G}(\mathbf{x}, t)$ at position **x** where the particle is at time t, and the value of $\mathcal{G}(\mathbf{x} + \mathbf{v}'\, dt, t + dt)$ at position $\mathbf{x} + \mathbf{v}'\, dt$ at time $t + dt$. If we expand the function $\mathcal{G}(\mathbf{x} + \mathbf{v}'\, dt, t + dt)$ in a multidimensional Taylor's series, we obtain

$$\mathcal{G}(\mathbf{x} + \mathbf{v}'\, dt, t + dt) = \mathcal{G}(\mathbf{x}, t) + (\mathbf{v}'\, dt) \cdot \nabla \mathcal{G} + dt \frac{\partial \mathcal{G}}{\partial t}$$

$$+ O(\mathbf{v}'\, dt, dt) \qquad (1.1.9)$$

In Eq. (1.1.9), ∇ is the vector gradient operator defined by

$$\nabla = \hat{e}_{(1)} \frac{\partial}{\partial x_1} + \hat{e}_{(2)} \frac{\partial}{\partial x_2} + \hat{e}_{(3)} \frac{\partial}{\partial x_3} \qquad (1.1.10)$$

where $\hat{e}_{(k)}(k = 1, 2, 3)$ is the set of orthogonal unit base vectors in our Cartesian coordinate system. Further, $\mathbf{v}' \cdot \nabla \mathcal{G}$ denotes the scalar product of the vectors $\mathbf{v}'$ and $\nabla \mathcal{G}$ and is equal to

$$\mathbf{v}' \cdot \nabla \mathcal{G} = v_1' \frac{\partial \mathcal{G}}{\partial x_1} + v_2' \frac{\partial \mathcal{G}}{\partial x_2} + v_3' \frac{\partial \mathcal{G}}{\partial x_3}$$

Finally, $O(\mathbf{v}'\, dt, dt)$ denotes terms of second and higher order in the expansion variables $\mathbf{v}'\, dt$ and dt.

Letting

$$d\mathcal{G} = \mathcal{G}(\mathbf{x} + \mathbf{v}'\, dt, t + dt) - \mathcal{G}(\mathbf{x}, t)$$

we can rewrite Eq. (1.1.9) as

$$\frac{d\mathcal{G}}{dt} = \frac{\partial \mathcal{G}}{\partial t} + \mathbf{v}' \cdot \nabla \mathcal{G} + \frac{1}{dt} O(\mathbf{v}'\, dt, dt)$$

Allowing $dt \to 0$, we can neglect terms of second order and higher to arrive at the desired result

$$\frac{d\mathcal{G}}{dt} = \frac{\partial \mathcal{G}}{\partial t} + \mathbf{v}' \cdot \nabla \mathcal{G} \qquad (1.1.11)$$

or, in component notation,

$$\frac{d\mathcal{G}}{dt} = \frac{\partial \mathcal{G}}{\partial t} + v_i' \frac{\partial \mathcal{G}}{\partial x_i} \qquad (1.1.12)$$

In Eq. (1.1.12) we have used the Einstein summation convention, wherein a repeated index, here i, is summed over. This convention will be used throughout the remainder of this book.

To summarize, $\partial \mathscr{G}/\partial t$ is the rate of change of $\mathscr{G}$ as observed at a fixed position x, whereas $d\mathscr{G}/dt$ is the rate of change of $\mathscr{G}$ as observed when moving with the fluid particle.

1.2. THE JACOBIAN

A fundamental result that we will need is how an elementary volume in one set of coordinates transforms to an elementary volume in a second set of coordinates. If x_1, x_2, x_3 are the coordinates in a Cartesian orthogonal coordinate system, then $dV = dx_1 \, dx_2 \, dx_3$ is an element of volume in this system. We need to ask what volume is to be associated with the three small changes $d\xi_1, d\xi_2, d\xi_3$ in a second Cartesian orthogonal system with coordinates ξ_1, ξ_2, ξ_3. We will show in this section that the two elementary volumes are related by $dV = J \, d\xi_1 \, d\xi_2 \, d\xi_3$, where J is the Jacobian of the transformation. Then we will need to determine dJ/dt, the time rate of change of the Jacobian. The time derivative of the Jacobian is needed in order to prove the Reynolds transport theorem which is derived in Section 1.3.

We begin with the following definitions and theorems.

Definition 1.2.1. Let $\hat{e}_{(k)}$ for $k = 1, 2, 3$ be a set of orthogonal base vectors for a Cartesian coordinate system. Let **a** and **b** be two arbitrary vectors. Then the vector product of **a** and **b** can be defined as

$$\mathbf{a} \times \mathbf{b} = \varepsilon_{ijk} a_i b_j \hat{e}_{(k)} \qquad (1.2.1)$$

where a_i and b_j are the components of **a** and **b**, respectively, and the permutation symbol ε_{ijk} is given by

$$\varepsilon_{ijk} \begin{aligned} &= 0, \text{ if any two of } i, j, k \text{ are the same} \\ &= 1, \text{ if } ijk \text{ is an even permutation of } 123 \\ &= -1, \text{ if } ijk \text{ is an odd permutation of } 123 \end{aligned} \qquad (1.2.2)$$

Let us show that this corresponds to the usual definition of the cross product of two vectors. Let $\mathbf{c} = \mathbf{a} \times \mathbf{b}$. Then from Eq. (1.2.1) we have

$$\mathbf{c} = \varepsilon_{ij_1} a_i b_j \hat{e}_{(1)} + \hat{e}_{ij_2} a_i b_j \hat{e}_{(2)} + \varepsilon_{ij_3} a_i b_j \hat{e}_{(3)}$$

The components of **c** are then the coefficients of the unit base vectors $\mathbf{e}_{(k)}$, so that we have

$$c_1 = \varepsilon_{ij_1} a_i b_j$$
$$c_2 = \varepsilon_{ij_2} a_i b_j$$
$$c_3 = \varepsilon_{ij_3} a_i b_j$$

6   THE FUNDAMENTAL EQUATIONS

On using the summation convention and Eq. (1.2.2), we have, for example,

$$c_1 = \varepsilon_{i11} a_i b_1 + \varepsilon_{i21} a_i b_2 + \varepsilon_{i31} a_i b_3$$

$$= \varepsilon_{111} a_1 b_1 + \varepsilon_{211} a_2 b_1 + \varepsilon_{311} a_3 b_1 + \varepsilon_{121} a_1 b_2 + \varepsilon_{221} a_2 b_2 + \varepsilon_{321} a_3 b_2$$

$$+ \varepsilon_{131} a_1 b_3 + \varepsilon_{231} a_2 b_3 + \varepsilon_{331} a_3 b_3$$

$$= a_2 b_3 - a_3 b_2$$

Similarly for the other components,

$$c_2 = a_3 b_1 - a_1 b_3$$

$$c_3 = a_1 b_2 - a_2 b_1$$

Thus we see that this agrees with the usual definition. We will need the following theorems.

Theorem 1.2.1. The area A of the parallelogram with vectors **a** and **b** forming adjacent edges is given by the relation

$$A = |\mathbf{a} \times \mathbf{b}|$$

Proof. The geometry is shown in Figure 1.2.1.

$$A = 2A_1 + A_2$$

$$= (b \sin \theta)(b \cos \theta) + (b \sin \theta)(a - b \cos \theta)$$

$$= ab \sin \theta = |\mathbf{a} \times \mathbf{b}|$$

Theorem 1.2.2. The volume V of the parallelepiped with the vectors **a**, **b**, and **c** forming adjacent edges is given by the relation

$$V = |\mathbf{a} \cdot (\mathbf{b} \times \mathbf{c})|$$

Proof. The geometry is shown in Figure 1.2.2. Let $\mathbf{d} = \mathbf{b} \times \mathbf{c}$ and $d = |\mathbf{b} \times \mathbf{c}|$. By Theorem 1.2.1, d is the area of the parallelogram forming the base

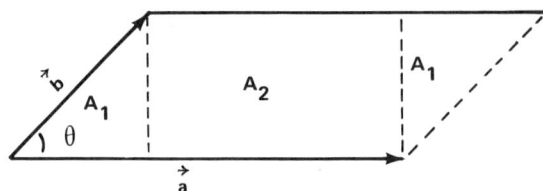

FIGURE 1.2.1.   Geometrical area as a vector cross product.

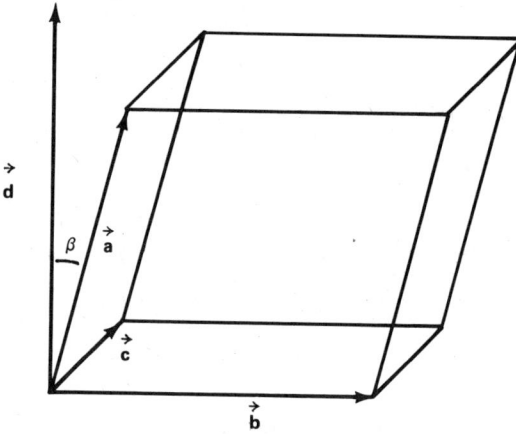

FIGURE 1.2.2. Geometrical volume as a vector triple scalar product.

of the parallelepiped. From geometry

$$V = dh$$

where h is the altitude of the parallelepiped. But

$$h = a|\cos \beta|$$

Thus

$$V = |ad \cos \beta|$$
$$= |\mathbf{a} \cdot \mathbf{d}|$$
$$= |\mathbf{a} \cdot (\mathbf{b} \times \mathbf{c})|$$

Definition 1.2.2. A 3×3 determinant can be defined with the aid of the permutation symbol ε_{ijk} in the following manner:

$$|a_{ij}| = \begin{vmatrix} a_{11} a_{12} a_{13} \\ a_{21} a_{22} a_{23} \\ a_{31} a_{32} a_{33} \end{vmatrix} = \varepsilon_{ijk} a_{i1} a_{j2} a_{k3} = \varepsilon_{ijk} a_{1i} a_{2j} a_{3k}$$

where i, j, and k are summed from 1 to 3.

A good introduction to the theory of determinants using the ε-notation is found in Lass [1.9].

8 THE FUNDAMENTAL EQUATIONS

Let us show that this is the usual definition of a determinant. Consider the first definition

$$|a_{ij}| = \varepsilon_{ijk} a_{i1} a_{j2} a_{k3}$$

$$= \varepsilon_{123} a_{11} a_{22} a_{33} + \varepsilon_{132} a_{11} a_{32} a_{23} + \varepsilon_{213} a_{21} a_{12} a_{33}$$
$$+ \varepsilon_{231} a_{21} a_{32} a_{13} + \varepsilon_{312} a_{31} a_{12} a_{23} + \varepsilon_{321} a_{31} a_{22} a_{13}$$

$$= a_{11} a_{22} a_{33} + a_{31} a_{12} a_{23} + a_{21} a_{32} a_{13}$$
$$- a_{11} a_{32} a_{23} - a_{21} a_{12} a_{33} - a_{31} a_{22} a_{13}$$

where only nonzero terms have been kept. It is seen that this last result is the usual definition of a 3 × 3 determinant.

Now in a Cartesian orthogonal coordinate system x_1, x_2, x_3 the element of volume is simply

$$dV = dx_1 \, dx_2 \, dx_3$$

Sometimes it is more convenient to describe the position by some other coordinates, say ξ_1, ξ_2, ξ_3. We may ask what volume is to be associated with the three small changes $d\xi_1$, $d\xi_2$, and $d\xi_3$. In order to calculate dV in the ξ_1, ξ_2, ξ_3 coordinate system, let us first review some fundamental concepts of coordinate systems. The change of coordinates is specified by a set of functions

$$x_i = x_i(\xi_j) \quad \text{for} \quad i, j = 1, 2, 3$$

which have the property of Theorem 1.1.1. Three surfaces pass through each point P in the ξ coordinate system, namely,

$$\xi_1 = \text{constant}, \quad \xi_2 = \text{constant}, \quad \xi_3 = \text{constant}$$

These surfaces are called the coordinate surfaces. On each coordinate surface, one coordinate is constant and two are variable. A surface is designated by the coordinate that is constant. Two surfaces intersect in a curve, called a coordinate curve, along which two coordinates are constant and one is variable. A coordinate curve is designated by the variable coordinate. The geometry is shown in Figure 1.2.3. Let **r** denote the vector from an arbitrary origin to a variable point $P(x_1, x_2, x_3)$. The point, and consequently its position vector **r**, may be considered a function of the coordinates ξ_i

$$\mathbf{r} = \mathbf{r}(\xi_i)$$

A differential change in **r** due to small displacements along the coordinate curves ξ_i is expressed by

$$d\mathbf{r} = \frac{\partial \mathbf{r}}{\partial \xi_1} d\xi_1 + \frac{\partial \mathbf{r}}{\partial \xi_2} d\xi_2 + \frac{\partial \mathbf{r}}{\partial \xi_3} d\xi_3$$

THE JACOBIAN

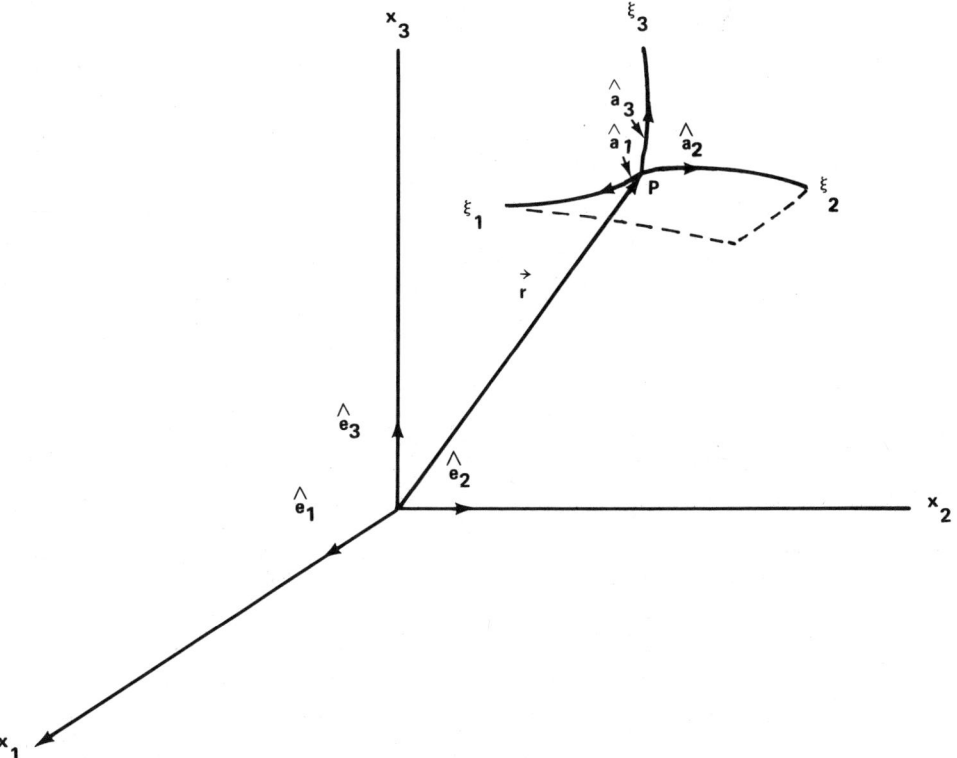

FIGURE 1.2.3. Geometry defining a general curvilinear coordinate system.

Now if one moves unit distance along the ξ_1 curve, the change in **r** is directed tangentially to this curve and is equal to $\partial \mathbf{r}/\partial \xi_1$. Thus the vectors

$$\mathbf{a}_1 = \frac{\partial \mathbf{r}}{\partial \xi_1}, \qquad \mathbf{a}_2 = \frac{\partial \mathbf{r}}{\partial \xi_2}, \qquad \mathbf{a}_3 = \frac{\partial \mathbf{r}}{\partial \xi_3}$$

are known as the unitary vectors associated with point P. These constitute a system of base vectors in the ξ coordinate system.

It must be noted that the unitary vectors are not necessarily of unit length. It is now a relatively simple matter to obtain expressions for the element of volume in the ξ coordinate system. Let $d\mathbf{s}_1$ be a small displacement at $P(\xi_1, \xi_2, \xi_3)$ along the ξ_1 curve, and similarly for $d\mathbf{s}_2$ and $d\mathbf{s}_3$. So

$$d\mathbf{s}_1 = \mathbf{a}_1 \, d\xi_1, \qquad d\mathbf{s}_2 = \mathbf{a}_2 \, d\xi_2, \qquad d\mathbf{s}_3 = \mathbf{a}_3 \, d\xi_3$$

By Theorem 1.2.2, a volume element bounded by the coordinate surface in the ξ frame is given by

$$dV = d\mathbf{s}_1 \cdot (d\mathbf{s}_2 \times d\mathbf{s}_3)$$

10 THE FUNDAMENTAL EQUATIONS

If $\hat{e}_1$, $\hat{e}_2$, and $\hat{e}_3$ are the base vectors in the x coordinate system, so that

$$\mathbf{r} = x_1\hat{e}_1 + x_2\hat{e}_2 + x_3\hat{e}_3$$

then

$$d\mathbf{s}_1 = \frac{\partial \mathbf{r}}{\partial \xi_1} d\xi_1 = \frac{\partial x_l}{\partial \xi_1} d\xi_1 \hat{e}_l$$

$$d\mathbf{s}_2 = \frac{\partial \mathbf{r}}{\partial \xi_2} d\xi_2 = \frac{\partial x_l}{\partial \xi_2} d\xi_2 \hat{e}_l$$

$$d\mathbf{s}_3 = \frac{\partial \mathbf{r}}{\partial \xi_3} d\xi_3 = \frac{\partial x_l}{\partial \xi_3} d\xi_3 \hat{e}_l$$

Using Definition 1.2.1 for the vector product, and using the usual definition of the inner product of two vectors, we get

$$\begin{aligned} dV &= d\mathbf{s}_1 \cdot (d\mathbf{s}_2 \times d\mathbf{s}_3) \\ &= \frac{\partial x_l}{\partial \xi_1} d\xi_1 \hat{e}_l \cdot \hat{e}_k \varepsilon_{ijk} \frac{\partial x_i}{\partial \xi_2} d\xi_2 \frac{\partial x_j}{\partial \xi_3} d\xi_3 \\ &= \varepsilon_{ijk} \frac{\partial x_i}{\partial \xi_1} \frac{\partial x_j}{\partial \xi_2} \frac{\partial x_l}{\partial \xi_3} d\xi_1 d\xi_2 d\xi_3 \, \delta_{kl} \\ &= \varepsilon_{ijk} \frac{\partial x_i}{\partial \xi_1} \frac{\partial x_j}{\partial \xi_2} \frac{\partial x_k}{\partial \xi_3} d\xi_1 d\xi_2 d\xi_3 \\ &= J \, d\xi_1 \, d\xi_2 \, d\xi_3 \end{aligned} \qquad (1.2.3)$$

where we have used the fact that

$$\hat{e}_l \cdot \hat{e}_k = \delta_{lk}$$

Here δ_{lk} is the Kronecker delta, defined by

$$\delta_{lk} = \begin{cases} 1, & \text{if } l = k \\ 0, & \text{if } l \neq k \end{cases} \qquad (1.2.4)$$

Also, we have set

$$J = \varepsilon_{ijk} \frac{\partial x_i}{\partial \xi_1} \frac{\partial x_j}{\partial \xi_2} \frac{\partial x_k}{\partial \xi_3}$$

J is the Jacobian of the transformation.

In Eq. (1.2.3) we let $\boldsymbol{\xi}$ be the material coordinates of the particle under consideration. Since the material coordinates are the Cartesian coordinates of the particle at time $t = 0$, the quantity $d\xi_1 d\xi_2 d\xi_3$ is the initial volume dV_0 about the given point $\boldsymbol{\xi}$. By the motion of the fluid this initial volume is convected and distorted, but since the motion is continuous, the volume dV_0 cannot break up. Hence at a time t later, the initial volume is at point $\mathbf{x}(\boldsymbol{\xi}, t)$ and its volume dV by Eq. (1.2.3) is

$$dV = J \, dV_0 \tag{1.2.5}$$

The assumption that Eq. (1.1.1) can be inverted to give Eq. (1.1.3) and vice versa is equivalent to requiring that neither J nor J^{-1} vanish. Thus

$$0 < J < \infty \tag{1.2.6}$$

We now wish to determine how the Jacobian changes with time; that is, we wish to determine the material derivative dJ/dt. Now

$$J = \begin{vmatrix} \dfrac{\partial x_1}{\partial \xi_1} & \dfrac{\partial x_1}{\partial \xi_2} & \dfrac{\partial x_1}{\partial \xi_3} \\ \dfrac{\partial x_2}{\partial \xi_1} & \dfrac{\partial x_2}{\partial \xi_2} & \dfrac{\partial x_2}{\partial \xi_3} \\ \dfrac{\partial x_3}{\partial \xi_1} & \dfrac{\partial x_3}{\partial \xi_2} & \dfrac{\partial x_3}{\partial \xi_3} \end{vmatrix} \tag{1.2.7}$$

Consequently

$$\frac{d}{dt}\left(\frac{\partial x_i}{\partial \xi_j}\right) = \frac{\partial}{\partial \xi_j}\left(\frac{dx_i}{dt}\right) = \frac{\partial v_i'}{\partial \xi_j} \tag{1.2.8}$$

If we regard v_i' as a function of x_j, we have

$$\frac{\partial v_i'}{\partial \xi_j} = \frac{\partial v_i'}{\partial x_1}\frac{\partial x_1}{\partial \xi_j} + \frac{\partial v_i'}{\partial x_2}\frac{\partial x_2}{\partial \xi_j} + \frac{\partial v_i'}{\partial x_3}\frac{\partial x_3}{\partial \xi_j}$$

$$\frac{\partial v_i'}{\partial \xi_j} = \frac{\partial v_i}{\partial x_l}\frac{\partial x_l}{\partial \xi_j} \tag{1.2.9}$$

From Definition 1.2.2 and Eq. (1.2.7) we have

$$\frac{dJ}{dt} = \frac{d}{dt}\left\{\varepsilon_{ijk}\frac{\partial x_1}{\partial \xi_i}\frac{\partial x_2}{\partial \xi_j}\frac{\partial x_3}{\partial \xi_k}\right\}$$

$$= \varepsilon_{ijk}\frac{d}{dt}\left(\frac{\partial x_1}{\partial \xi_i}\right)\frac{\partial x_2}{\partial \xi_j}\frac{\partial x_3}{\partial \xi_k} + \varepsilon_{ijk}\frac{\partial x_1}{\partial \xi_i}\frac{d}{dt}\left(\frac{\partial x_2}{\partial \xi_j}\right)\frac{\partial x_3}{\partial \xi_k}$$

$$+ \varepsilon_{ijk}\frac{\partial x_1}{\partial \xi_i}\frac{\partial x_2}{\partial \xi_j}\frac{d}{dt}\left(\frac{\partial x_3}{\partial \xi_k}\right)$$

$$= \varepsilon_{ijk}\frac{\partial v_1'}{\partial \xi_i}\frac{\partial x_2}{\partial \xi_j}\frac{\partial x_3}{\partial \xi_k} + \varepsilon_{ijk}\frac{\partial x_1}{\partial \xi_i}\frac{\partial v_2'}{\partial \xi_j}\frac{\partial x_3}{\partial \xi_k} + \varepsilon_{ijk}\frac{\partial x_1}{\partial \xi_i}\frac{\partial x_2}{\partial \xi_j}\frac{\partial v_3'}{\partial \xi_k}$$

$$\tag{1.2.10}$$

12 THE FUNDAMENTAL EQUATIONS

where Eq. (1.2.8) was used to obtain the last result. For convenience, let

$$\frac{dJ}{dt} = J_1 + J_2 + J_3 \tag{1.2.11}$$

Consider J_1. Using Eq. (1.2.9), we get

$$J_1 = \varepsilon_{ijk} \frac{\partial v'_1}{\partial \xi_i} \frac{\partial x_2}{\partial \xi_j} \frac{\partial x_3}{\partial \xi_k}$$

$$= \varepsilon_{ijk} \frac{\partial v'_1}{\partial x_l} \frac{\partial x_l}{\partial \xi_i} \frac{\partial x_2}{\partial \xi_j} \frac{\partial x_3}{\partial \xi_k}$$

$$= \varepsilon_{ijk} \frac{\partial v'_1}{\partial x_1} \frac{\partial x_1}{\partial \xi_i} \frac{\partial x_2}{\partial \xi_j} \frac{\partial x_3}{\partial \xi_k} + \varepsilon_{ijk} \frac{\partial v'_1}{\partial x_2} \frac{\partial x_2}{\partial \xi_i} \frac{\partial x_2}{\partial \xi_j} \frac{\partial x_3}{\partial \xi_k}$$

$$+ \varepsilon_{ijk} \frac{\partial v'_1}{\partial x_3} \frac{\partial x_3}{\partial \xi_i} \frac{\partial x_2}{\partial \xi_j} \frac{\partial x_3}{\partial \xi_k} \tag{1.2.12}$$

The last two terms in Eq. (1.2.12) vanish because they represent determinants with two rows identical. Hence,

$$J_1 = \frac{\partial v'_1}{\partial x_1} J \tag{1.2.13}$$

Similarly,

$$J_2 = \frac{\partial v'_2}{\partial x_2} J \tag{1.2.14}$$

$$J_3 = \frac{\partial v'_3}{\partial x_3} J \tag{1.2.15}$$

Using Eqs. (1.2.13), (1.2.14), and (1.2.15), Eq. (1.2.11) becomes

$$\frac{dJ}{dt} = \left(\frac{\partial v'_1}{\partial x_1} + \frac{\partial v'_2}{\partial x_2} + \frac{\partial v'_3}{\partial x_3} \right) J$$

$$\frac{dJ}{dt} = \frac{\partial v'_k}{\partial x_k} J \tag{1.2.16}$$

or, in vector notation,

$$\frac{dJ}{dt} = (\nabla \cdot \mathbf{v}') J \tag{1.2.17}$$

This is the desired result. We will need this to prove the Reynolds transport theorem in the next section.

1.3. REYNOLDS TRANSPORT THEOREM

This theorem, attributed to Reynolds, concerns the rate of change of any volume integral. It is necessary for the derivation of the conservation of mass equation and the conservation of momentum equation.

Let $\mathscr{G}(\mathbf{x}, t)$ be an arbitrary function describing some physical property of the fluid, and $V(t)$ be a closed volume moving with the fluid, that is, consisting of the same fluid particles. Let

$$G(t) = \iiint_{V(t)} \mathscr{G}(\mathbf{x}, t) \, dV \qquad (1.3.1)$$

We wish to determine dG/dt. We cannot interchange the time derivative and the volume integration because the limits of integration depend on the time t. To circumvent this difficulty, we use the transformation Eq. (1.1.1) to transform the variables of integration from $\mathbf{x}$ to $\boldsymbol{\xi}$. Since $\boldsymbol{\xi}$ is the initial position at time $t = 0$, the volume of integration described in the $\boldsymbol{\xi}$ coordinates is independent of time and hence the operations of differentiation and integration can be interchanged. Now let us carry out this prescription mathematically. We use the transformation given by Eqs. (1.1.1) and (1.2.3) to give

$$\frac{d}{dt} \iiint_{V(t)} \mathscr{G}(\mathbf{x}, t) \, dV = \frac{d}{dt} \iiint_{V_0} \mathscr{G}[\mathbf{x}(\boldsymbol{\xi}, t), t] J \, dV_0$$

$$= \iiint_{V_0} \left(\frac{d\mathscr{G}}{dt} J + \mathscr{G} \frac{dJ}{dt} \right) dV_0$$

$$= \iiint_{V_0} \left\{ \frac{d\mathscr{G}}{dt} + \mathscr{G}(\nabla \cdot \mathbf{v}') \right\} J \, dV_0$$

$$= \iiint_{V} \left\{ \frac{d\mathscr{G}}{dt} + \mathscr{G}(\nabla \cdot \mathbf{v}') \right\} dV \qquad (1.3.2)$$

Eq. (1.2.17) was used to derive Eq. (1.3.2), the Reynolds transport equation.

1.4. CONSERVATION OF MASS

Let $V(t)$ be a material volume, and let $\rho'(\mathbf{x}, t)$ be the density or mass per unit volume of the fluid. Then the mass in the volume $V(t)$ is

$$m = \iiint_{V(t)} \rho'(\mathbf{x}, t) \, dV \qquad (1.4.1)$$

14 THE FUNDAMENTAL EQUATIONS

Let Q' be the rate of creation of mass per unit volume within $V(t)$. Q' has the dimension of mass/[(length)^3 × time]. Then the time rate of change of mass must be equal to the rate that it is being created, that is,

$$\frac{dm}{dt} = \iiint_V Q' \, dV \tag{1.4.2}$$

Using the Reynolds transport theorem, Eq. (1.3.2), Eq. (1.4.1) becomes

$$\frac{dm}{dt} = \iiint_{V(t)} \left\{ \frac{d\rho'}{dt} + \rho'(\nabla \cdot \mathbf{v}') \right\} dV \tag{1.4.3}$$

Combining Eqs. (1.4.2) and (1.4.3) results in

$$\iiint_{V(t)} \left\{ \frac{d\rho'}{dt} + \rho'(\nabla \cdot \mathbf{v}') - Q' \right\} dV = 0 \tag{1.4.4}$$

Now this is true for an arbitrary volume and hence the integrand must vanish everywhere. If we suppose it did not vanish at some point P but were positive there, then, since the integrand is continuous, it would have to be positive for some neighborhood of P. We could then take V to be entirely within this neighborhood and for this V the integral would not vanish. It follows that

$$\frac{d\rho'}{dt} + \rho'(\nabla \cdot \mathbf{v}') = Q' \tag{1.4.5}$$

Eq. (1.4.5) is called the equation of continuity and expresses the conservation of mass. Next let us combine the equation of continuity with Reynolds transport theorem to derive another useful result.

Let $\mathcal{G} = \rho' H$, where H is an arbitrary function. Using Eqs. (1.3.2) and (1.4.5), we get

$$\frac{d}{dt} \iiint_V \rho' H \, dV = \iiint_V \left\{ \frac{d}{dt}(\rho' H) + \rho' H(\nabla \cdot \mathbf{v}') \right\} dV$$

$$= \iiint_V \left\{ \rho' \frac{dH}{dt} + H\left(\frac{d\rho'}{dt} + \rho' \nabla \cdot \mathbf{v}'\right) \right\} dV$$

$$= \iiint_V \left\{ \rho' \frac{dH}{dt} + HQ' \right\} dV \tag{1.4.6}$$

1.5. CONSERVATION OF MOMENTUM FOR A PERFECT FLUID

The conservation of momentum is expressed by Newton's Second Law, which states that the total force $\mathbf{F}$ acting on a particle of mass m is equal to the time rate of change of its momentum, that is,

$$\mathbf{F} = \frac{d}{dt}(m\mathbf{v}') \tag{1.5.1}$$

CONSERVATION OF MOMENTUM FOR A PERFECT FLUID

We shall consider only perfect fluids. A perfect fluid is one in which there is no viscous stress. In a perfect fluid an element of area always experiences a stress normal to itself, and this stress is independent of the orientation of the surface element. The stress in a perfect fluid is simply the pressure p', which is defined as a force per unit area.

Let $V(t)$ be a material volume of fluid and let S be the area enclosing V. Let dS be an infinitesimal element of area of S. Let p' be the pressure of the fluid external to V on the fluid within V. The force associated with this pressure, say $\mathbf{F}_I$, is directed toward the interior of V and is normal to the surface S. The geometry is illustrated in Figure 1.5.1. Let $\hat{n}$ be the outward unit normal to S. Then

$$\frac{d\mathbf{F}_I}{dS} = -p'\hat{n} \qquad (1.5.2)$$

The total force on V due to the pressure is then

$$\mathbf{F}_I = -\iint_S p'\hat{n}\, dS \qquad (1.5.3)$$

Let dm be the mass of the infinitesimal volume dV, and let ρ' be its density, so that

$$dm = \rho'\, dV$$

The momentum of this volume element is then

$$dm\,\mathbf{v}' = (\rho'\, dV)\mathbf{v}'$$

where $\mathbf{v}'$ is the velocity of the fluid. The total momentum over the finite volume V is given by the relation

$$\text{Total momentum} = \iiint_V \rho'\mathbf{v}''\, dv \qquad (1.5.4)$$

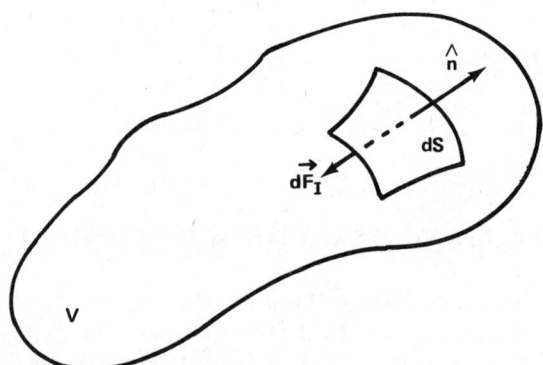

FIGURE 1.5.1. Force exerted on a volume V due to fluid external to V.

16 THE FUNDAMENTAL EQUATIONS

Let **f** be the gravitational force per unit mass acting on the fluid element V. The gravitational force is an external force that can be considered to be reaching into the medium and acting throughout the volume. In contrast, force $\mathbf{F}_I$ due to pressure p' is an internal force that can be considered to be acting on an element of volume through its bounding surface.

Newton's Second Law, Eq. (1.5.1), becomes, upon using Eqs. (1.5.4) and (1.5.3) and taking the gravitational force into account,

$$\frac{d}{dt} \iiint_V \rho' \mathbf{v}' \, dV = - \iint_S p' \hat{n} \, dS + \iiint_V \mathbf{f} \rho' \, dV \qquad (1.5.5)$$

This is the integral form of the conservation of momentum. We require a differential form. Using Eq. (1.4.6), we have

$$\frac{d}{dt} \iiint_V \rho' \mathbf{v}' \, dV = \iiint_V \left\{ \rho' \frac{d\mathbf{v}'}{dt} + Q' \mathbf{v}' \right\} dV \qquad (1.5.6)$$

Using a variant of Gauss' theorem, which is proven in Coburn [1.10] or Hay [1.11], we obtain

$$\iint_S p' \hat{n} \, dS = \iiint_V \nabla p' \, dV \qquad (1.5.7)$$

Substituting the results of Eqs. (1.5.6) and (1.5.7) into Eq. (1.5.5) gives

$$\iiint_V \left\{ \rho' \frac{d\mathbf{v}'}{dt} + Q' \mathbf{v}' \right\} dV = - \iiint_V \nabla p' \, dV + \iiint_V \rho' \mathbf{f} \, dV$$

or

$$\iiint_V \left\{ \rho' \frac{d\mathbf{v}'}{dt} + Q' \mathbf{v}' + \nabla p' - \rho' \mathbf{f} \right\} dV = 0 \qquad (1.5.8)$$

Since V is arbitrary, we have the differential form of the conservation of momentum, namely,

$$\rho' \frac{d\mathbf{v}'}{dt} = - \nabla p' + \rho' \mathbf{f} - Q' \mathbf{v}' \qquad (1.5.9)$$

1.6. CONSERVATION OF ENERGY FOR A PERFECT FLUID

At this point in the description of fluid motion we have five variables; ρ', p', and $\mathbf{v}'$; and four equations, namely, the conservation of mass equation given by Eq. (1.4.5) and the conservation of momentum equations given by Eq. (1.5.9).

CONSERVATION OF ENERGY FOR A PERFECT FLUID

From thermodynamic considerations we know that p' and ρ' are related through an equation of state and that an equation of state actually involves a relation between three thermodynamic variables. We shall take as the third thermodynamic variable the entropy per unit mass S' and denote the equation of state as

$$p' = p'(\rho', S') \tag{1.6.1}$$

While we have gained one more equation to make a total of five, we have also gained one more variable. So we still have one more unknown variable than the number of equations. This remaining gap of one equation is filled by the conservation of energy equation, which we now proceed to derive. For the derivation of the energy equation we will assume for simplicity that we are always working in a region outside the source so that we can take $Q' = 0$.

Recall from Eq. (1.5.2) and the accompanying figure that $d\mathbf{F}_I$ is the force exerted on the volume V through the bounding surface element dS. The rate at which this force does work is just $d\mathbf{F}_I \cdot \mathbf{v}'$, and hence the total power for the closed surface S is

$$\int \mathbf{v}' \cdot d\mathbf{F}_I = -\iint p'\mathbf{v}' \cdot \hat{n}\, dS \tag{1.6.2}$$

Using Gauss' theorem (Coburn [1.10]) and the conservation of momentum, Eq. (1.5.9), we can reformulate the expression for the power expended by doing work on the closed surface S given in Eq. (1.6.2) as

$$-\iint_S p'\mathbf{v}' \cdot \hat{n}\, dS = -\iiint \nabla \cdot (p'\mathbf{v}')\, dV$$

$$= -\iiint (\mathbf{v}' \cdot \nabla p' + p' \nabla \cdot \mathbf{v}')\, dV$$

$$= \iiint \left(\mathbf{v}' \cdot \rho' \frac{d\mathbf{v}'}{dt} - \rho'\mathbf{v}' \cdot \mathbf{f} - p' \nabla \cdot \mathbf{v}'\right) dV$$

Rearranging this last expression and using Eq. (1.4.6) on the first term on the right gives

$$\frac{d}{dt} \iiint_V \frac{1}{2}\rho' v'^2\, dV = \iiint_V \rho' \mathbf{f} \cdot \mathbf{v}'\, dV + \iiint_V p' \nabla \cdot \mathbf{v}'\, dV$$

$$- \iint_S p'\mathbf{v}' \cdot \hat{n}\, dS \tag{1.6.3}$$

Eq. (1.6.3) says that the rate of change of kinetic energy of a material volume is

18 THE FUNDAMENTAL EQUATIONS

the sum of three parts: (1) the rate at which the body forces do work, (2) the rate at which the internal stresses do work, and (3) the rate at which the surface stresses do work.

We now need to invoke the first law of thermodynamics, which is simply the conservation of energy law. This law states that the increase of total energy density (i.e., energy per unit volume) E' in a material volume is the sum of the heat transferred and the work done on the volume. Let $\mathbf{q}$ denote the heat flux vector; then, since $\hat{n}$ is the outward normal to the surface, $-\mathbf{q} \cdot \hat{n}$ is the heat flux into the volume. The balance expressed by the first law of thermodynamics is then

$$\frac{d}{dt} \iiint_V E' \, dV = \iiint_V \rho' \mathbf{f} \cdot \mathbf{v}' \, dV - \iint_S p' \mathbf{v}' \cdot \hat{n} \, dS - \iint_S \mathbf{q} \cdot \hat{n} \, dS \quad (1.6.4)$$

where we have seen above that the first term on the right is the rate at which the body forces do work and the second term on the right is the rate at which the surface stresses do work. Now the total energy density E' can be decomposed into a sum of kinetic energy density and internal energy density

$$E' = \tfrac{1}{2} \rho' v'^2 + \rho' U' \quad (1.6.5)$$

where U' is the internal energy per unit mass. From a macroscopic point of view, Eq. (1.6.5) can be considered as the definition of internal energy. That is, the internal energy is that energy left over after the kinetic energy is subtracted from the total energy. Further, we will consider only fluids for which the heat conductivity vanishes, that is, for which $\mathbf{q} = 0$. With this last assumption and using Eq. (1.6.5), Eq. (1.6.4) becomes

$$\frac{d}{dt} \iiint_V \left(\frac{1}{2} \rho' v'^2 + \rho' U' \right) dV = \iiint_V \rho' \mathbf{f} \cdot \mathbf{v}' \, dV - \iint_S p' \mathbf{v}' \cdot \hat{n} \, dS \quad (1.6.6)$$

Eq. (1.6.6) may be simplified by subtracting Eq. (1.6.3) from it:

$$\frac{d}{dt} \iiint_V \rho' U' \, dV = - \iiint_V p' \nabla \cdot \mathbf{v}' \, dV$$

Using Eq. (1.4.6), this last equation becomes

$$\iiint_V \left(\rho' \frac{dU'}{dt} + p' \nabla \cdot \mathbf{v}' \right) dV = 0 \quad (1.6.7)$$

and arguing as before, we have the differential form of the energy conservation law

$$\rho' \frac{dU'}{dt} = -p' \nabla \cdot \mathbf{v}' \quad (1.6.8)$$

Recalling from Eqs. (1.2.17) and (1.2.3) that $\nabla \cdot \mathbf{v}'$ is related to the change in a material volume, we see from Eq. (1.6.8) that the internal energy increases with the compression of the fluid. Using Eq. (1.4.5) and remembering $Q' = 0$, Eq. (1.6.8) can be written

$$\rho' \frac{dU'}{dt} - \frac{p'}{\rho'} \frac{d\rho'}{dt} = 0 \qquad (1.6.9)$$

Next we would like to express Eq. (1.6.9) in terms of the entropy S' instead of the internal energy U'. Thus we need a relationship between U' and S'. The most straightforward way to obtain the desired relationship is to follow the postulated approach of Callen [1.1]. For completeness we will state the postulates. Reference [1.1] should be consulted for more details.

First we need to define some basic concepts. We will restrict our attention to simple systems. The restriction is not a basic limitation on the generality of thermodynamic theory but is adopted merely for simplicity of exposition. Simple systems are systems that are macroscopically homogeneous, isotropic, uncharged, and chemically inert; that are sufficiently large that surface effects can be neglected; and that are unaffected by electric, magnetic, or gravitational fields. A simple system has a definite chemical composition that must be described by an appropriate set of parameters. One set of convenient parameters is the mole numbers defined as the actual number of each type of molecule divided by Avogadro's number (6.024×10^{23}).

Postulate I. There exist particular states (called equilibrium states) of simple systems that, macroscopically, are characterized completely by the internal energy $\mathscr{U}$, the volume $\mathscr{V}$, and the mole numbers $N_1, \ldots, N_r$ of the chemical components.

Only differences in energy, rather than absolute values of the energy, have physical significance, either at the atomic level or in macroscopic systems. It is conventional, therefore, to adopt some particular state of a system as a reference state, the energy of which is arbitrarily taken as zero. The energy of a system in any other state, relative to the energy of the system in the reference state, is then called the thermodynamic internal energy, $\mathscr{U}$, of the system in that state.

A composite system is just two or more simple systems. The parameters $\mathscr{U}$, $\mathscr{V}$, and $N_1, \ldots, N_r$ have values in a composite system that are equal to the sum of their values in each of the subsystems. Such parameters are called extensive parameters. A description of a thermodynamic system requires the specification of the "walls" that separate it from its surroundings and provide its boundary conditions. It is by manipulation of the walls that the extensive parameters of the system are altered and the processes are initiated. The processes arising by manipulation of the walls generally are associated with a redistribution of some quantity among various systems or among various

20 THE FUNDAMENTAL EQUATIONS

portions of a single system. A wall that constrains an extensive parameter of a system to have a definite and particular value is said to be restrictive with respect to that parameter.

A composite system is called closed if it is surrounded by a wall that is restrictive with respect to the total energy, the total volume, and the total mole numbers of each component of the composite system. The individual simple systems within a closed composite system need not be closed. Constraints that prevent the flow of energy, volume, or matter among the simple systems constituting the composite system are known as internal constraints. If a closed composite system is in equilibrium with respect to certain internal constraints, and if some of these constraints are then removed, the system eventually comes to a new equilibrium state. The basic problem of thermodynamics is the determination of the equilibrium state that eventually results after the removal of internal constraints in a closed composite system.

Postulate II. There exists a function, called the entropy $\mathscr{S}$, of the extensive parameters of any composite system, defined for all equilibrium states and having the following property: the values assumed by the extensive parameters in the absence of an internal constraint are those that maximize the entropy over the manifold of constrained equilibrium states.

In the absence of a constraint, the system is free to select any one of a number of states, each of which might also be realized in the presence of a suitable constraint. The entropy of each of these constrained equilibrium states is definite, and the entropy is largest in some particular state of the set. In the absence of the constraint, it is the state of maximum entropy that is selected by the system. The relation that gives the entropy as a function of the extensive parameters is known as a fundamental relation. It follows that if the fundamental relation of a particular system is known, all conceivable thermodynamic information about the system is ascertainable from it. The fundamental relation is not to be confused with the equation of state of the system. The relationship between the equation of state and the fundamental relation will be made clear later.

Postulate III. The entropy of a composite system is additive over the constituent subsystems. The entropy is continuous with continuous first partial derivatives and is a monotonically increasing function of the energy.

The additive property when applied to conceptually distinct (rather than actually physically distinct) subsystems requires the following property: the entropy of a simple system is a homogeneous, first order function of the extensive parameters; that is, if all the extensive parameters of a system are multiplied by a constant λ, the entropy is multiplied by this same constant. Or

$$\mathscr{S}(\lambda \mathscr{U}, \lambda \mathscr{V}, \lambda N_1, \ldots, \lambda N_r) = \lambda \mathscr{S}(\mathscr{U}, \mathscr{V}, N_1, \ldots, N_r) \qquad (1.6.10)$$

Because of Postulate III, the entropy function can be inverted with respect to

CONSERVATION OF ENERGY FOR A PERFECT FLUID

the energy, and the energy is a single valued, continuous function with continuous first partial derivatives of $\mathscr{S}, \mathscr{V}, N_1, \ldots, N_r$ (compare Theorem 1.1.1).

The function

$$\mathscr{S} = \mathscr{S}(\mathscr{U}, \mathscr{V}, N_1, \ldots, N_r) \qquad (1.6.11)$$

can be solved uniquely for $\mathscr{U}$ in the form

$$\mathscr{U} = \mathscr{U}(\mathscr{S}, \mathscr{V}, N_1, \ldots, N_r) \qquad (1.6.12)$$

Eqs. (1.6.11) and (1.6.12) are alternate forms of the fundamental relation, and each contains all thermodynamic information about the system. However, it can be shown that the energy $\mathscr{U}$ must be minimized rather than maximized as the entropy.

Now let us restrict our attention to a thermodynamic system with a single component. Eq. (1.6.11) becomes

$$\mathscr{S} = \mathscr{S}(\mathscr{U}, \mathscr{V}, N) \qquad (1.6.13)$$

If in Eq. (1.6.10) we take the scale factor $\lambda = 1/N$, we can write Eq. (1.6.13) as

$$S' = S'(U', V') \qquad (1.6.14)$$

when we have put

$$U' = \frac{\mathscr{U}}{N}$$

$$V' = \frac{\mathscr{V}}{N} \qquad (1.6.15)$$

$$S' = \frac{\mathscr{S}}{N}$$

Solving Eq. (1.6.14) for U' gives

$$U' = U'(S', V') \qquad (1.6.16)$$

and taking the total differential of U' gives

$$dU' = \left(\frac{\partial U'}{\partial S'}\right)_{V'} dS' + \left(\frac{\partial U'}{\partial V'}\right)_{S'} dV' \qquad (1.6.17)$$

The following definitions are now made. We define the temperature T' and the thermodynamic pressure p' as

$$T' = \left(\frac{\partial U'}{\partial S'}\right)_{V'} \qquad (1.6.18)$$

$$p' = -\left(\frac{\partial U'}{\partial V'}\right)_{S'} \qquad (1.6.19)$$

22   THE FUNDAMENTAL EQUATIONS

The quantities T' and p' are called intensive parameters. With this notation, Eq. (1.6.17) becomes

$$dU' = T' \, dS' - p' \, dV' \qquad (1.6.20)$$

Now in Eq. (1.6.13) we characterized the chemical composition by the mole number. We also could have used the mass of the chemical constituent. Then in Eq. (1.6.15) U' would be the energy per unit mass, V' would be the volume per unit mass, and S' would be the entropy per unit mass.

Consequently, the mass density $\rho' = 1/V'$ and Eq. (1.6.20) can be expressed in terms of ρ' as

$$dU' = T' \, dS' + \frac{p'}{\rho'^2} d\rho' \qquad (1.6.21)$$

Combining Eq. (1.6.21) with Eq. (1.6.9), we get our desired result that

$$\frac{dS'}{dt} = 0 \qquad (1.6.22)$$

for a perfect fluid with no heat conductivity.

Before leaving the subject of thermodynamics, some remarks should be made concerning the relationship of the fundamental relation with the equation or, actually. equations of state. We will confine our remarks to a simple component system so that the fundamental relation is given by Eq. (1.6.16). We see from Eqs. (1.6.18) and (1.6.19) that the temperature T' and pressure p' are functions of S' and V' because the partial derivatives are functions of S' and V'. We thus have a set of functional relationships

$$T' = T'(S', V') \qquad (1.6.23)$$

$$p' = p'(S', V') \qquad (1.6.24)$$

Relationships that express intensive parameters in terms of the independent extensive parameters are called "equations of state."

Knowledge of a single equation of state does not constitute complete knowledge of the thermodynamic properties of a system. It can be shown that knowledge of all equations of state of a system is equivalent to knowledge of the fundamental equation.

1.7. THE ACOUSTIC WAVE EQUATION

The basic hydrodynamic equations derived above for a perfect fluid with no heat conductivity are as follows:

Conservation of mass:

$$\frac{d\rho'}{dt} + \rho' \nabla \cdot \mathbf{v}' = Q' \qquad (1.7.1)$$

Conservation of momentum:

$$\rho' \frac{d\mathbf{v}'}{dt} = -\nabla p' + \rho' \mathbf{f} - Q' \mathbf{v}' \qquad (1.7.2)$$

Conservation of energy:

$$\frac{dS'}{dt} = 0 \qquad (1.7.3)$$

Equation of state:

$$p' = p'(\rho', S') \qquad (1.7.4)$$

These equations constitute a system of six equations for the six unknown field variables ρ', $\mathbf{v}'$, p', and S'. For the approximation that we will consider in this work, we will assume that the ocean is described by these equations. Because the remainder of this work will deal exclusively with acoustic propagation in the ocean, it is convenient at this point to introduce a Cartesian orthogonal coordinate system that will find extensive use. We will take the xy plane to coincide with the sea surface, and we will take the z-axis as positive downward.

Let $\hat{e}_1$, $\hat{e}_2$, and $\hat{e}_3$ be the unit base vectors along the x, y, and z axes, respectively. From time to time we will interchangeably use the notation x_1, x_2, x_3 for x, y, z when it is convenient to do so. We also will assume that the gravitation force $\mathbf{f}$ is constant and is given by the equation

$$\mathbf{f} = g\hat{e}_3 \qquad (1.7.5)$$

where $g = 9.8$ m/sec^2 is the acceleration due to gravity.

In the absence of a sound wave we will assume the undisturbed fluid variables have the following values:

$$\rho' = \rho_0 \qquad (1.7.6a)$$

$$p' = p_0 \qquad (1.7.6b)$$

$$\mathbf{v}' = \mathbf{v}_0 = 0 \qquad (1.7.6c)$$

$$S' = S_0 \qquad (1.7.6d)$$

$$Q' = Q_0 = 0 \qquad (1.7.6e)$$

Eq. (1.7.6c) states that the fluid is at rest. This is the only case we treat. The hydrodynamic equations describing the undisturbed fluid then becomes

$$\frac{\partial \rho_0}{\partial t} = 0 \qquad (1.7.7a)$$

$$\nabla p_0 = \rho_0 \mathbf{f} \qquad (1.7.7b)$$

$$\frac{\partial S_0}{\partial t} = 0 \qquad (1.7.7c)$$

$$p_0 = p_0(S_0, \rho_0) \qquad (1.7.7d)$$

24 THE FUNDAMENTAL EQUATIONS

Eqs. (1.1.11) and (1.7.6c) were used in deriving these equations. Note that Eqs. (1.7.7a) and (1.7.7c) imply that p_0 and S_0 are independent of time. Combining this result with Eq. (1.7.7d) implies that p_0 is also independent of time:

$$\frac{\partial p_0}{\partial t} = 0 \tag{1.7.8}$$

Suppose now that an acoustic source is turned on with strength Q_1. By an acoustic source is meant a vibrating object whose amplitude and velocity of vibration are so small that they cause only a small perturbation of the undisturbed fluid variables. Let these small perturbations be denoted by ρ_1, p_1, $\mathbf{v}_1$, and S_1. Note that Q_1 is of the same order of magnitude as the perturbed fluid variables. These perturbed variables are called acoustic variables. Now p_0 and ρ_0 are positive quantities because they represent the pressure and density of the undisturbed fluid. However, p_1 and ρ_1 can be positive or negative quantities since they represent excursions about the undisturbed state p_0 and ρ_0, respectively.

Thus, the total density in the fluid ρ', the total pressure p', the total velocity $\mathbf{v}'$, the total entropy S', and the total source strength Q' can be represented in the form

$$\rho' = \rho_0 + \rho_1 \tag{1.7.9a}$$

$$p' = p_0 + p_1 \tag{1.7.9b}$$

$$\mathbf{v}' = \mathbf{v}_0 + \mathbf{v}_1 = \mathbf{v}_1 \tag{1.7.9c}$$

$$S' = S_0 + S_1 \tag{1.7.9d}$$

$$Q' = Q_0 + Q_1 = Q_1 \tag{1.7.9e}$$

where the acoustic variables ρ_1, p_1, $\mathbf{v}_1$, S_1, and Q_1 are to be treated as small quantities. The acoustic wave equation will be derived under the assumption that squares or higher order terms involving the acoustic variables can be neglected. We now proceed to derive the acoustic wave equation.

Now substitute Eqs. (1.7.9a) and (1.7.9c) into Eq. (1.7.1), and using Eq. (1.1.11) gives

$$\frac{d\rho'}{dt} + \rho' \nabla \cdot \mathbf{v}' = Q'$$

$$\frac{\partial \rho'}{\partial t} + \mathbf{v}' \cdot \nabla \rho' + \rho' \nabla \cdot \mathbf{v}' = Q'$$

$$\frac{\partial \rho_0}{\partial t} + \frac{\partial \rho_1}{\partial t} + \mathbf{v}_1 \cdot \nabla \rho_0 + \mathbf{v}_1 \cdot \nabla \rho_1 + \rho_0 \nabla \cdot \mathbf{v}_1 + \rho_1 \nabla \cdot \mathbf{v}_1 = Q_1$$

THE ACOUSTIC WAVE EQUATION

Using Eq. (1.7.7a) and neglecting second and higher order terms in the acoustic variables gives

$$\frac{\partial \rho_1}{\partial t} + \mathbf{v}_1 \cdot \nabla \rho_0 + \rho_0 \nabla \cdot \mathbf{v}_1 = Q_1 \qquad (1.7.10)$$

Substituting Eqs. (1.7.6a), (1.7.6b), and (1.7.6c) into Eq. (1.7.2) and using Eq. (1.1.11) results in

$$\rho' \frac{d\mathbf{v}'}{dt} = -\nabla p' + \rho' \mathbf{f} - Q'\mathbf{v}'$$

$$\rho' \frac{\partial \mathbf{v}'}{\partial t} + \rho' \mathbf{v}' \cdot \nabla \mathbf{v}' = -\nabla p' + \rho' \mathbf{f} - Q'\mathbf{v}'$$

$$(\rho_0 + \rho_1)\frac{\partial \mathbf{v}_1}{\partial t} + (\rho_0 + \rho_1)\mathbf{v}_1 \cdot \nabla \mathbf{v}_1 = -\nabla p_0 - \nabla p_1 + (\rho_0 + \rho_1)\mathbf{f} - Q_1 \mathbf{v}_1$$

Using Eq. (1.7.7b) and neglecting second and higher order terms in the acoustic variables gives

$$\rho_0 \frac{\partial \mathbf{v}_1}{\partial t} = -\nabla p_1 + \rho_1 \mathbf{f} \qquad (1.7.11)$$

Since we assumed Q_1 is the source term for the acoustic wave, it is of the same order of magnitude as $\mathbf{v}_1$; thus the product $Q_1 \mathbf{v}_1$ is neglected.

Next we want to eliminate the three components of the particle velocity $\mathbf{v}_1$ between Eqs. (1.7.10) and (1.7.11). This will result in an equation containing only the two variables ρ_1 and p_1.

Differentiating Eq. (1.7.10) with respect to time and taking the divergence of Eq. (1.7.11) gives

$$\frac{\partial^2 \rho_1}{\partial t^2} + \frac{\partial \mathbf{v}_1}{\partial t} \cdot \nabla \rho_0 + \rho_0 \nabla \cdot \frac{\partial \mathbf{v}_1}{\partial t} = \frac{\partial Q_1}{\partial t} \qquad (1.7.12)$$

$$\rho_0 \nabla \cdot \frac{\partial \mathbf{v}_1}{\partial t} + \frac{\partial \mathbf{v}_1}{\partial t} \cdot \nabla \rho_0 = -\nabla^2 p_1 + \nabla \cdot (\rho_1 \mathbf{f}) \qquad (1.7.13)$$

Eq. (1.7.7a) was used in the derivation of Eq. (1.7.12). Subtracting the second equation from the first yields

$$\frac{\partial^2 \rho_1}{\partial t^2} = \nabla^2 p_1 - \nabla \cdot (\rho_1 f) + \frac{\partial Q_1}{\partial t} \qquad (1.7.14)$$

We now have an equation in the two variables ρ_1 and p_1. To eliminate ρ_1 and obtain an equation for p_1 alone, the equation needed is one of the equations of state of the system. This is given by Eq. (1.6.1), which was

$$p' = p'(\rho', S') \tag{1.7.15}$$

where S' is the entropy per unit mass. This equation can be linearized by expanding p' about the equilibrium state and neglecting second and higher order terms in the acoustic variables:

$$p'(\rho_0 + \rho_1, S_0 + S_1) = p'(\rho_0, S_0) + \left(\frac{\partial p'}{\partial \rho'}\right)_{\substack{S' \\ \rho' = \rho_0}} \rho_1 + \left(\frac{\partial p'}{\partial S'}\right)_{\substack{\rho' \\ S' = S_0}} S_1 \tag{1.7.16}$$

Note that

$$p_1 = p'(\rho_0 + \rho_1, S_0 + S_1) - p'(\rho_0, S_0) \tag{1.7.17}$$

Let

$$c^2 = \left(\frac{\partial p'}{\partial \rho'}\right)_{\substack{S' = \text{const} \\ \rho' = \rho_0}} \tag{1.7.18}$$

and

$$A = \left(\frac{\partial p'}{\partial S'}\right)_{\substack{\rho' = \text{const} \\ S' = S_0}} \tag{1.7.19}$$

At this point c has no physical interpretation. It is simply an abbreviation for the partial derivative appearing on the right side of Eq. (1.7.18).

Using Eqs. (1.7.17), (1.7.18), and (1.7.19), we can write Eq. (1.7.16) in the form

$$p_1 = c^2 \rho_1 + A S_1 \tag{1.7.20}$$

For an adiabatic compression of the fluid, the linearized equation of state becomes

$$p_1 = c^2 \rho_1 \tag{1.7.21}$$

To eliminate ρ_1 from Eq. (1.7.14) we need $\partial \rho_1/\partial t$.

However, we cannot simply take the partial derivative of Eq. (1.7.21). The equation of state holds for a given small volume of the fluid and not for a fixed point in space. Consequently, we must go back to Eq. (1.7.15) and take the material derivative since this describes the change in the fluid variables for a fixed fluid element, that is, the material derivative gives the rate of change of

THE ACOUSTIC WAVE EQUATION 27

the fluid variables for a given fluid element as observed when moving with that element. So we have

$$\frac{dp'}{dt} = \frac{\partial p'}{\partial \rho'}\frac{d\rho'}{dt} + \frac{\partial p'}{\partial S'}\frac{dS'}{dt} \tag{1.7.22}$$

Now for a perfect fluid with no heat conductivity or viscosity (which we are assuming) we have shown in Eq. (1.6.22) that

$$\frac{dS'}{dt} = 0 \tag{1.7.23}$$

Using Eqs. (1.7.18) and (1.7.23), Eq. (1.7.22) becomes

$$\frac{dp'}{dt} = c^2 \frac{d\rho'}{dt} \tag{1.7.24}$$

or

$$\frac{\partial p'}{\partial t} + \mathbf{v}' \cdot \nabla p' = c^2 \left\{ \frac{\partial \rho'}{\partial t} + \mathbf{v}' \cdot \nabla \rho' \right\} \tag{1.7.25}$$

Using Eqs. (1.7.9a), (1.7.9b), and (1.7.9c) and making the acoustic approximation gives

$$\frac{\partial}{\partial t}(p_0 + p_1) + \mathbf{v}_1 \cdot \nabla(p_0 + p_1) = c^2 \left\{ \frac{\partial}{\partial t}(\rho_0 + \rho_1) + \mathbf{v}_1 \cdot \nabla(\rho_0 + \rho_1) \right\}$$

$$\frac{\partial p_1}{\partial t} + \mathbf{v}_1 \cdot \nabla p_0 = c^2 \left\{ \frac{\partial \rho_1}{\partial t} + \mathbf{v}_1 \cdot \nabla \rho_0 \right\} \tag{1.7.26}$$

Eqs. (1.7.7a) and (1.7.8) were used to derive Eq. (1.7.26).
Taking the time derivative of Eq. (1.7.26) gives

$$\frac{\partial^2 p_1}{\partial t^2} + \frac{\partial \mathbf{v}_1}{\partial t} \cdot \nabla p_0 = c^2 \left\{ \frac{\partial^2 \rho_1}{\partial t^2} + \frac{\partial \mathbf{v}_1}{\partial t} \cdot \nabla \rho_0 \right\} \tag{1.7.27}$$

Note that we have assumed c to be independent of time.
Using Eq. (1.7.11), Eq. (1.7.27) becomes

$$\frac{\partial^2 p_1}{\partial t^2} - \frac{1}{\rho_0} \nabla p_1 \cdot \nabla p_0 + \frac{\rho_1}{\rho_0} \mathbf{f} \cdot \nabla p_0$$

$$= c^2 \left\{ \frac{\partial^2 \rho_1}{\partial t^2} - \frac{1}{\rho_0} \nabla p_1 \cdot \nabla \rho_0 + \frac{\rho_1}{\rho_0} \nabla \rho_0 \cdot \mathbf{f} \right\}$$

28 THE FUNDAMENTAL EQUATIONS

or

$$\frac{\partial^2 p_1}{\partial t^2} = \frac{1}{c^2}\left\{\frac{\partial^2 p_1}{\partial t^2} - \frac{1}{\rho_0}\nabla p_0 \cdot \nabla p_1 + \frac{p_1}{\rho_0}\mathbf{f}\cdot\nabla p_0\right\}$$

$$+ \frac{1}{\rho_0}\nabla \rho_0 \cdot \nabla p_1 - \frac{p_1}{\rho_0}\nabla \rho_0 \cdot \mathbf{f} \qquad (1.7.28)$$

Making use of the last equation, Eq. (1.7.14) becomes

$$\frac{1}{c^2}\left\{\frac{\partial^2 p_1}{\partial t^2} - \frac{1}{\rho_0}\nabla p_0 \cdot \nabla p_1 + \frac{p_1}{\rho_0}\mathbf{f}\cdot\nabla p_0\right\} + \frac{1}{\rho_0}\nabla \rho_0 \cdot \nabla p_1 - \frac{p_1}{\rho_0}\nabla \rho_0 \cdot \mathbf{f}$$

$$= \nabla^2 p_1 - \nabla \cdot (\rho_1 \mathbf{f}) + \frac{\partial Q_1}{\partial t}$$

or, upon using Eqs. (1.7.21) and (1.7.7b),

$$\nabla^2 p_1 - \frac{1}{\rho_0}\nabla \rho_0 \cdot \nabla p_1 - \frac{1}{c^2}\frac{\partial^2 p_1}{\partial t^2} = -\frac{\partial Q_1}{\partial t} + \frac{1}{\rho_0}\nabla p_0$$

$$\cdot \left\{\frac{1}{\rho_0 c^4}\nabla p_0 - \frac{2}{c^3}\nabla c - \frac{1}{\rho_0 c^2}\nabla \rho_0\right\} p_1 \qquad (1.7.29)$$

Next, let us estimate the relative order of magnitude of these terms (except the source term, which we will ignore for obvious reasons). We will assume that ρ_0, p_0, and c are a function of depth z only. For simplicity, in order to get an order of magnitude estimate for the acoustic pressure p_1, assume that p_1 has the form of a plane wave (see Kinsler and Frey [1.7]), namely,

$$p_1 = \cos(\mathbf{k}\cdot\mathbf{r} - \omega t)$$

where
$$\mathbf{r} = x\hat{e}_1 + y\hat{e}_2 + z\hat{e}_3$$
$$\mathbf{k} = k_x\hat{e}_1 + k_y\hat{e}_2 + k_z\hat{e}_3$$
$$\omega = 2\pi f, \quad f \text{ being the frequency.}$$

Here $\mathbf{k}$ is the wave vector of magnitude $k = 2\pi/\lambda$, λ being the wavelength. The quantities ω and k are related by the well-known relation $\omega = kc$. Then

$$\nabla^2 p_1 \sim k^2$$
$$\nabla p_1 \sim k$$
$$p_1 \sim 1$$
$$\frac{1}{c^2}\frac{\partial^2 p_1}{\partial t^2} = -\frac{\omega^2}{c^2}p_1 = -k^2 p_1 \sim -k^2$$

where the cosine factor is omitted from the estimate because it is common to all terms in the wave equation (having dropped the $\partial Q_1/\partial t$ term).

Now let us estimate the three terms on the right side of Eq. (1.7.29) that constitute the coefficient of p_1. Using typical values for the ocean we get

$$\frac{1}{\rho_0^2 c^4} \nabla p_0 \cdot \nabla p_0 = \frac{g^2}{c^4} = \frac{(9.8 \text{ m/sec}^2)^2}{(1500 \text{ m/sec})^4} = 2 \times 10^{-11} \text{ m}^{-2}$$

$$\frac{2}{\rho_0 c^3} \nabla p_0 \cdot \nabla c = \frac{2g}{c^3} \frac{dc}{dz} = \frac{2(9.8 \text{ m/sec}^2)}{(1500 \text{ m/sec})^3} (0.013 \text{ sec}^{-1})$$

$$= 8 \times 10^{-11} \text{ m}^{-2}$$

$$\frac{1}{\rho_0^2 c^2} \nabla p_0 \cdot \nabla \rho_0 = \frac{g}{\rho_0 c^2} \frac{d\rho_0}{dz} = \frac{(9.8 \text{ m/sec}^2)}{(1025 \text{ kg/m}^3)(1500 \text{ m/sec})^2}$$

$$\cdot \frac{(1046 \text{ kg/m}^3 - 1025 \text{ kg/m}^3)}{(4000 \text{ m} - 0 \text{ m})}$$

$$= 2 \times 10^{-11} \text{ m}^{-2}$$

Eqs. (1.7.5) and (1.7.7b) were used to get these results.

Finally, let us estimate the coefficient of ∇p_1 on the left side of Eq. (1.7.29).

$$\frac{1}{\rho_0} \frac{d\rho_0}{dz} = \frac{1}{(1025 \text{ kg/m}^2)} \frac{(1046 \text{ kg/m}^3 - 1025 \text{ kg/m}^3)}{(4000 \text{ m} - 0 \text{ m})}$$

$$= 5 \times 10^{-6} \text{ m}^{-1}$$

If we now put all these results together, Eq. (1.7.29) becomes in terms of relative orders of magnitude

$$k^2 - (5 \times 10^{-6})k + k^2 = \{(2 \times 10^{-11}) - (8 \times 10^{-11}) - (2 \times 10^{-11})\}$$

Now typical values of k of interest to us can vary from approximately 0.004 at a frequency of 1 Hz to approximately 21 at a frequency of 5000 Hz. Thus as a good approximation, at least for cases of interest to us, we can drop the three terms involving ∇p_0 on the right side of Eq. (1.7.29). The wave equation that we will work with is then given by

$$\nabla^2 p_1 - \frac{1}{\rho_0} \nabla p_0 \cdot \nabla p_1 - \frac{1}{c^2} \frac{\partial^2 p_1}{\partial t^2} = \frac{-\partial Q_1}{\partial t} \qquad (1.7.30)$$

30 THE FUNDAMENTAL EQUATIONS

1.8. CONSERVATION OF ENERGY IN THE ACOUSTIC APPROXIMATION

The linearized acoustic equations have been derived in Section 1.7. For reference they are

$$\frac{\partial \rho_1}{\partial t} + \mathbf{v}_1 \cdot \nabla \rho_0 + \rho_0 \nabla \cdot \mathbf{v}_1 = Q_1 \qquad (1.7.10)$$

$$\rho_0 \frac{\partial \mathbf{v}_1}{\partial t} = -\nabla p_1 + \frac{p_1}{\rho_0} \nabla p_0 \qquad (1.7.11)$$

In component notation these equations become

$$\frac{\partial \rho_1}{\partial t} + v_{1_i} \frac{\partial \rho_0}{\partial x_i} + \rho_0 \frac{\partial v_{1_i}}{\partial x_i} = Q_1$$

$$\rho_0 \frac{\partial v_{1_i}}{\partial t} = -\frac{\partial p_1}{\partial x_i} + \frac{p_1}{\rho_0} \frac{\partial p_0}{\partial x_i}$$

Taking the scalar product of v_{1_i} with Eq. (1.7.11) gives

$$\rho_0 v_{1_i} \frac{\partial v_{1_i}}{\partial t} = -v_{1_i} \frac{\partial p_1}{\partial x_i} + \frac{p_1}{\rho_0} v_{1_i} \frac{\partial p_0}{\partial x_i}$$

or

$$\rho_0 v_{1_i} \frac{\partial v_{1_i}}{\partial t} + \frac{\partial}{\partial x_i} (p_1 v_{1_i}) - p_1 \frac{\partial v_{1_i}}{\partial x_i} - \frac{p_1}{\rho_0} v_{1_i} \frac{\partial p_0}{\partial x_i} = 0 \qquad (1.8.1)$$

Solving Eq. (1.7.10) for $\partial v_{1_i}/\partial x_i$ and substituting this expression into Eq. (1.8.1) gives

$$\rho_0 v_{1_i} \frac{\partial v_{1_i}}{\partial t} + \frac{\partial}{\partial x_i}(p_1 v_{1_i}) + \frac{p_1}{\rho_0} \frac{\partial \rho_1}{\partial t} + \frac{p_1}{\rho_0} v_{1_i} \frac{\partial \rho_0}{\partial x_i} - \frac{p_1}{\rho_0} Q_1 - \frac{p_1}{\rho_0} v_{1_i} \frac{\partial p_0}{\partial x_i} = 0$$

$$(1.8.2)$$

If we solve Eq. (1.7.26) for $\partial \rho_1/\partial t$, we obtain

$$\frac{\partial \rho_1}{\partial t} = \frac{1}{c^2} \left\{ \frac{\partial p_1}{\partial t} + v_{1_i} \frac{\partial p_0}{\partial x_i} \right\} - v_{1_i} \frac{\partial \rho_0}{\partial x_i} \qquad (1.8.3)$$

CONSERVATION OF ENERGY IN THE ACOUSTIC APPROXIMATION

Substituting the expression in Eq. (1.8.3) for $\partial \rho_1/\partial t$ into Eq. (1.8.2) yields

$$\rho_0 v_{1_i} \frac{\partial v_{1_i}}{\partial t} + \frac{\partial}{\partial x_i}(p_1 v_{1_i}) + \overset{\text{①}}{\frac{p_1}{\rho_0} v_{1_i} \frac{\partial \rho_0}{\partial x_i}} - \frac{p_1}{\rho_0} Q_1 - \overset{\text{②}}{\frac{p_1}{\rho_0} v_{1_i} \frac{\partial \rho_0}{\partial x_i}}$$

$$+ \frac{p_1}{\rho_0 c^2} \frac{\partial p_1}{\partial t} + \overset{\text{②}}{\frac{p_1}{\rho_0 c^2} v_{1_i} \frac{\partial p_0}{\partial x_i}} - \overset{\text{①}}{\frac{p_1}{\rho_0} v_{1_i} \frac{\partial \rho_0}{\partial x_i}} = 0$$

where Eq. (1.7.21) was used to cancel the two terms designated by ②. Thus we have the final result

$$\frac{\partial}{\partial t}\left(\frac{1}{2}\rho_0 v_1^2 + \frac{p_1^2}{2\rho_0 c^2}\right) + \frac{\partial}{\partial x_i}(p_1 v_{1_i}) = \frac{p_1}{\rho_0} Q_1 \qquad (1.8.4)$$

Let

$$e = \frac{1}{2}\rho_0 v_1^2 + \frac{p_1^2}{2\rho_0 c^2} \qquad (1.8.5)$$

$$\mathbf{J} = p_1 \mathbf{v}_1 \qquad (1.8.6)$$

Then Eq. (1.8.4) can be written in vector form

$$\frac{\partial e}{\partial t} + \nabla \cdot \mathbf{J} = \frac{p_1 Q_1}{\rho_0} \qquad (1.8.7)$$

Eq. (1.8.7) has the form of a continuity equation. It is the differential expression for the conservation of energy in an acoustic radiation field.

Now let us examine in some detail the physical significance of the terms in Eq. (1.8.7). First, let us consider the expression for the quantity denoted e and defined by Eq. (1.8.5). The term $\rho_0 v_1^2/2$ is clearly the kinetic energy per unit volume, or the kinetic energy density. It will be shown that the term $p_1^2/(2\rho_0 c^2)$ is the potential energy per unit volume, or the potential energy density. Thus e is the sum of the kinetic energy density and the potential energy density, or e is the total energy density due to the passage of the sound wave. The vector $\mathbf{J}$ has the dimensions of instantaneous power per unit area. Thus $\mathbf{J}$ is the instantaneous acoustic intensity or energy flux. The term $p_1 Q_1/\rho_0$ is, of course, the source of energy for the radiation field.

The interpretation of Eq. (1.8.7) is analogous to the continuity equation for mass. Consider a volume V. Eq. (1.8.7) states that the time rate of change of the energy per unit volume inside V plus the rate that it is being created is equal to the decrease of the divergence of the energy flux $\mathbf{J}$ through the surface S of V.

32   THE FUNDAMENTAL EQUATIONS

Thus, in general, the instantaneous intensity is p_1v_1. Only in special cases such as a plane wave or a spherical wave is $v_1 \propto p_1$ and the intensity is proportional to p_1^2.

Now let us show that $p_1^2/(2\rho_0 c^2)$ is the potential energy density in the medium due to the passage of the sound wave. For simplicity we will assume $Q_1 = 0$ and $\rho_0 =$ constant.

Consider a small volume V_0 in the undisturbed fluid. The potential energy of this volume element is the work done on it by the acoustic wave to compress it. Since we are considering a small volume V_0, we can assume that the portion of an arbitrary wave front acting on it is planar. Let us take the x-axis to be in the direction of propagation of the wave front.

Consider a small cross-sectional area of wave front S and the volume $V_0 = S\,dx$ of the undisturbed fluid. The mass of this element is $m_0 = \rho_0 S\,dx$. Let F be the total force exerted on this volume at x due to the acoustic wave. Then the plane originally at x is displaced a distance ξ to the right.

The plane at $x + dx$ is displaced a distance $\xi + (\partial \xi/\partial x)\,dx$. The new volume is $V = S\,d\tau$, where $d\tau$ is shown in Figure 1.8.1. Figure 1.8.1 shows that

$$d\tau = x + dx + \xi + \frac{\partial \xi}{\partial x}dx - (x + \xi)$$

$$= \left(1 + \frac{\partial \xi}{\partial x}\right)dx \qquad (1.8.8)$$

So the new volume is

$$V = S\left(1 + \frac{\partial \xi}{\partial x}\right)dx$$

$$V = \left(1 + \frac{\partial \xi}{\partial x}\right)V_0 \qquad (1.8.9)$$

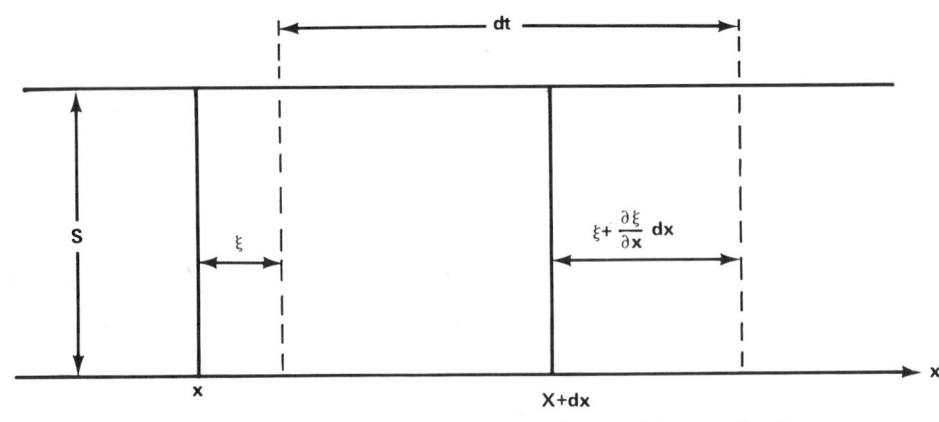

FIGURE 1.8.1. Geometry used to calculate the potential energy density.

CONSERVATION OF ENERGY IN THE ACOUSTIC APPROXIMATION

Now the potential energy, PE, of the volume V_0 due to compression by the acoustic wave is

$$PE = -\int p_1 \, dV \tag{1.8.10}$$

where the minus sign is required so that the potential energy will increase as work is done on the volume element when its volume is decreased by action of a positive pressure p_1.

In order to carry out the integration, it is necessary to express V as a function of p_1.

We will consider only time harmonic waves with a time factor $\cos(-i\omega t)$. Thus Eq. (1.7.10) can be written for the x-coordinate (remembering that ρ_0 is constant and Q_1 vanishes) as

$$\rho_1 + \rho_0 \frac{\partial \xi}{\partial x} = 0 \tag{1.8.11}$$

and using Eq. (1.7.21),

$$\frac{p_1}{c^2} + \rho_0 \frac{\partial \xi}{\partial x} = 0$$

or

$$\frac{\partial \xi}{\partial x} = -\frac{p_1}{\rho_0 c^2} \tag{1.8.12}$$

Putting this into Eq. (1.8.9) gives

$$V = \left(1 - \frac{p_1}{\rho_0 c^2}\right) V_0 \tag{1.8.13}$$

So

$$dV = -\frac{V_0}{\rho_0 c^2} \, dp_1 \tag{1.8.14}$$

Substituting the expression for dV given by Eq. (1.8.14) into Eq. (1.8.10) gives for the integral

$$PE = \frac{V_0}{\rho_0 c^2} \int_0^{p_1} p_1 \, dp_1$$

$$= \frac{V_0}{2\rho_0 c^2} p_1^2$$

34 THE FUNDAMENTAL EQUATIONS

Thus, the potential energy density PE/V_0 is

$$\frac{PE}{V_0} = \frac{p_1^2}{2\rho_0 c^2} \qquad (1.8.15)$$

In the development for the remainder of this section we will consider those regions outside the source, so that Q_1 will be taken to be zero.

Now the physical meaningful quantity is the time average of the intensity vector **J**. Let us take the time average of Eq. (1.8.7) over one period, T, of the sound wave.

$$\frac{1}{T}\int_0^T \frac{\partial e}{\partial t} dt + \frac{1}{T}\int_0^T \nabla \cdot \mathbf{J}\, dt = 0$$

$$\frac{1}{T}[e(T) - e(0)] + \nabla \cdot \langle \mathbf{J} \rangle = 0 \qquad (1.8.16)$$

where

$$\langle \mathbf{J} \rangle = \frac{1}{T}\int_0^T \mathbf{J}\, dt \qquad (1.8.17)$$

Since we are assuming that $e(t)$ is a periodic function of t of period T, we have

$$e(T) - e(0) = 0$$

Thus, we get for the average intensity

$$\nabla \cdot \langle \mathbf{J} \rangle = 0 \qquad (1.8.18)$$

In general, it is convenient to represent the field variable by complex quantities. The real part of the expression is understood to represent the physical quantity in question. For example, for time harmonic waves we can write

$$p_1 = \text{Re}\{\mathscr{P}(\mathbf{r}, t)\} = \text{Re}\{p(\mathbf{r})e^{-i\omega t}\} \qquad (1.8.19)$$

$$\mathbf{v}_1 = \text{Re}\{\mathscr{V}(\mathbf{r}, t)\} = \text{Re}\{\mathbf{V}(\mathbf{r})e^{-i\omega t}\} \qquad (1.8.20)$$

where p and **V** are in general complex and Re denotes the real part of a complex number.

In Stratton [1.12], it is shown that

$$\langle \mathbf{J} \rangle = \frac{1}{T}\int_0^T \mathbf{J}\, dt = \frac{1}{T}\int_0^T p_1 \mathbf{v}_1\, dt$$

$$= \frac{1}{T}\int_0^T (\text{Re}\,\mathscr{P})(\text{Re}\,\mathscr{V})\, dt$$

$$= \tfrac{1}{2}\text{Re}(\mathscr{P}\mathscr{V}^*) \qquad (1.8.21)$$

1.9. BOUNDARY CONDITIONS

Consider the case when there are no external body forces and no sources present in the volume of fluid under consideration. The linearized hydrodynamic equations or the system of first order differential equations for the acoustic radiation field are given by Eqs. (1.7.10) and (1.7.11), namely,

$$\frac{\partial \rho_1}{\partial t} + \nabla \cdot (\rho_0 \mathbf{v}_1) = 0 \qquad (1.9.1)$$

$$\rho_0 \frac{\partial \mathbf{v}_1}{\partial t} = -\nabla p_1 \qquad (1.9.2)$$

The validity of these equations is only postulated for points in whose neighborhood the physical properties of the medium vary continuously. However, across a surface that bounds one medium from another there can occur sharp changes in the parameters ρ_0 and c.

Imagine that the surface S bounding medium a from medium b has been replaced by a very thin transition layer in which ρ_0 and c vary rapidly but continuously from their values near S in medium a to their values near S in medium b. Within this layer, as within media a and b, the acoustic variables ρ_1, p_1, and $\mathbf{v}_1$ and their first derivatives are continuous and bounded functions of position and time. Through the layer we now draw a small right cylinder as shown in Figure 1.9.1.

Now let us integrate Eq. (1.9.2) over the volume of the cylinder V

$$\iiint_V \rho_0 \frac{\partial \mathbf{v}_1}{\partial t} dV = -\iiint \nabla p_1 \, dV$$

Using Eq. (1.5.7), we obtain

$$\iiint_V \rho_0 \frac{\partial \mathbf{v}_1}{\partial t} dV = -\iint p_1 \hat{n} \, dS \qquad (1.9.3)$$

If the base of the cylinder whose area is ΔA is made sufficiently small, it may be assumed that ρ_0, $\partial \mathbf{v}_1/\partial t$, and p_1 have constant values over each end. Eq. (1.9.3) then becomes

$$\rho_0 \frac{\partial \mathbf{v}_1}{\partial t} \Delta A \, \Delta w = -(p_1 \hat{n}_a + p_1 \hat{n}_b) \Delta A + \text{contributions of the walls}$$

$$(1.9.4)$$

The contribution of the walls to the surface integral is directly proportional to Δw. Now let the transition layer shrink to the surface S. In the limit, as

36 THE FUNDAMENTAL EQUATIONS

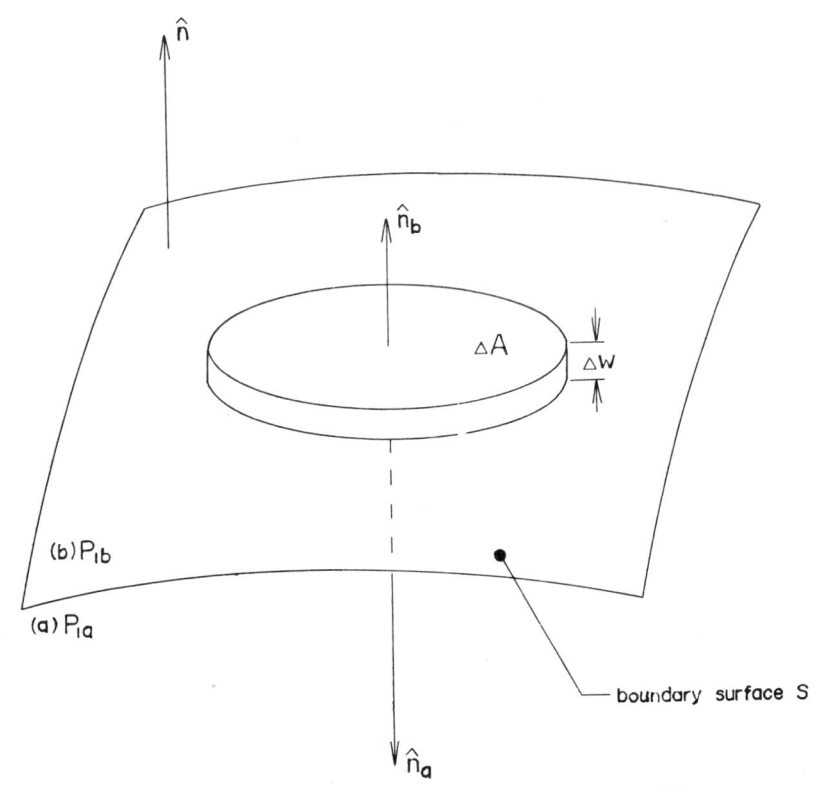

FIGURE 1.9.1. Geometry for determination of boundary conditions.

$\Delta w \to 0$, the contribution of the walls becomes vanishingly small. The value of p_1 at a point on S in medium a will be denoted by p_{1_a} and the corresponding value of p_1 across the surface in medium b will be denoted p_{1_b}. On side b, $\hat{n}_b = \hat{n}$ and on side a, $\hat{n}_a = -\hat{n}$.

Then as $\Delta w \to 0$ and $\Delta A \to 0$, Eq. (1.9.4) becomes

$$p_{1_b} - p_{1_a} = 0 \tag{1.9.5}$$

That is, the acoustic pressure field is continuous across any interface, even if it has a discontinuity in ρ_0 or c.

Now let us integrate Eq. (1.9.1) over the volume of the cylinder and use Green's theorem (or the divergence theorem):

$$\iiint \frac{\partial \rho_1}{\partial t} dV + \iiint \nabla \cdot (\rho_0 \mathbf{v}_1) \, dV = 0$$

$$\iiint \frac{\partial \rho_1}{\partial t} dV + \iint \rho_0 \mathbf{v}_1 \cdot \hat{n} \, dS = 0 \tag{1.9.6}$$

Following this same reasoning, we get

$$\frac{\partial \rho_1}{\partial t} \Delta A \, \Delta w + (\rho_0 \mathbf{v}_1 \cdot \hat{n}_a + \rho_0 \mathbf{v}_1 \cdot \hat{n}_b) \Delta A$$

$$+ \text{contributions of the walls} = 0 \qquad (1.9.7)$$

In the limit as $\Delta A \to 0$ and $\Delta w \to 0$, Eq. (1.9.7) becomes

$$\rho_{0_b} \mathbf{v}_{1_b} \cdot \hat{n} - \rho_{0_a} \mathbf{v}_{1_a} \cdot \hat{n} = 0 \qquad (1.9.8)$$

When the external body force **f** vanishes, the acoustic pressure p_1 and acoustic particle velocity $\mathbf{v}_1$ are related by Eq. (1.7.11).

$$\rho_0 \frac{\partial \mathbf{v}_1}{\partial t} = -\nabla p_1 \qquad (1.9.9)$$

In this book we shall only consider acoustic fields with a simple harmonic time variation of the form $e^{-i\omega t}$. Consequently, Eq. (1.9.9) becomes for the time harmonic case

$$i\omega \rho_0 \mathbf{v}_1 = \nabla p_1 \qquad (1.9.10)$$

If we combine Eqs. (1.9.8) and (1.9.10), we get

$$\nabla p_{1_b} \cdot \hat{n} - \nabla p_{1_a} \cdot \hat{n} = 0$$

Now, from vector calculus, we have

$$\hat{n} \cdot \nabla = \frac{\partial}{\partial n}$$

where $\partial/\partial n$ denotes the normal derivative in the direction of the unit normal $\hat{n}$. Consequently, we can write this in the form

$$\frac{\partial p_{1_b}}{\partial n} - \frac{\partial p_{1_a}}{\partial n} = 0 \qquad (1.9.11)$$

This states that the normal derivative of the acoustic pressure is continuous across the interface.

REFERENCES

1.1. H. B. Callen, *Thermodynamics* (Wiley, New York, 1963), Chapters 1–4.
1.2. R. Courant and F. John, *Introduction to Calculus and Analysis* (Wiley, New York, 1974).
1.3. W. E. Boyce and R. C. Di Prima, *Elementary Differential Equations, and Boundary Value Problems* (Wiley, New York, 1977).

38 THE FUNDAMENTAL EQUATIONS

1.4. R. V. Churchhill, J. W. Brown, and R. F. Verhey, *Complex Variables and Applications* (McGraw-Hill, New York, 1976).
1.5. N. Coburn, *Vector and Tensor Analysis* (Macmillan, New York, 1955), Chapters 1-4.
1.6. G. E. Hay, *Vector and Tensor Analysis* (Dover Publications, New York, 1953), Chapters 1-5.
1.7. L. E. Kinsler and A. R. Frey, *Fundamentals of Acoustics* (Wiley, New York, 1962), Chapters 1, 5, 6, 7, 15.
1.8. See Courant and John, pp. 218-277.
1.9. H. Lass, *Pure and Applied Mathematics* (McGraw-Hill, New York, 1957), Chapters 2, 3.
1.10. See Coburn, pp. 73-75.
1.11. See Hay, pp. 149-151.
1.12. J. A. Stratton, *Electromagnetic Theory* (McGraw-Hill, New York, 1941), pp. 135-136.

2 MATHEMATICAL REVIEW

2.0. INTRODUCTION

This chapter is intended to be a comprehensive review of those aspects of the theory of differential equations that we will need to solve the boundary value problem associated with acoustic propagation in an oceanic waveguide. Since there is no single source that the reader can be referred to, the material in this chapter was collected from multiple sources, put into a form that suits our needs, and appears here as a convenience to the reader. The multiple sources from which this material was taken will be referenced as we go along.

2.1. SEPARATION OF VARIABLES

A powerful technique for solving partial differential equations is the method of separation of variables due to Bernoulli. By this method, the original partial differential equation in several independent variables is separated into a set of ordinary differential equations, each involving just one independent variable.

There are limitations on the applicability of the method of separation of variables. The following theorem proven in Sagan [2.1] clarifies the situation.

Theorem 2.1.1. Consider the most general form of a linear homogeneous partial differential equation of the second order in two independent variables.

$$a_{11}\frac{\partial^2 u}{\partial \xi^2} + 2a_{12}\frac{\partial^2 u}{\partial \xi \partial \eta} + a_{22}\frac{\partial^2 u}{\partial \eta^2} + a_{10}\frac{\partial u}{\partial \xi} + a_{01}\frac{\partial u}{\partial \eta} + a_{00}u = 0$$

where the a_{ik} are functions of ξ and η. This partial differential equation can be

39

reduced to two ordinary differential equations of second order, both containing an arbitrary parameter λ by Bernoulli's separation method if there exists a transformation

$$x = x(\xi, \eta)$$
$$y = y(\xi, \eta)$$

whose Jacobian does not vanish, so, the resulting equation

$$A_{11}\frac{\partial^2 u}{\partial x^2} + A_{22}\frac{\partial^2 u}{\partial y^2} + A_{10}\frac{\partial u}{\partial x} + A_{01}\frac{\partial u}{\partial y} + A_{00}u = 0$$

does not contain a term of the type $\partial^2 u/\partial x \partial y$ and, if there exists a function $B(x, y)$ so that $A_{11}/B = C_{11}$, $A_{10}/B = C_{10}$ are functions of x only; and if $A_{22}/B = C_{22}$, $A_{01}/B = C_{01}$ are functions of y only; and if A_{00}/B can be split up into a function of x and a function of y, as $A_{00}/B = C_{00}^{(1)}(x) + C_{00}^{(2)}(y)$.

In order to illustrate how the technique of separation of variables works, we will apply it to the wave equation given by Eq. (1.7.30). We will assume that the density ρ_0 is a constant and the source term $\partial Q_1/\partial t$ vanishes. Under these conditions, the wave equation becomes

$$\nabla^2 \mathscr{P} = \frac{1}{c^2}\frac{\partial^2 \mathscr{P}}{\partial t^2} \qquad (2.1.1)$$

In the remainder of this book we will denote the solution of the wave equation by the complex quantity $\mathscr{P}$, discussed in Eq. (1.8.19), instead of the quantity p_1.

As we saw in Theorem 2.1.1, there are limitations on the applicability of the method of separation of variables. First, there are only 11 coordinate systems in which the wave equation will separate. These 11 coordinate systems are discussed in Morse and Feshbach [2.2]. Consequently, separation of variables can only be applied to the wave equation if the boundary surface of the problem coincides with the coordinate surface of one of these 11 coordinate systems.

In addition, as stated in Theorem 2.1.1, the speed of sound must either be a function of a single coordinate or $1/c^2$ must be the sum of functions, each function depending on only one variable. For the case of a realistic ocean with a rough sea surface and a sound speed c, which depends on the two coordinates representing horizontal range and depth in the ocean, separation of variables cannot be used. In Chapter 7 we will discuss another technique for solving such problems.

For the present case, however, we will consider only the case of a horizontally stratified medium. That is, we will assume the speed of sound to be a

SEPARATION OF VARIABLES

function of a single variable only. We take a Cartesian orthogonal coordinate system oriented so that the positive z-axis is vertical downward and the xy plane coincides with the flat sea surface. Consequently, the speed of sound will be a function of depth z only. The geometry is shown in Figure 2.1.1.

First, let us assume a solution of the form

$$\mathscr{P}(\mathbf{r}, t) = p(\mathbf{r})T(t) \qquad (2.1.2)$$

Here $p(\mathbf{r})$ is a function only of the spatial coordinates x, y, and z where the position vector $\mathbf{r} = x\hat{e}_{(1)} + y\hat{e}_{(2)} + z\hat{e}_{(3)}$, and $T(t)$ is a function only of the time t. We are also going to let $p(\mathbf{r})$ and $T(t)$ be complex numbers.

Now let us substitute the expression for $\mathscr{P}$ given by Eq. (2.1.2) back into Eq. (2.1.1) to obtain

$$T\nabla^2 p = \frac{1}{c^2(z)} p \frac{d^2 T}{dt^2}$$

or

$$\frac{c^2(z)}{p} \nabla^2 p = \frac{1}{T} \frac{d^2 T}{dt^2} \qquad (2.1.3)$$

Now the left-hand side of Eq. (2.1.3) is a function of $\mathbf{r}$ only and the right-hand side is a function of t only. This can only be true if both sides are equal to the same constant, say $-\omega^2$. Thus

$$\frac{c^2(z)}{p} \nabla^2 p = \frac{1}{T} \frac{d^2 T}{dt^2} = -\omega^2$$

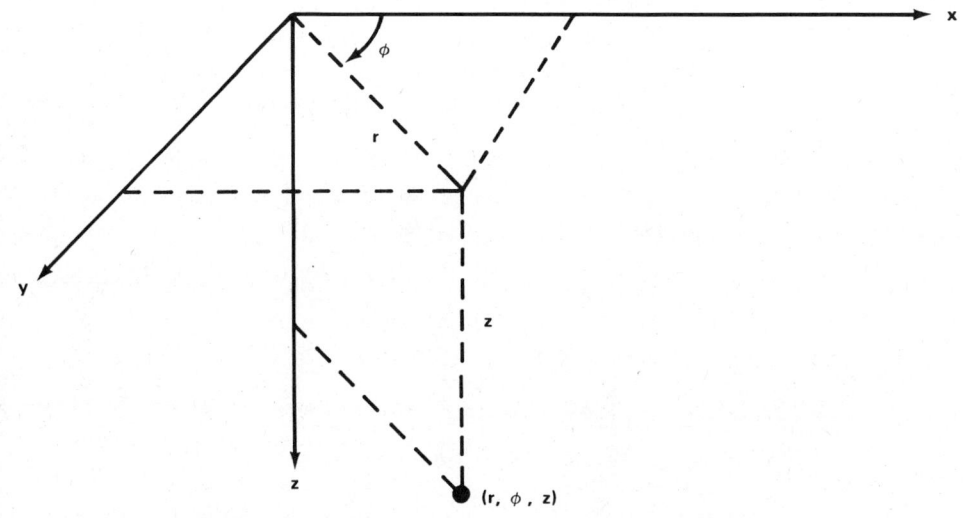

FIGURE 2.1.1. Coordinate geometry used to describe an acoustic waveguide.

or

$$\nabla^2 p + \frac{\omega^2}{c^2(z)} p = 0 \qquad (2.1.4)$$

$$\frac{d^2 T}{dt^2} + \omega^2 T = 0 \qquad (2.1.5)$$

The arbitrary constant ω is called a separation constant. The minus sign was inserted for later convenience.

Eq. (2.1.5) is a well-known differential equation that has as its two independent solutions

$$T = A e^{\pm i\omega t}$$

It is simply a matter of convenience which of the two solutions is chosen. In this book we will choose

$$T = A e^{-i\omega t} \qquad (2.1.6)$$

Thus the solution of Eq. (2.1.1), which is represented by Eq. (2.1.2), becomes

$$\mathscr{P}(\mathbf{r}, t) = p(\mathbf{r}) e^{-i\omega t} \qquad (2.1.7)$$

where we have absorbed the arbitrary constant A into $p(\mathbf{r})$. The constant ω has a simple interpretation from Eq. (2.1.7). The function $\mathscr{P}(\mathbf{r}, t)$ executes simple harmonic motion in time with a frequency $\omega/2\pi$.

We now need to solve Eq. (2.1.4). This equation is called the Helmholtz equation. We repeat the reasoning that led from the wave equation given by Eq. (2.1.1), to Eqs. (2.1.4) and (2.1.5). In order to separate the Helmholtz equation, we need to specify the coordinate systems. It is shown in Morse and Feshbach [2.2] that the Helmholtz equation, is separable into both Cartesian and cylindrical coordinates.

Let c_0 be a constant reference sound speed. This can be chosen to be any convenient sound speed. We then define the index of refraction n and a reference wave number k_0 by

$$n(z) = \frac{c_0}{c(z)} \qquad (2.1.8)$$

$$k_0 = \frac{\omega}{c_0} \qquad (2.1.9)$$

Then Eq. (2.1.4) can be written in the form

$$\nabla^2 p + k_0^2 n^2 p = 0 \qquad (2.1.10)$$

SEPARATION OF VARIABLES

In Cartesian coordinates x, y, and z, Eq. (2.1.10) takes the form

$$\frac{\partial^2 p}{\partial x^2} + \frac{\partial^2 p}{\partial y^2} + \frac{\partial^2 p}{\partial z^2} + k_0^2 n^2(z) p = 0 \qquad (2.1.11)$$

However, for oceanic propagation it is convenient to work in cylindrical coordinates r, ϕ, and z, defined by the transformation

$$x = r \cos \phi$$
$$y = r \sin \phi$$
$$z = z \qquad (2.1.12)$$

The relationship between the Cartesian orthogonal coordinate system and the cylindrical coordinate system is shown in Figure 2.1.1. Except for the material in Sections 7.5 and 7.6, we will assume that the acoustic pressure $p(\mathbf{r})$ is independent of the angle ϕ; that is, we will assume the ocean is cylindrically symmetric. With this assumption the Helmholtz equation, Eq. (2.1.11), becomes in polar coordinates r and z

$$\frac{\partial^2 p}{\partial r^2} + \frac{1}{r}\frac{\partial p}{\partial r} + \frac{\partial^2 p}{\partial z^2} + k_0^2 n^2(z) p = 0 \qquad (2.1.13)$$

We assume a solution of the form

$$p = R(r)\psi(z) \qquad (2.1.14)$$

and substitute this back into Eq. (2.1.13) to get

$$\psi R'' + \frac{1}{r}\psi R' + R\psi'' + k_0^2 n^2 R\psi = 0$$

where a prime denotes differentiation with respect to an argument. Dividing through this equation by $R\psi$ and rearranging gives

$$\frac{R''}{R} + \frac{R'}{rR} = -\frac{\psi''}{\psi} - k_0^2 n^2 \qquad (2.1.15)$$

Now the left side of this equation is a function of r only, and the right side is a function of z only. This can be true only if both sides are equal to the same constant, say $(-k_0^2 \xi^2)$, that is,

$$\frac{R''}{R} + \frac{R'}{rR} = -\frac{\psi''}{\psi} - k_0^2 n^2 = -k_0^2 \xi^2 \qquad (2.1.16)$$

The quality ξ, called a separation constant, is an arbitrary quantity at this point. The separation constant is written in the form $(-k_0^2\xi^2)$ instead of a single arbitrary number only for later convenience. Eq. (2.1.16) is equivalent to the two ordinary differential equations:

$$\frac{d^2R}{dr^2} + \frac{1}{r}\frac{dR}{dr} + k_0^2\xi^2 R = 0 \qquad (2.1.17)$$

$$\frac{d^2\psi}{dz^2} + k_0^2(n^2 - \xi^2)\psi = 0 \qquad (2.1.18)$$

Thus, instead of having to solve a partial differential equation [Eq. (2.1.13)], we have the simpler problem of solving two ordinary differential equations, Eqs. (2.1.17) and (2.1.18).

In the next section, we will study properties of the solutions of these separated equations.

2.2. STURM-LIOUVILLE THEORY

The ordinary differential equations arising from the separation of the time-independent wave equation (the Helmholtz equation),

$$\nabla^2 p + k^2 p = 0$$

can be written in the form

$$-\frac{d}{dz}\left[p(z)\frac{d\psi}{dz}\right] + [q(z) - \lambda w(z)]\psi = 0 \qquad (2.2.1)$$

This equation is called the Sturm-Liouville equation. The constant parameter λ is the separation constant. Each of the functions $p(z)$, $q(z)$, and $w(z)$ is characteristic of the coordinates used in the separation. We will assume that the real functions $p(z)$, $q(z)$, and $w(z)$ are continuous, $p(z)$ differentiable, and $p(z) > 0$, $w(z) > 0$ on the finite interval $a \le z \le b$ (also denoted $[a, b]$). The function $p(z)$ that appears in Eq. (2.2.1) is of course not the pressure p that appears in the wave equation. There should be no confusion.

The problem we wish to consider is the problem of determining the dependence of the general behavior of ψ on the parameter λ and the dependence of the values of λ on the boundary conditions imposed on ψ. We shall take the boundary conditions to be of the form

$$A_1 \frac{d\psi(a)}{dz} + B_1 \psi(a) = 0 \qquad (2.2.2)$$

$$A_2 \frac{d\psi(b)}{dz} + B_2 \psi(b) = 0 \qquad (2.2.3)$$

The differential equation (2.2.1), along with boundary conditions (2.2.2) and (2.2.3) and the conditions on $p(z)$, $q(z)$, and $w(z)$, will be called the Sturm-Liouville problem.

Given two differentiable functions ψ_1 and ψ_2 on some open interval $a < z < b$, the function $(\psi_1\psi_2' - \psi_1'\psi_2)$ is referred to as the Wronskian of ψ_1 and ψ_2 and is usually written as

$$W(\psi_1, \psi_2) = \begin{vmatrix} \psi_1 & \psi_2 \\ \psi_1' & \psi_2' \end{vmatrix} = \psi_1\psi_2' - \psi_2\psi_1' \qquad (2.2.4)$$

Here $\psi' = d\psi(z)/dz$.

We can then state one of the fundamental theorems for the solution of differential equations.

Theorem 2.2.1. If the functions $P(z)$ and $Q(z)$ are continuous on the open interval $a < z < b$, and if ψ_1 and ψ_2 are solutions of the differential equation

$$\psi'' + P(z)\psi' + Q(z)\psi = 0$$

on the interval $a < z < b$, and if there is at least one point in $a < z < b$ where $W(\psi_1, \psi_2)$ is not zero, then every solution $\psi = f(z)$ of the differential equation can be expressed in the form

$$f(z) = C_1\psi_1(z) + C_2\psi_2(z)$$

If the Wronskian does not vanish for two solutions ψ_1 and ψ_2, they are said to be linearly independent.

Suppose ψ_1 and ψ_2 are two linearly independent solutions of Eq. (2.2.1). Then the general solution of Eq. (2.2.1) can be written in the form

$$\psi(z) = C_1\psi_1(z, \lambda) + C_2\psi_2(z, \lambda) \qquad (2.2.5)$$

where we have indicated the explicit dependence on the parameter λ. Substituting for ψ in the boundary conditions Eqs. (2.2.2) and (2.2.3) yields the equations

$$C_1[A_1\psi_1'(a, \lambda) + B_1\psi_1(a, \lambda)] + C_2[A_1\psi_2'(a, \lambda) + B_1\psi_2(a, \lambda)] = 0$$
$$C_1[A_2\psi_1'(b, \lambda) + B_2\psi_1(b, \lambda)] + C_2[A_2\psi_2'(b, \lambda) + B_2\psi_2(b, \lambda)] = 0$$
$$(2.2.6)$$

This set of linear homogeneous equations has nontrivial solutions if and only if the determinant of coefficients vanishes, that is, if and only if

$$\begin{vmatrix} A_1\psi_1'(a, \lambda) + B_1\psi_1(a, \lambda) & A_1\psi_2'(a, \lambda) + B_1\psi_2(a, \lambda) \\ A_2\psi_1'(b, \lambda) + B_2\psi_1(b, \lambda) & A_2\psi_2'(b, \lambda) + B_2\psi_2(b, \lambda) \end{vmatrix} = 0 \qquad (2.2.7)$$

46 MATHEMATICAL REVIEW

This is called the characteristic equation. Values of λ, if any, satisfying this determinantal equation are the eigenvalues of the boundary value problem. Corresponding to each eigenvalue there is at least one nontrivial solution, called an eigenfunction, that is determined at most to within an arbitrary multiplicative constant. We can label the eigenvalues λ_n and corresponding eigenfunction ψ_n. The totality of eigenvalues is called the spectrum of eigenvalues.

Next let us derive an expression for the Wronskian. Let ψ_1 and ψ_2 be two independent solutions of Eq. (2.2.1) corresponding to the same value of λ.

$$\frac{d}{dz}(p\psi_1') - (q - \lambda w)\psi_1 = 0$$

$$\frac{d}{dz}(p\psi_2') - (q - \lambda w)\psi_2 = 0$$

Multiply the first equation by ψ_2, the second by ψ_1, and subtract.

$$\psi_2 \frac{d}{dz}(p\psi_1') - \psi_2(q - \lambda w)\psi_1 - \psi_1 \frac{d}{dz}(p\psi_2') + \psi_1(q - \lambda w)\psi_2 = 0$$

$$\psi_2 \frac{d}{dz}(p\psi_i') - \psi_1 \frac{d}{dz}(p\psi_2') = 0$$

or

$$\frac{d}{dz}\left[p\psi_1'\psi_2 - p\psi_1\psi_2'\right] = 0$$

or

$$\frac{d}{dz} pW(\psi_1, \psi_2) = 0$$

$$W(\psi_1, \psi_2) = \frac{\text{Constant}}{p} \tag{2.2.8}$$

If p is a constant, the Wronskian is independent of z.

Next let us show that the set of eigenfunctions derived from the Sturm-Liouville problem is orthogonal. Let $\psi_n(z)$ and $\psi_m(z)$ be eigenfunctions of the Sturm-Liouville problem belonging to eigenvalues λ_n and λ_m, respectively. Then these eigenfunctions satisfy the Sturm equation:

$$-\frac{d}{dz}(p\psi_n') + (q - \lambda_n w)\psi_n = 0 \tag{2.2.9}$$

$$-\frac{d}{dz}(p\psi_m') + (q - \lambda_m w)\psi_m = 0 \tag{2.2.10}$$

STURM-LIOUVILLE THEORY

Now multiply Eq. (2.2.9) by ψ_m and Eq. (2.2.10) by ψ_n and subtract

$$-\psi_m \frac{d}{dz}(p\psi_n') + \psi_m q \psi_n - \psi_m \lambda_n w \psi_n$$

$$+ \psi_n \frac{d}{dz}(p\psi_m') - \psi_n q \psi_m + \psi_n \lambda_m w \psi_m = 0$$

Rearranging

$$(\lambda_n - \lambda_m) w \psi_n \psi_m = \psi_n \frac{d}{dz}(p\psi_m') - \psi_m \frac{d}{dz}(p\psi_n')$$

$$= \frac{d}{dz}(p\psi_n \psi_m' - p\psi_m \psi_n') \qquad (2.2.11)$$

Now let us integrate Eq. (2.2.11) over the interval $a \leq z \leq b$.

$$(\lambda_n - \lambda_m) \int_a^b w \psi_n \psi_m \, dz = \left[p\psi_n \psi_m' - p\psi_m \psi_n' \right]_a^b \qquad (2.2.12)$$

Let us apply the boundary condition at $z = b$ given by Eq. (2.2.3) to ψ_m and ψ_n.

$$A_2 \psi_m'(b) + B_2 \psi_m(b) = 0$$

$$A_2 \psi_n'(b) + B_2 \psi_n(b) = 0$$

or

$$\psi_m'(b) = -\frac{B_2}{A_2} \psi_m(b)$$

$$\psi_n'(b) = -\frac{B_2}{A_2} \psi_n(b)$$

So

$$\psi_n(b)\psi_m'(b) - \psi_m(b)\psi_n'(b) = -\psi_n(b)\frac{B_2}{A_2}\psi_m(b) + \psi_m(b)\frac{B_2}{A_2}\psi_n(b) = 0$$

Thus the integrated term on the right side of Eq. (2.2.12) vanishes for $z = b$. It similarly vanishes at the lower limit of integration for $z = a$. Hence, if $\lambda_n \neq \lambda_m$, Eq. (2.2.12) becomes

$$(\lambda_n - \lambda_m) \int_a^b w(z) \psi_n(z) \psi_m(z) \, dz = 0 \qquad (2.2.13)$$

Thus the eigenfunctions that satisfy the Sturm-Liouville problem are orthogonal. It is important to note that orthogonality depends on the boundary conditions.

We have noted in the solution of Eqs. (2.2.6) that the eigenfunctions are determined only to within an arbitrary constant. We can use this freedom to normalize the eigenfunctions. That is, we will always choose this arbitrary constant so that the eigenfunctions satisfy the normalization condition

$$\int_a^b w(z)\psi_n^2(z)\,dz = 1 \qquad (2.2.14)$$

Eqs. (2.2.13) and (2.2.14) can be combined into a single equation with the Kronecker delta:

$$\int_a^b w(z)\psi_n(z)\psi_m(z)\,dz = \delta_{nm} \qquad (2.2.15)$$

A set of functions that satisfy Eq. (2.2.15) is called orthonormal.

Next let us state the fundamental theorem on the Sturm-Liouville problem. We will prove parts of it, but a rigorous proof of the entire theorem would take us too far afield. The theorem is proved in the works of Ince [2.3], Sagan [2.4], and Courant and Hilbert [2.5].

Theorem 2.2.2. Consider the Sturm-Liouville system

$$\frac{d}{dz}\left[p\frac{d\psi}{dz}\right] - (q - \lambda w)\psi = 0$$

$$\alpha'\psi(a) - \alpha\psi'(a) = 0$$

$$\beta'\psi(b) + \beta\psi(b) = 0$$

where it will be assumed that p, q, and w are real continuous functions with p differentiable when $a \leq z \leq b$; are independent of λ; and are such that $p > 0$, $w > 0$. The coefficients α, α', β, β' are also independent of λ. Then there exists an infinite set of real numbers $\lambda_1, \lambda_2, \ldots$ that have no limit point except $\lambda = +\infty$. If the corresponding eigenfunctions are $\psi_1, \psi_2, \ldots$, then ψ_m has exactly $m - 1$ zeros in the interval $a < z < b$.

For the conditions cited in the theorem, the eigenvalues form a discrete set of numbers. We say that the boundary value problem has a discrete spectrum.

Let us show that the eigenvalues of the Sturm-Liouville problem are real. Let λ be a complex eigenvalue and ϕ the corresponding eigenfunction. Since λ is complex, ϕ must be a complex function, so we can write $\phi = u(z) + iv(z)$ where $u(z)$ and $v(z)$ are real functions of the real variable z. If we take the complex conjugate of Eq. (2.2.1), remembering that p, q, and w are real functions, we find that the complex conjugate of λ, say λ^*, is an eigenvalue and ϕ^* is the corresponding eigenfunction. Thus we have

$$-\frac{d}{dz}\left(p\frac{d\phi}{dz}\right) + (q - \lambda w)\phi = 0$$

$$-\frac{d}{dz}\left(p\frac{d\phi^*}{dz}\right) + (q - \lambda^* w)\phi^* = 0$$

Multiplying the first equation by ϕ^* and the second by ϕ, subtracting, integrating over z from $z = a$ to $z = b$, and applying the boundary conditions gives

$$(\lambda - \lambda^*)\int_a^b w\phi\phi^*\,dz = 0$$

or

$$(\lambda - \lambda^*)\int_a^b w(u^2 + v^2)\,dz = 0$$

Since $\lambda \neq \lambda^*$, we have

$$\int_a^b w(u^2 + v^2)\,dz = 0$$

Since we have assumed $w > 0$ on the interval $a \leq z \leq b$, the integral is positive and therefore cannot vanish. This contradicts the assumption that λ can be complex. Therefore λ must be real.

It is important to remember that all the results we have obtained apply only to a finite interval $[a, b]$, to the particular boundary conditions expressed by Eqs. (2.2.2) and (2.2.3), and to the particular condition imposed on p, q, and w.

2.3. CALCULUS OF VARIATIONS

In this section we present the fundamental ideas of the calculus of variations that we will need in this book. Since there is no single source that the reader can be referred to, the material was collected from multiple sources, put into a form that suits our needs, and appears here as a convenience to the reader.

We need only consider problems in one-dimensional form. The basic problem is that we wish to find a path $y = y(\tau)$ between two values a and b, such that the line integral

$$J = \int_a^b f(y, \dot{y}, \tau)\,d\tau \qquad (2.3.1)$$

of some function $f(y, \dot{y}, \tau)$, where $\dot{y} = dy/d\tau$ is an extremum (local maximum, local minimum, or saddle point). We will consider only such varied paths for which $y(a) = y_1$ and $y(b) = y_2$. We will put the problem in the form that enables us to use the familiar apparatus of the differential calculus for obtaining an extremum. So suppose that $y(\tau)$ has continuous second derivatives, that $f(y, \dot{y}, \tau)$ is continuous, and that the first and second partial derivatives of f exist and are continuous.

We can label all possible curves $y(\tau)$ under examination with different values of a parameter α, such that for some value of α, say $\alpha = 0$, the curve would coincide with the path giving an extremum for the integral. The quantity y would then be a function of both τ and α. For example, we can represent y by

$$y(\tau, \alpha) = y(\tau, 0), + \alpha \eta(\tau) \tag{2.3.2}$$

where $\eta(\tau)$ is any function of τ that vanishes at $\tau = a$ and $\tau = b$. Then J in Eq. (2.3.1) is also a function of α:

$$J(\alpha) = \int_a^b f[y(\tau, \alpha), \dot{y}(\tau, \alpha), \tau] \, d\tau \tag{2.3.3}$$

and the condition for obtaining an extremum is the familiar one that

$$\left(\frac{\partial J}{\partial \alpha} \right)_{\alpha = 0} = 0 \tag{2.3.4}$$

By the usual method of differentiating under the integral sign one finds that

$$\frac{\partial J}{\partial \alpha} = \int_a^b \left\{ \frac{\partial f}{\partial y} \frac{\partial y}{\partial \alpha} + \frac{\partial f}{\partial \dot{y}} \frac{\partial \dot{y}}{\partial \alpha} \right\} d\tau \tag{2.3.5}$$

Consider the second of these integrals:

$$\int_a^b \frac{\partial f}{\partial \dot{y}} \frac{\partial \dot{y}}{\partial \alpha} d\tau = \int_a^b \frac{\partial f}{\partial \dot{y}} \frac{\partial^2 y}{\partial \tau \partial \alpha} d\tau$$

Integrating by parts, the integral becomes

$$\int_a^b \frac{\partial f}{\partial \dot{y}} \frac{\partial^2 y}{\partial \tau \partial \alpha} d\tau = \frac{\partial f}{\partial \dot{y}} \frac{\partial y}{\partial \alpha} \bigg|_a^b - \int_a^b \frac{d}{d\tau} \left(\frac{\partial f}{\partial \dot{y}} \right) \frac{\partial y}{\partial \alpha} d\tau \tag{2.3.6}$$

The conditions on all the varied curves are that they pass through points $(a, y_1), (b, y_2)$ and hence $\partial y / \partial \alpha$ must vanish at a and b. Therefore, the first term of Eq. (2.3.6) vanishes and Eq. (2.3.5) reduces to

$$\frac{\partial J}{\partial \alpha} = \int_a^b \left\{ \frac{\partial f}{\partial y} - \frac{d}{d\tau} \left(\frac{\partial f}{\partial \dot{y}} \right) \right\} \frac{\partial y}{\partial \alpha} d\tau$$

To obtain the extremum condition, we multiply through by a differential $d\alpha$ and evaluate the derivative at $\alpha = 0$, resulting in

$$\left(\frac{\partial J}{\partial \alpha}\right)_0 d\alpha = \int_a^b \left\{\frac{\partial f}{\partial y} - \frac{d}{d\tau}\left(\frac{\partial f}{\partial \dot{y}}\right)\right\}\left(\frac{\partial y}{\partial \alpha}\right)_0 d\alpha\, d\tau \qquad (2.3.7)$$

We shall call

$$\left(\frac{\partial J}{\partial \alpha}\right)_0 d\alpha = \delta J \qquad (2.3.8)$$

the variation of J. Similarly,

$$\left(\frac{\partial y}{\partial \alpha}\right)_0 d\alpha = \delta y \qquad (2.3.9)$$

Here, δy represents some arbitrary variation of $y(\tau)$, obtained by variation of the arbitrary parameter α about its zero value. Thus we can write Eq. (2.3.7) in the form

$$\delta J = \int_a^b \left\{\frac{\partial f}{\partial y} - \frac{d}{d\tau}\frac{\partial f}{\partial \dot{y}}\right\} \delta y\, d\tau \qquad (2.3.10)$$

or, using Eqs. (2.3.2) and (2.3.9),

$$\frac{\partial J}{\partial \alpha} = \int_a^b \left\{\frac{\partial f}{\partial y} - \frac{d}{d\tau}\frac{\partial f}{\partial \dot{y}}\right\} \eta(\tau)\, d\tau \qquad (2.3.11)$$

Now Eq. (2.3.11) holds for every $\eta(\tau)$, and it is therefore reasonable to suspect that the quantity in the braces vanishes identically. We have in fact the following lemma, which is proved in Wilf [2.6].

Lemma 2.3.1. Let $h(\tau)$ be a continuous function, and suppose that

$$\int_a^b h(\tau)\eta(\tau)\, d\tau = 0$$

for every function $\eta(\tau)$ that has a continuous derivative and that vanishes at a and b. Then $h(\tau) = 0$ on (a, b).

We have therefore proved the following theorem.

Theorem 2.3.1. If $y(\tau)$ has continuous second derivatives and gives to the integral

$$J = \int_a^b f(y, \dot{y}, \tau)\, d\tau$$

an extreme value (local maximum, local minimum, or saddle point) in the class of all such functions that assume given values at $\tau = a$, $\tau = b$, and if in addition the function $f(y, \dot{y}, \tau)$ has continuous second partial derivatives, then we have

$$\frac{d}{d\tau}\left(\frac{\partial f}{\partial \dot{y}}\right) - \frac{\partial f}{\partial y} = 0 \qquad (2.3.12)$$

for $a \leq \tau \leq b$.

Therefore J is an extremum only for curves of $y(\tau)$ such that f satisfies the differential equation, Eq. (2.3.12). Eq. (2.3.12) is called the Euler-Lagrange equation.

In many cases the Lagrange integral is to be minimized subject to additional requirements further restricting the independent variables. In some cases we can use the method of Lagrange multipliers. Let us show how they work by using an example given in Morse and Feshbach [2.7].

Suppose a function $f(y, z)$ is to be maximized (or minimized). If there are no auxiliary conditions, we solve the two equations

$$\frac{\partial f}{\partial z} = 0; \quad \frac{\partial f}{\partial y} = 0 \qquad (2.3.13)$$

simultaneously. The resulting pair of values of y and z (z_0, y_0) specify the point at which f has a maximum, minimum, or saddle point, and the value $f(y_0, z_0)$ is the value of f at this maximum or minimum. A typical case is pictured in Figure 2.3.1 in which f is depicted in terms of contour lines. But suppose we wish to find the maximum of $f(y, z)$ along the line given by the auxiliary equation $y = y_a(z)$. This line usually does not run through point (z_0, y_0) so the solution cannot be the same. However, there may be one or more points along the line where $f(y, z)$ has an extremum value, such as point (z_1, y_1) shown in Figure 2.3.1. These points may be computed by inserting the expression for y in terms of z into the form for f, which gives the value of f along the line as a function of z. We then differentiate with respect to z to find the extremum:

$$\frac{d}{dz}f[y_a(z), z] = \frac{\partial f}{\partial z} + \frac{\partial f}{\partial y}\frac{d}{dz}[y_a(z)] = 0 \qquad (2.3.14)$$

The position of the extremum is then the solution z_1 of the equation and the related value $y_1 = y_a(z_1)$. However, we can solve the same problem by a method which at first appears to be different from and more complicated than that above, but in more complex cases turns out to be the easier of the two methods. Suppose the auxiliary equation is

$$g(y, z) = 0$$

CALCULUS OF VARIATIONS

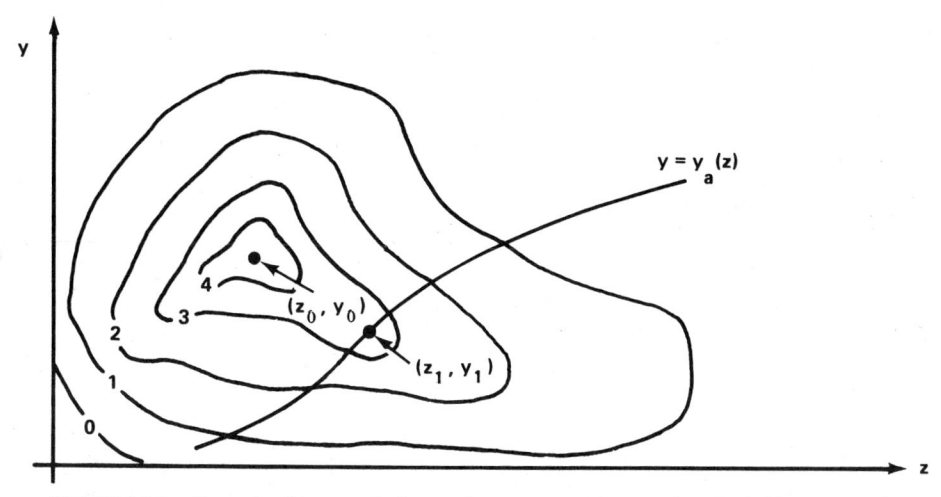

FIGURE 2.3.1. Example of how a relative maximum on a surface can be affected by a constraint.

We first introduce a third unknown λ and then minimize the new function $(f - \lambda g)$ subject to the relation $g = 0$. Note that since λ is completely arbitrary at this point, we could have chosen to work with the combination $(f + \lambda g)$. Both choices lead to the same result. To be consistent with our choice of signs later, we will choose the form $(f - \lambda g)$. We now must solve the three equations

$$\frac{\partial f}{\partial z} - \lambda \frac{\partial g}{\partial z} = 0 \qquad (2.3.15)$$

$$\frac{\partial f}{\partial y} - \lambda \frac{\partial g}{\partial y} = 0 \qquad (2.3.16)$$

$$g = 0 \qquad (2.3.17)$$

simultaneously to determine the proper values for z, y, and λ. It is not immediately apparent that the solution of Eqs. (2.3.15) through (2.3.17) is identical with the solution of Eq. (2.3.14), but the connection becomes clearer when we write the auxiliary equation $g(y, z) = 0$ in the form used above, $y_a(z) - y = 0$. Then Eq. (2.3.17) becomes

$$g = y_a(z) - y = 0$$

and Eqs. (2.3.15) and (2.3.16) become

$$\frac{\partial f}{\partial z} - \lambda \frac{dy_a}{dz} = 0$$

$$\frac{\partial f}{\partial y} + \lambda = 0$$

54 MATHEMATICAL REVIEW

Substituting the second of these into the first gives us

$$\frac{\partial f}{\partial z} + \frac{\partial f}{\partial y}\frac{dy_a}{dz} = 0$$

which is just Eq. (2.3.14). This is true in the general case of more than two variables.

A second type of restriction occurs when the auxiliary condition is in the form of an integral condition. The problem is to find the extreme values of the integral

$$J = \int_a^b f(\tau, y, \dot{y}) \, d\tau \qquad (2.3.18)$$

subject to the condition

$$y(a) = y_1, \; y(b) = y_2$$

and

$$\int_a^b g(\tau, y, \dot{y}) \, d\tau = K \qquad (2.3.19)$$

where K is a given constant. We will follow the method given in Wilf [2.8].

Suppose we consider a one-parameter family of curves such as those given by Eq. (2.3.2):

$$y(\tau, \alpha) = y(\tau, 0) + \alpha \eta(\tau)$$

Suppose we have found a function, say $y(\tau, 0)$, that gives Eq. (2.3.18) a stationary value for the value of $\alpha = 0$. In general this function will not satisfy the auxiliary condition, Eq. (2.3.19). Thus we must consider a two-parameter family of curves. Let

$$y(\tau, \alpha_1, \alpha_2) = y(\tau, 0, 0) + \alpha_1 \eta_1(\tau) + \alpha_2 \eta_2(\tau) \qquad (2.3.20)$$

where $\eta_1(\tau)$ and $\eta_2(\tau)$ each vanish at the end points and the parameters α_1, α_2 are not independent, but are connected by the condition

$$\psi(\alpha_1, \alpha_2) = \int_a^b g(\tau, y + \alpha_1\eta_1 + \alpha_2\eta_2, \dot{y} + \alpha_1\dot{\eta}_1 + \alpha_2\dot{\eta}_2) \, d\tau = K$$

$$(2.3.21)$$

Our problem then is that the function

$$\phi(\alpha_1, \alpha_2) = \int_a^b f(\tau, y + \alpha_1\eta_1 + \alpha_2\eta_2, \dot{y} + \alpha_1\dot{\eta}_1 + \alpha_2\dot{\eta}_2) \, d\tau \qquad (2.3.22)$$

CALCULUS OF VARIATIONS   55

is to be stationary at $\alpha_1 = \alpha_2 = 0$ with respect to values of α_1, α_2, which satisfy Eq. (2.3.21).

The problem is now reduced to one of extremizing a function of two real variables with a side condition. We already know from the theory of Lagrange multiplier that for some number λ we have

$$\frac{\partial}{\partial \alpha_1}\{\phi(\alpha_1, \alpha_2) - \lambda \psi(\alpha_1, \alpha_2)\}_{\alpha_1=\alpha_2=0} = 0 \qquad (2.3.23)$$

$$\frac{\partial}{\partial \alpha_2}\{\phi(\alpha_1, \alpha_2) - \lambda \psi(\alpha_1, \alpha_2)\}_{\alpha_1=\alpha_2=0} = 0 \qquad (2.3.24)$$

Eq. (2.3.23) gives

$$\int_a^b \left\{ \frac{\partial f}{\partial y}\eta_1 + \frac{\partial f}{\partial \dot{y}}\dot{\eta}_1 - \lambda \frac{\partial g}{\partial y}\eta_1 - \lambda \frac{\partial g}{\partial \dot{y}}\dot{\eta}_1 \right\} d\tau = 0$$

and Eq. (2.3.24) gives

$$\int_a^b \left\{ \frac{\partial f}{\partial y}\eta_2 + \frac{\partial f}{\partial \dot{y}}\dot{\eta}_2 - \lambda \frac{\partial g}{\partial y}\eta_2 - \lambda \frac{\partial g}{\partial \dot{y}}\dot{\eta}_2 \right\} d\tau = 0$$

Integrating by parts, as before, yields

$$\int_a^b \left\{ \left[\frac{\partial f}{\partial y} - \frac{d}{d\tau}\left(\frac{\partial f}{\partial \dot{y}}\right) \right] - \lambda \left[\frac{\partial g}{\partial y} - \frac{d}{d\tau}\left(\frac{\partial g}{\partial \dot{y}}\right) \right] \right\} \eta_1 \, d\tau = 0 \qquad (2.3.25)$$

$$\int_a^b \left\{ \left[\frac{\partial f}{\partial y} - \frac{d}{d\tau}\left(\frac{\partial f}{\partial \dot{y}}\right) \right] - \lambda \left[\frac{\partial g}{\partial y} - \frac{d}{d\tau}\left(\frac{\partial g}{\partial \dot{y}}\right) \right] \right\} \eta_2 \, d\tau = 0 \qquad (2.3.26)$$

If $\partial g/\partial y \neq d/d\tau(\partial g/\partial \dot{y})$, then Eq. (2.3.25) states that λ is independent of η_2 and Eq. (2.3.26) shows that

$$\frac{\partial f}{\partial y} - \frac{d}{d\tau}\left(\frac{\partial f}{\partial \dot{y}}\right) - \lambda\left\{\frac{\partial g}{\partial y} - \frac{d}{d\tau}\left(\frac{\partial g}{\partial \dot{y}}\right)\right\} = 0$$

or

$$\frac{\partial}{\partial y}(f - \lambda g) - \frac{d}{d\tau}\left\{\frac{\partial}{\partial \dot{y}}(f - \lambda g)\right\} = 0 \qquad (2.3.27)$$

Eq. (2.3.27) is the Euler equation of the problem. Eq. (2.3.27), along with the auxiliary condition Eq. (2.3.19), determines y and the Lagrange multipler λ.

The fundamental problem of the calculus of variation is easily generalized to the case in which f is a function of N independent variables y_i and their

56 MATHEMATICAL REVIEW

derivatives $\dot{y}_i = dy_i/d\tau$. All the y_i are considered functions of the parameter τ. Then the variation of the integral J

$$\delta J = \delta \int_a^b f[y_1(\tau), y_2(\tau), \ldots, \dot{y}_1(\tau), \dot{y}_2(\tau), \ldots, \tau] \, d\tau \quad (2.3.28)$$

is obtained as before by considering J as a function of a parameter α, which labels all positive curves $y_i(\tau, \alpha)$. Thus we may introduce α by setting

$$y_1(\tau, \alpha) = y_1(\tau, 0) + \alpha \eta_1(\tau)$$
$$y_2(\tau, \alpha) = y_2(\tau, 0) + \alpha \eta_2(\tau)$$
$$\vdots \quad (2.3.29)$$

where $y_1(\tau, 0)$, $y_2(\tau, 0)$, and so on, are the solutions of the extremum problem and η_1, η_2, and so forth, are completely arbitrary functions of τ, except that

$$\eta_i(a) = \eta_i(b) = 0, \quad i = 1, \ldots, N$$

The calculation proceeds as before. The variation of J is

$$\frac{\partial J}{\partial \alpha} d\alpha = \int_a^b \left\{ \frac{\partial f}{\partial y_i} \frac{\partial y_i}{\partial \alpha} d\alpha + \frac{\partial f}{\partial \dot{y}_i} \frac{\partial \dot{y}_i}{\partial \alpha} d\alpha \right\} d\tau \quad (2.3.30)$$

where the repeated subscripts indicate a summation over i from $i = 1$ to N. Again we integrate by parts the integral involved in the second sum of Eq. (2.3.30):

$$\int_a^b \frac{\partial f}{\partial \dot{y}_i} \frac{\partial^2 y_i}{\partial \alpha \partial \tau} d\tau = \left. \frac{\partial f}{\partial \dot{y}_i} \frac{\partial y_i}{\partial \alpha} \right|_a^b - \int_a^b \frac{\partial y_i}{\partial \alpha} \frac{d}{d\tau}\left(\frac{\partial f}{\partial \dot{y}_i} \right) d\tau$$

where the first term vanishes because, by hypothesis, all curves pass through the fixed end points. Substituting this last result into Eq. (2.3.30), δJ becomes

$$\delta J = \int_a^b \left(\frac{\partial f}{\partial y_i} - \frac{d}{d\tau} \frac{\partial f}{\partial \dot{y}_i} \right) \delta y_i \, d\tau \quad (2.3.31)$$

where, in analogy with Eq. (2.3.9), the variation δy_i is

$$\delta y_i = \frac{\partial y_i}{\partial \alpha} d\alpha$$

Since the y variables are independent, the variations δy_i are independent. Hence $\delta J = 0$ if and only if the coefficients of the δy_i vanish separately

$$\frac{\partial f}{\partial y_i} - \frac{d}{d\tau}\left(\frac{\partial f}{\partial \dot{y}_i} \right) = 0, \quad i = 1, \ldots, N \quad (2.3.32)$$

which represent the appropriate generalization of Eq. (2.3.12) to several variables. The solutions of the differential equation (2.3.32) represent those curves for which the variation of an integral of the form given in Eq. (2.3.28) vanishes.

2.4. EIGENFUNCTIONS AND THE VARIATIONAL PRINCIPLE

Following Morse and Feshbach [2.9], we will show that the solution of the Sturm-Liouville equation (2.2.1) is the same as minimizing the integral

$$\Omega(\psi) = \int_a^b \left[p\left(\frac{d\psi}{dz}\right)^2 + q\psi^2 \right] dz \qquad (2.4.1)$$

subject to the auxiliary condition

$$\int_a^b w\psi^2 \, dz = 1 \qquad (2.4.2)$$

Comparing Eqs. (2.4.1) and (2.4.2) with Eqs. (2.3.18) and (2.3.19), we see that

$$J = \Omega$$

$$f = p\left(\frac{d\psi}{dz}\right)^2 + q\psi^2$$

$$\tau = z$$

$$y = \psi$$

$$\dot{y} = \frac{d\psi}{dz}$$

$$K = 1$$

$$g = w\psi^2$$

The Euler-Lagrange equation for this variational problem is given by Eq. (2.3.27), which becomes for the Sturm-Liouville problem

$$\frac{\partial}{\partial \psi}\left\{(p\psi'^2 + q\psi^2) - \lambda w\psi^2\right\} - \frac{d}{dz}\left\{\frac{\partial}{\partial \psi'}\left[(p\psi'^2 + q\psi^2) - \lambda w\psi^2\right]\right\} = 0$$

$$2q\psi - 2\lambda w\psi - \frac{d}{dz}(2p\psi') = 0$$

$$-\frac{d}{dz}(p\psi') + (q - \lambda w)\psi = 0$$

which is just Sturm's equation.

At this point we are going to specialize our boundary conditions even further than those given by Eqs. (2.2.2) and (2.2.3). First, it is easier to prove completeness of the set of eigenfunctions of the Sturm-Liouville problem for these restricted boundary conditions, and second, these restricted boundary conditions will be the actual boundary conditions we will use in our model of the ocean. So, for the remainder of our work, we will take the boundary conditions:

$$\psi(a) = 0 \qquad (2.4.3)$$

$$\frac{d\psi}{dz}(b) = 0 \qquad (2.4.4)$$

Now let us examine the variational statement of the Sturm-Liouville problem in somewhat greater detail. We require that the trial functions ψ be continuous and have a piecewise continuous first derivative in the closed interval $[a, b]$. A function $f(z)$ is said to be piecewise continuous in the interval $a \leq z \leq b$ if there are only finitely many discontinuities $z_1, z_2, \ldots, z_n$ of $f(z)$ in $a \leq z \leq b$ and if

$$\lim_{z \to z_i^-} f(z) = f(z_i^-), \quad \lim_{z \to z_i^+} f(z) = f(z_i^+)$$

exist for all $i = 1, \ldots, n$. The notation $z \to z_i^-$ and $z \to z_i^+$ means that z approaches z_i from the left and from the right, respectively. We further require that ψ satisfy the boundary conditions, Eqs. (2.4.3) and (2.4.4), and that it be normalized according to Eq. (2.4.2).

We then compute the integral $\Omega(\psi)$ given by Eq. (2.4.1) and vary the trial function in every possible way (subject, of course, to the above restrictions that we have imposed on it) until we find the ψ that results in the lowest value of Ω. The fact that the Sturm-Liouville equation corresponds to the variational equation ensures that the ψ for which Ω is the lowest possible is ψ_1, the lowest eigenfunction. Let us determine the corresponding value of $\Omega(\psi_1)$:

$$\Omega(\psi_1) = \int_a^b (p\psi_1'^2 + q\psi_1^2)\, dz$$

$$= \left[p\psi_1 \frac{d\psi_1}{dz} \right]_a^b + \int_a^b \left\{ -\psi_1 \frac{d}{dz}(p\psi_1') + q\psi_1^2 \right\} dz$$

where we have used integration by parts. Now, using the boundary conditions and Eqs. (2.4.3) and (2.4.4), the integrated term vanishes and, using Sturm's equation (2.2.1), we get

$$\Omega(\psi_1) = \int_a^b \psi_1 \lambda_1 w \psi_1 \, dz$$

Since we assume ψ_1 satisfied Eq. (2.4.2), we have that

$$\Omega(\psi_1) = \lambda_1 \tag{2.4.5}$$

is the lowest eigenvalue.

To find the next eigenfunction and eigenvalue, we minimize Ω for a normalized trial function ψ subject to the boundary conditions at a and b, which in addition, is orthogonal to ψ_1, found above. This minimum value of Ω is λ_2 and the corresponding trial function is ψ_2. To find ψ_n and λ_n we compute Ω for a trial function satisfying the boundary conditions at a and b, which is normalized and which is orthogonal $\psi_1, \ldots, \psi_{n-1}$, which have all previously been computed. The function that then makes Ω a minimum is ψ_n and Ω is equal to λ_n.

To prove completeness, it is convenient to take $\lambda_1 = 0$. There is no loss in generality in doing so. In the expression $(q - \lambda w)$ we can always subtract $\lambda_1 w$ from q and add $\lambda_1 w$ to $-\lambda w$, producing $[(q - \lambda_1 w) - (\lambda - \lambda_1)w]$. We now define a new q equal to $q - \lambda_1 w$ and a new λ equal to $\lambda - \lambda_1$. The lowest eigenvalue for the new equation will be zero.

2.5. COMPLETENESS OF A SET OF EIGENFUNCTIONS

The problem that we are faced with is that of solving the Helmholtz equation for quite general boundary conditions. We saw in Section 2.1 that in cylindrical coordinates (assuming cylindrical symmetry) the Helmholtz equation is

$$\frac{\partial^2 p}{\partial r^2} + \frac{1}{r}\frac{\partial p}{\partial r} + \frac{\partial^2 p}{\partial z^2} + k_0^2 n^2(z) p = 0$$

We also saw in Section 2.1 that the separated equations arising from the Helmholtz equation by separation of variables are

$$\frac{d^2 R}{dr^2} + \frac{1}{r}\frac{dR}{dr} + k_0^2 \xi^2 R = 0$$

$$\frac{d^2 \psi}{dz^2} + k_0^2(n^2 - \xi^2)\psi = 0$$

where the acoustic pressure p is related to the solution of the separated equation by

$$p(r, z) = R(r)\psi(z)$$

Now, if we impose boundary conditions of the type given by Eqs. (2.2.2) and (2.2.3), which we will in our model of the ocean, then the depth equation along with these boundary conditions are a Sturm-Liouville system. Thus, there is a set of eigenfunctions, say ψ_n, and corresponding eigenvalues ξ_n that are the solutions of this Sturm-Liouville problem. Since the range equation is a function of the already determined eigenvalues ξ_n, there will be a discrete set of solutions for it, say $R_n(r)$. An individual term of the separated solution $R_n \psi_n$ is called a mode. While a single mode is a solution of the Helmholtz equation because of the way it was chosen, it may not satisfy all the boundary conditions imposed on the acoustic pressure $p(r, z)$. Thus we need a more general solution than a single mode in order to satisfy the boundary conditions on the Helmholtz equation.

Let $p(r, z)$ be the exact solution of the Helmholtz equation that satisfies all the boundary conditions. Since the Helmholtz equation is a linear partial differential equation, a linear combination of solutions is also a solution. To obtain a more general solution than a single mode, we might try to represent the solution $p(r, z)$ as a finite sum of modes, say

$$p(r, z) = \sum_{n=1}^{N} R_n(r)\psi_n(z)$$

While the sum will certainly be a solution of the wave equation, it will not be able to satisfy the boundary conditions in general. The next generalization would be to try and represent the solution $p(r, z)$ by an infinite series of eigenfunctions, say

$$p(r, z) = \sum_{n=1}^{\infty} R_n(r)\psi_n(z)$$

This seemingly slight generalization raises many questions. First, we must define in what sense we want the eigenfunction $\psi_n(z)$ to represent the function $p(r, z)$, and then determine what class of functions $p(r, z)$ can be represented in this manner. There are further questions on the convergence, integration, and differentiation of an infinite series of functions.

Let us first take up the question of defining what we mean when we say that the set of functions $\psi_n(z)$ represents a given function $f(z)$. Suppose we are given the set of functions $\psi_1, \psi_2, \ldots, \psi_n$, which are continuous on the interval $a \le z \le b$, and satisfy the orthonormality condition, Eq. (2.2.13), namely,

$$\int_a^b w(z)\psi_n(z)\psi_m(z)\,dz = \delta_{nm} \tag{2.5.1}$$

We will follow the discussion in Boyce and DiPrima [2.10] to illustrate various possible methods of representing a function. Suppose that we wish to approximate a given function $f(z)$, defined on $a \le z \le b$, by a finite linear combination of $\psi_1, \ldots, \psi_n$. That is, if

$$S_n(z) = \sum_{i=1}^{n} C_i \psi_i(z) \tag{2.5.2}$$

we wish to choose the coefficients $C_1, \ldots, C_n$ so that the function S_n will best approximate f on $a \le z \le b$. The first problem that we must face in doing this is to state precisely what we mean by "best approximation to f on $a \le z \le b$." There are several reasonable meanings that can be attached to this phrase. Later we will generalize the method we finally choose to represent a function to the case of an infinite linear combination of eigenfunctions.

1. We can choose n points $z_1, \ldots, z_n$ in the interval $a \le z \le b$ and require that $S_n(z)$ have the same value as $f(z)$ at each of these points. The coefficients are found by solving the set of linear algebraic equations

$$\sum_{i=1}^{n} C_i \psi_i(z_j) = f(z_j) \tag{2.5.3}$$

This method has the advantage that it is easy to write down Eq. (2.5.3); one needs only to evaluate the functions involved at points $z_1, \ldots, z_n$. If these points are well chosen, and if n is fairly large, then presumably $S_n(z)$ will not only be equal to $f(z)$ at the chosen points but will be reasonably close to it at other points as well. However, this method has several deficiencies. One is that the coefficients C_i are found by solving a set of linear equations. This is normally done on a computer, and it may not be a simple job if n is very large. A more important disadvantage is that if one more base function ψ_{n+1} is added, then one more point is required, and *all* of the coefficients must be recomputed. Further, the C_i depends on the location of points $z_1, \ldots, z_n$ and it is not clear how best to select these points.

2. Alternately, we can consider the difference $|f(z) - S_n(z)|$ and try to make it as small as possible. The trouble here is that $|f(z) - S_n(z)|$ is a function of z as well as the coefficients $C_1, \ldots, C_n$, and it is not obvious what criterion should be used in selecting the C_i. The choice of C_i that makes $|f(z) - S_n(z)|$ small at one point may make it large at another. One way to proceed is to consider instead the least upper bound (l.u.b.) of $|f(z) - S_n(z)|$ for z in $a \le z \le b$, and then to choose $C_1, \ldots, C_n$ so as to make this quantity as small as possible. That is, if

$$E_n(C_1, \ldots, C_n) = \underset{a \le z \le b}{\text{l.u.b.}} |f(z) - S_n(z)|$$

then choose $C_1, \ldots, C_n$ so as to minimize E_n. This approach is intuitively appealing and often used in theoretical calculations. However, in practice, it is usually very hard, if not impossible, to write an explicit formula for E_n. Further, this procedure shares one of the disadvantages of method A; namely, on adding an additional term to S_n, one must recompute all of the preceding coefficients.

3. Another way to proceed is to consider

$$I_n(C_1, \ldots, C_n) = \int_a^b w |f(z) - S_n(z)| \, dz \tag{2.5.4}$$

62 MATHEMATICAL REVIEW

If $w = 1$, then I_n is the area between the graphs of $y = f(z)$ and $y = S_n(z)$. We can then determine the coefficients C_i so as to minimize I_n. To avoid the complications resulting from calculations with absolute values, it is more convenient to consider instead

$$R_n(C_1,\ldots,C_n) = \int_a^b w[f(z) - S_n(z)]^2 \, dz \tag{2.5.5}$$

as our measure of the quality of approximation of the linear combination $S_n(z)$ to $f(z)$. While R_n is clearly similar in some ways to I_n, it lacks the simple geometric interpretation of the latter. Nevertheless, it is much easier mathematically to deal with R_n than with I_n. The quantity R_n is called the mean square error of the approximation S_n to f. If $C_1,\ldots,C_n$ are chosen to minimize R_n, then S_n is said to best approximate f in the mean square sense.

In order to choose $C_1,\ldots,C_n$ so as to minimize R_n, we must satisfy the necessary conditions

$$\frac{\partial R_n}{\partial C_i} = 0, \quad i = 1,\ldots,n \tag{2.5.6}$$

We obtain from Eq. (2.5.5)

$$\frac{\partial R_n}{\partial C_i} = -2\int_a^b w[f(z) - S_n(z)]\psi_i(z) \, dz = 0 \tag{2.5.7}$$

Substituting for $S_n(z)$ from Eq. (2.5.2) and making use of Eq. (2.5.1), we obtain

$$\int_a^b w(f - S_n)\psi_i \, dz = \int_a^b w\left(f - \sum_j C_j\psi_j\right)\psi_i \, dz$$

$$= \int_a^b wf\psi_i \, dz - \int_a^b w\sum_j C_j\psi_j\psi_i \, dz$$

$$= \int_a^b wf\psi_i \, dz - C_i$$

So

$$C_i = \int_a^b wf\psi_i \, dz \tag{2.5.8}$$

The coefficients defined by Eq. (2.5.8) are called the Fourier coefficients of f with respect to the orthonormal set $\psi_1,\ldots,\psi_n$ and the weight function w. Since the condition Eq. (2.5.6) is only necessary and not sufficient for R_n to be a minimum, a separate argument is required to show that R_n is actually minimized if the C_i are chosen by Eq. (2.5.8). Eq. (2.5.8) is noteworthy in two other important respects. In the first place, it gives a formula for each C_i separately. This is due to the orthogonality of the set $\psi_1, \ldots, \psi_n$. Further, the formula for C_i is independent of n, the number of terms in S_n. The practical significance of this is as follows. Suppose that, in order to obtain a better approximation to f, we desire to use an approximation with more terms, say k terms, when $k > n$. It is then unnecessary to recompute the first n coefficients in S_n.

The most practical way to approximate a function then is in the mean square sense. We now must generalize this notion to an infinite set of eigenfunctions. First, suppose by some means we have already shown that a certain function $f(z)$ can be expanded in a series of eigenfunctions, that is, suppose we know

$$f(z) = \sum_{n=1}^{\infty} C_n \psi_n(z) \qquad (2.5.9)$$

Multiplying Eq. (2.5.9) by $w\psi_m$ and using the orthonormality condition Eq. (2.5.1), we see that the C_n's are determined by the condition

$$C_m = \int_a^b w f \psi_m \, dz \qquad (2.5.10)$$

But these are just the Fourier coefficients given by Eq. (2.5.8). We are now ready to define the concept of completeness for a set of eigenfunctions that satisfy the Sturm-Liouville problem.

Let $\mathscr{S}$ be a set of admissible functions and let $f \in \mathscr{S}$. Let $\psi_1(z), \ldots, \psi_n(z), \ldots$ be an orthonormal set of eigenfunctions which are solutions of the Sturm equation (2.2.1) and which satisfy the boundary condition, Eqs. (2.4.3) and (2.4.4). Then we say that the series

$$\sum_{n=1}^{\infty} C_n \psi_n(z)$$

where the C_n are the Fourier coefficients given by

$$C_n = \int_a^b w(z) f(z) \psi_n(z) \, dz \qquad (2.5.11)$$

converge in the mean to f if

$$\lim_{n \to \infty} \int_a^b w \left[f - \sum_{i=1}^n C_i \psi_i \right]^2 dz = 0 \qquad (2.5.12)$$

and we write

$$f(z) \doteq \sum_{n=1}^{\infty} C_n \psi_n(z) \qquad (2.5.13)$$

Again, let $\mathscr{S}$ be an admissible set of functions defined on the interval $a \le z \le b$. Let $\psi_1(z), \ldots, \psi_n(z), \ldots$ be a set of orthonormal eigenfunctions satisfying the Sturm-Liouville system, Eqs. (2.2.1), (2.4.3), and (2.4.4). Let

$$S_n(z) = \sum_{i=1}^{n} C_i \psi_i(z)$$

where C_i is given by Eq. (2.5.11). Then the orthonormal set $\psi_1, \ldots, \psi_n, \ldots$ is said to be complete for $\mathscr{S}$ if for every $f \in \mathscr{S}$, we have

$$\lim_{n \to \infty} \int_a^b w(f - S_n)^2 \, dz = 0$$

Completeness of the set of eigenfunctions of the Sturm-Liouville problem can be proven for a very general class of functions $\mathscr{S}$. We will not need such a general result. Therefore we will prove completeness under much more restrictive conditions. We will now prove the following theorem which can be found in Morse and Feshbach [2.11].

Let $\psi_1, \ldots, \psi_n, \ldots$ be an orthonormal set of eigenfunctions that satisfy the Sturm-Liouville equation (2.2.1) and the boundary conditions (2.4.3) and (2.4.4). Let $\mathscr{S}$ be a set of continuous functions with a piecewise continuous first derivative on the interval $[a, b]$. Further assume that every $f \in \mathscr{S}$ satisfies the boundary conditions (2.4.3) and (2.4.4). Then the set of eigenfunctions are complete on the set $\mathscr{S}$.

We can reword our conclusions of Section 2.4 as follows. Suppose that F is any function of the set $\mathscr{S}$ satisfying the boundary conditions and normalized in the interval $[a, b]$; then the integral $\Omega(F)$, defined in Eq. (2.4.1), is never smaller than λ_1. Since we can adjust q and λ so $\lambda_1 = 0$, we have that $\Omega(F) \ge 0$.

Likewise, if F_n is a function such that

$$\int_a^b w F_n^2 \, dz = 1, \qquad \int_a^b F_n \psi_m w \, dz = 0, \qquad m = 1, \ldots, n \qquad (2.5.14)$$

then the integral $\Omega(F_n)$ is never smaller than λ_{n+1}. The series

$$\sum_{m=1}^{\infty} C_m \psi_m(z); \quad C_m = \int_a^b f(z) \psi_m(z) w(z) \, dz \qquad (2.5.15)$$

is supposed to equal (in the mean square sense) the function $f(z) \in \mathscr{S}$ over the range $a \le z \le b$. If it is to do so, the difference between the function f and the first n terms of the series

$$f_n(z) = f(z) - \sum_{m=1}^{n} C_m \psi_m(z) \qquad (2.5.16)$$

COMPLETENESS OF A SET OF EIGENFUNCTIONS

must approach zero (in the mean square sense) as $n \to \infty$. In other words, if the series is to be a good least squares fit, we must show, according to Eq. (2.5.12) and using Eq. (2.5.16), that

$$\lim_{n \to \infty} A_n^2 \equiv \lim_{n \to \infty} \int_a^b f_n^2(z) w(z) \, dz = 0 \qquad (2.5.17)$$

But

$$A_n^2 = \int_a^b w \left(f - \sum_{m=1}^n C_m \psi_m \right)^2 dz$$

$$= \int_a^b w \left(f^2 - 2f \sum_{m=1}^n C_m \psi_m + \sum_{m=1}^n \sum_{l=1}^n C_m \psi_m C_l \psi_l \right) dz$$

$$= \int_a^b w f^2 \, dz - 2 \sum_{m=1}^n C_m \int_a^b w f \psi_m \, dz + \sum_{l=1}^n \sum_{m=1}^n C_m C_l \int \psi_m \psi_l w \, dz$$

$$= \int_a^b w f^2 \, dz - \sum_{m=1}^n C_m^2 \qquad (2.5.18)$$

where use was made of Eqs. (2.5.1) and (2.5.8).

We can now apply our variational argument to obtain a measure of the size of A_n, for the function

$$F_n = \frac{f_n(z)}{A_n}$$

has the properties specified above for F_n.

$$\int_a^b F_n^2 w \, dz = \int_a^b \frac{f_n^2}{A_n^2} w \, dz = 1$$

by the definition of A_n, given by Eq. (2.5.17):

$$\int_a^b F_n \psi_m w \, dz = \int_a^b \frac{f_n}{A_n} \psi_m w \, dz$$

$$= \frac{1}{A_n} \int_a^b \psi_m w \left(f - \sum_{l=1}^n C_l \psi_l \right) dz$$

$$= \left\{ \frac{1}{A_n} \int_a^b f w \psi_m \, dz - \sum_{l=1}^n C_l \int w \psi_m \psi_l \, dz \right\}$$

$$= \frac{1}{A_n} \left(C_m - \sum_{l=1}^n C_l \delta_{lm} \right)$$

$$= \frac{1}{A_n} (C_m - C_m) = 0, \quad \text{for} \quad m \le n$$

66 MATHEMATICAL REVIEW

Therefore, according to Section 2.4,

$$\Omega(F_n) = \int_a^b \left[p\left(\frac{dF_n}{dz}\right)^2 + qF_n^2 \right] dz$$

$$= \frac{1}{A_n^2} \int_a^b \left[p\left(\frac{df_n}{dz}\right)^2 + qf_n^2 \right] dz$$

$$= \frac{1}{A_n^2} \int_a^b \left\{ p\left[\frac{df}{dz} - \sum_{m=1}^n C_m \frac{d\psi_m}{dz}\right]^2 + q\left[f - \sum_{m=1}^n C_m \psi_m\right]^2 \right\} dz$$

$$= \frac{1}{A_n^2} \int_a^b \left[p\left(\frac{df}{dz}\right)^2 + qf^2 \, dz \right] - \frac{2}{A_n^2} \int_a^b \sum_{m=1}^n C_m \left[p\frac{df}{dz}\frac{d\psi_m}{dz} + qf\psi_m \right] dz$$

$$+ \frac{1}{A_n^2} \sum_{m=1}^n C_m \int_a^b \sum_{l=1}^n C_l \left[p\frac{d\psi_m}{dz}\frac{d\psi_l}{dz} + q\psi_m\psi_l \right] dz$$

$$\geq \lambda_{n+1} \qquad (2.5.19)$$

The first term in Eq. (2.5.19) is $\Omega(f)/A_n^2$, which we have just seen is never negative. The first term in the second integral can be integrated by parts and, using the fact that ψ_m satisfies the Liouville equation and that both f and the ψ's satisfy the boundary conditions, we obtain

$$\frac{2}{A_n^2} \int_a^b f \sum_{m=1}^n C_m \left[\frac{-d}{dz}\left(p\frac{d\psi_m}{dz} \right) + q\psi_m \right] dz + \frac{2}{A_n^2} \left[\sum_{m=1}^n C_m pf \frac{d\psi_m}{dz} \right]_a^b$$

$$= \frac{2}{A_n^2} \int_a^b f \sum_{m=1}^n C_m \lambda_m w \psi_m \, dz$$

$$= \frac{2}{A_n^2} \sum_{m=1}^n C_m^2 \lambda_m$$

Similarly, the third integral in Eq. (2.5.19) equals

$$\frac{1}{A_n^2} \sum_{m=1}^n C_m \int_a^b \sum_{l=1}^n C_l \left[-\frac{d}{dz}\left(p\frac{d\psi_m}{dz} \right) \psi_l + q\psi_m\psi_l \right] dz$$

$$= \frac{1}{A_n^2} \sum_{m=1}^n C_m \int_a^b \sum_{l=1}^n C_l \lambda_m w \psi_m \psi_l \, dz$$

$$= \frac{1}{A_n^2} \sum_{m=1}^n C_m^2 \lambda_m$$

Combining all of these results, Eq. (2.5.19) becomes

$$\Omega(F_n) = \frac{1}{A_n^2}\left[\Omega(f) - \sum_{m=1}^{n} C_m^2 \lambda_m\right] \geq \lambda_{n+1}$$

Since $\sum_{m=1}^{n} C_m^2 \lambda_m$ is positive, we see

$$\Omega(f) > A_n^2 \lambda_{n+1}$$

or

$$A_n^2 < \frac{\Omega(f)}{\lambda_{n+1}}$$

But $\Omega(f) > 0$, independent of n and we have stated that $\lambda_{n+1} \to \infty$ as $n \to \infty$. Therefore,

$$\lim_{n \to \infty} A_n^2 = 0 \qquad \text{Q.E.D.}$$

There are two other important types of convergence in addition to convergence in the mean. These are pointwise convergence and uniform convergence.

Let

$$f_1(z) + f_2(z) + \cdots + f_n(z) + \cdots$$

be a series of functions defined on a set A and let

$$F_n(z) = \sum_{i=1}^{n} f_i(z) \qquad (2.5.20)$$

We say that this series of functions converges on A in case the sequence $F_n(z)$, $n = 1, 2, \ldots$ converges on A. In pointwise convergence, let $F_1(z), F_2(z), \ldots, F_n(z)$, be a sequence of functions given by Eq. (2.5.20) and defined on a set A. We say that this sequence of functions converges pointwise on A in case, for every fixed $z \in A$, the sequence of constants $F_n(z)$, $n = 1, 2, \ldots$ converges. Assume the sequence $F_n(z)$ converges on A and define

$$F(z) = \lim_{n \to \infty} F_n(z)$$

Then the rapidity with which $F_n(z)$ approaches $F(z)$ can be expected to depend rather heavily on z. More precisely, the sequence of functions $F_n(z)$ converges pointwise to the function $F(z)$ if, corresponding to any point $z \in A$ and any $\varepsilon > 0$, there exists a number $N = N(z, \varepsilon)$ such that $n > N(z, \varepsilon)$ implies

$$|F_n(z) - f(z)| < \varepsilon$$

68 MATHEMATICAL REVIEW

In uniform convergence, a sequence of functions $F_n(z)$, $n = 1, 2, \ldots$, defined on a set A, converges uniformly on A to a function $f(x)$ defined on A if and only if corresponding to $\varepsilon > 0$ there exists a number $N = N(\varepsilon)$ depending on ε alone and not on the point z such that $n > N$ implies

$$|F_n(z) - f(z)| < \varepsilon$$

for every $z \in A$.

Figure 2.5.1 illustrates in general the concept of uniform convergence. If we assign an arbitrary measure of accuracy ε, then from a certain index N onward all the function $F_n(z)$ should lie between $f(z) - \varepsilon$ and $f(z) + \varepsilon$ for all values of z, so that their graphs $y = F_n(z)$ lie entirely in the strip shown in Figure 2.5.1.

Let us look at a simple example of pointwise and uniform convergence.

We restrict our attention to the interval $0 \leq z \leq 1$, and make the definition for $n \geq 2$

$$F_n(z) = zn^\alpha \quad \text{for} \quad 0 \leq z \leq \frac{1}{n}$$

$$F_n(z) = \left(\frac{2}{n} - z\right)n^\alpha \quad \text{for} \quad \frac{1}{n} \leq z \leq \frac{2}{n}$$

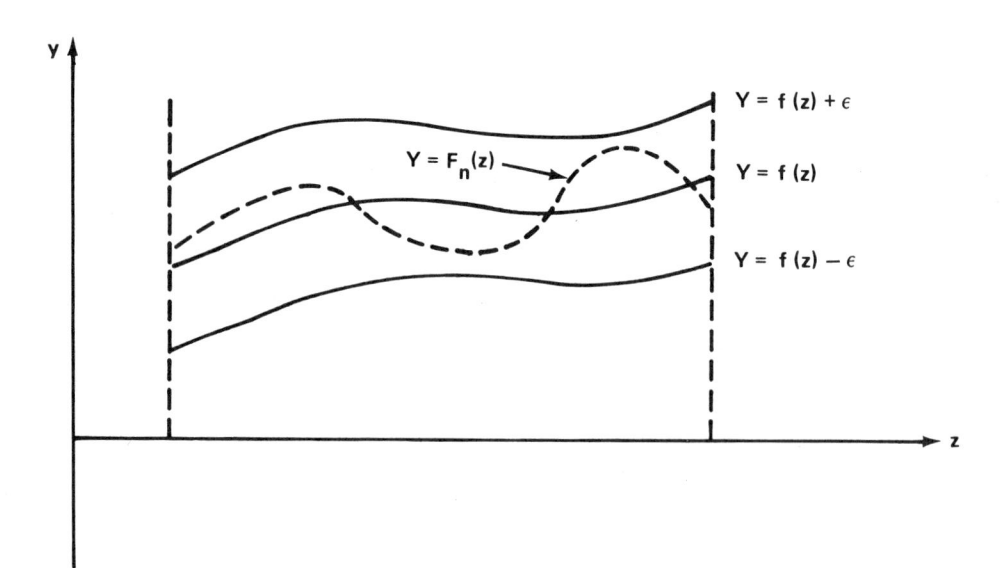

$$F_n(z) = 0 \quad \text{for} \quad \frac{2}{n} \leq z \leq 1$$

FIGURE 2.5.1. General concept of uniform convergence.

COMPLETENESS OF A SET OF EIGENFUNCTIONS   69

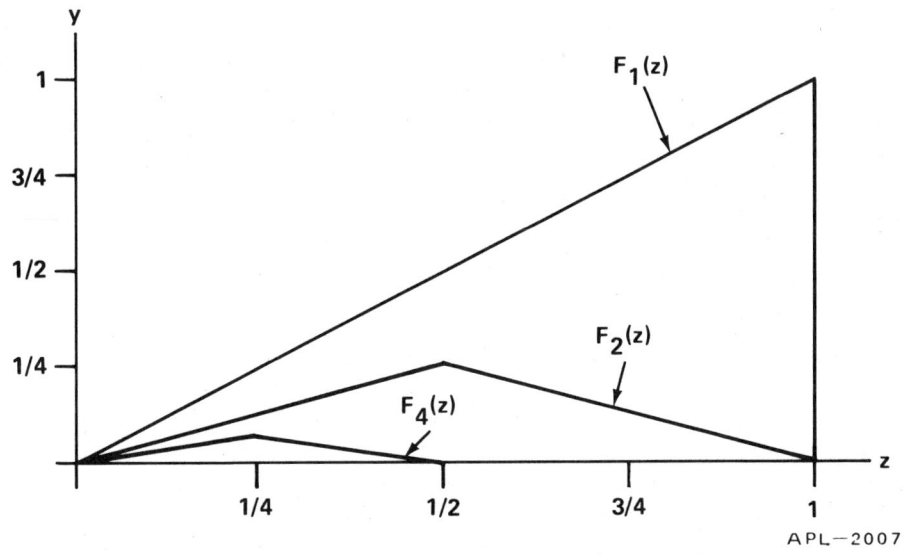

FIGURE 2.5.2. Example of a function that converges uniformly.

Consider first the case $\alpha = -1$. This is a case of uniform convergence. Figure 2.5.2 illustrates several functions of the sequence $F_n(z)$.

For this case, the altitude of the highest point of the graph, which has in general the value $(1/n^2)$, will tend to zero as n increases. The curves will then tend toward the z-axis and the functions $F_n(z)$ will converge uniformly to the limit function $f(z) = 0$.

Now consider the case $\alpha = 2$. This is a case of pointwise convergence. Figure 2.5.3 illustrates several functions of the sequence $F_n(z)$. In this case, the

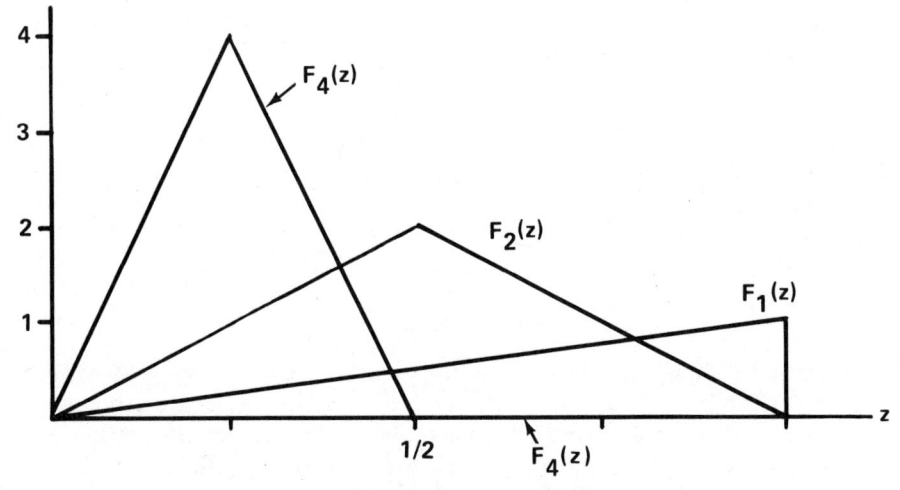

FIGURE 2.5.3. Example of a function that converges pointwise.

height of the peak will increase beyond all bounds as n increases. However, this sequence still tends to the limit function $f(z) = 0$, for if $z > 0$ we have $z > 2/n$ for all sufficiently large values of n so that z is not under the roof-shaped part of the graph and $F_n(z) = 0$; for $z = 0$, all the functional values $F_n(z)$ are equal to zero, so in either case, limit $F_n(z) = 0$ as $n \to \infty$.

The following theorems show the importance of uniform convergence.

Theorem 2.5.1. If $f(z) = \sum_{n=1}^{\infty} f_n(z)$ where the series converges uniformly on a set A, and if every term of the series is continuous on A, then $f(z)$ is continuous on A.

Theorem 2.5.2. If $f(z) = \sum_{n=1}^{\infty} f_n(z)$ where the series converges uniformly on $[a, b]$, and if every term of the series is integrable on $[a, b]$, then $f(z)$ is integrable on $[a, b]$ and the series can be integrated term by term

$$\int_a^b f(z)\, dz = \int_a^b \sum_{n=1}^{\infty} f_n(z)\, dz = \sum_{n=1}^{\infty} \int f_n(z)\, dz$$

Theorem 2.5.3. If $\sum_{n=1}^{\infty} f_n(z)$ is a series of differentiable functions on $[a, b]$, convergent at one point of $[a, b]$, and if the derived series $\sum_{n=1}^{\infty} f_n'(z)$ converges uniformly on $[a, b]$, then the original series converges uniformly on $[a, b]$ to a differentiable function $f(z)$ whose derivatives are represented on $[a, b]$ by

$$f'(z) = \sum_{n=1}^{\infty} f_n'(z)$$

It is very important to note that completeness of an orthonormal system $\psi_1(z), \ldots, \psi_n(z), \ldots,$ for an admissable set of functions $\mathscr{S}$ does not imply that if $f \in \mathscr{S}$, the orthonormal system converges pointwise or uniformly to f. That is why we wrote in Eq. (2.5.13)

$$f(z) \doteq \sum_{n=1}^{\infty} C_n \psi_n$$

instead of

$$f = \sum_{n=1}^{\infty} C_n \psi_n$$

It must be shown separately that the complete orthonormal system converges uniformly to f, and it is uniform convergence rather than pointwise convergence that we are most concerned about. This is because we need to show that $\sum_{n=1}^{\infty} C_n \psi_n$ is a solution of our differential equation and to do this we need Theorems 2.5.1 and 2.5.3.

An orthonormal system can be complete for certain general classes of functions for which we will not have uniform convergence. However, for the function of interest for our problem of acoustic propagation in an oceanic waveguide, the following theorem guarantees uniform convergence.

Theorem 2.5.4. Let $\mathscr{S}$ be the class of functions $f(z)$ that are continuous, and that have a piecewise continuous first derivative on an interval $a \leq z \leq b$. Let $\psi_1(z), \ldots, \psi_n(z), \ldots$ be a set of orthonormal eigenfunctions of the Sturm-Liouville system, Eqs. (2.2.1), (2.4.3), and (2.4.4). Then every function $f \in \mathscr{S}$ can be expanded in a uniformly convergent series

$$f = \sum_{n=1}^{\infty} C_n \psi_n(z), \qquad C_n = \int_a^b w f \psi_n \, dz$$

We can state this last result in another way. If $\psi_1, \ldots, \psi_n$ is a complete orthonormal system, then it is possible to expand any continuous function $f(z)$ that has a piecewise continuous derivative and satisfies the boundary conditions, Eqs. (2.4.3) and (2.4.4), into a uniformly convergent series.

2.6. THE CONTINUOUS SPECTRUM

The eigenvalues of the problems considered thus far form a denumerably infinite sequence. However, if the coefficients of the differential equation are singular at the boundary points of the fundamental interval $[a, b]$ or, in particular, if the fundamental interval is infinite, the spectrum, or totality of eigenvalues, may behave quite differently. In particular, continuous spectra may occur.

Let us do a simple problem to show the effect of boundary conditions on the spectra. Let us calculate the normal modes of a rectangular box. Figure 2.6.1 shows a diagram of the box with sides of length L_x, L_y, and L_z.

The Helmholtz equation in rectangular coordinates is

$$\frac{\partial^2 p}{\partial x^2} + \frac{\partial^2 p}{\partial y^2} + \frac{\partial^2 p}{\partial z^2} + k^2 p = 0 \qquad (2.6.1)$$

where

$$k = \frac{\omega}{c}$$

ω being the angular frequency and c the constant speed of sound in the box. Using separation of variables, we assume a solution of the form

$$P = X(x) Y(y) Z(z) \qquad (2.6.2)$$

Substituting Eq. (2.6.2) into the reduced wave equation (2.6.1) and dividing by

72 MATHEMATICAL REVIEW

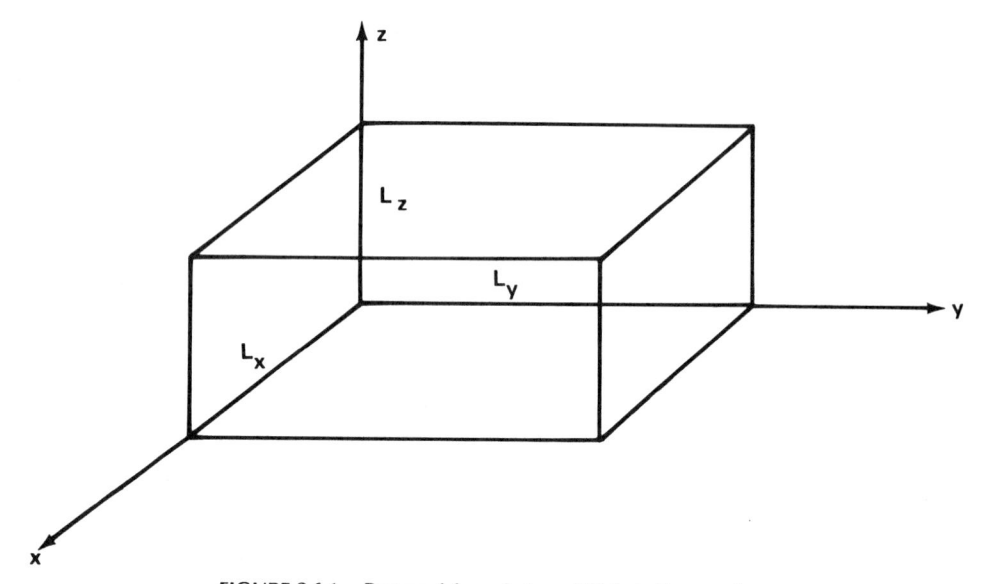

FIGURE 2.6.1. Box used for solution of Helmholtz equation.

XYZ gives

$$\frac{1}{X}\frac{d^2X}{dx^2} + \frac{1}{Y}\frac{d^2Y}{dy^2} + \frac{1}{Z}\frac{d^2Z}{dz^2} + k^2 = 0$$

Rearranging,

$$\frac{1}{Y}\frac{d^2Y}{dy^2} + \frac{1}{Z}\frac{d^2Z}{dz^2} + k^2 = -\frac{1}{X}\frac{d^2X}{dx^2} = k_1^2$$

where k_1^2 is the first separation constant. Regrouping terms, we have

$$\frac{1}{Z}\frac{d^2Z}{dz^2} + k^2 - k_1^2 = -\frac{1}{Y}\frac{d^2Y}{dy^2} = k_2^2$$

and

$$\frac{d^2X}{dx^2} + k_1^2 X = 0$$

where k_2^2 is the second separation constant.

THE CONTINUOUS SPECTRUM

Letting

$$k_3^2 \equiv k^2 - k_1^2 - k_2^2 \qquad (2.6.3)$$

the separated equations become

$$\frac{d^2X}{dx^2} + k_1^2 X = 0$$

$$\frac{d^2Y}{dy^2} + k_2^2 Y = 0$$

$$\frac{d^2Z}{dz^2} + k_3^2 Z = 0 \qquad (2.6.4)$$

where, from Eq. (2.6.3), $\qquad k^2 = k_1^2 + k_2^2 + k_3^2 \qquad (2.6.5)$

The solutions of Eq. (2.6.4) are

$$X(x) = A_1 e^{-ik_1 x} + B_1 e^{ik_1 x}$$

$$Y(y) = A_2 e^{-ik_2 y} + B_2 e^{ik_2 y}$$

$$Z(z) = A_3 e^{-ik_3 z} + B_3 e^{ik_3 z} \qquad (2.6.6)$$

Let us assume that the walls of the box satisfy the pressure release boundary condition, that is, we assume

$$p = 0 \qquad (2.6.7)$$

on all the walls. If we apply this boundary condition to the solutions in Eqs. (2.6.6), we arrive very easily at the following system of equations

$$A_1 + B_1 = 0$$

$$A_2 + B_2 = 0$$

$$A_3 + B_3 = 0$$

$$A_1 e^{-ik_1 L_x} + B_1 e^{ik_1 L_x} = 0$$

$$A_2 e^{-ik_2 L_y} + B_2 e^{ik_2 L_y} = 0$$

$$A_3 e^{-ik_3 L_z} + B_3 e^{ik_3 L_z} = 0 \qquad (2.6.8)$$

It is trivial to solve this system of equations. We get

$$\sin k_1 L_x = 0$$
$$\sin k_2 L_y = 0$$
$$\sin k_3 L_z = 0 \tag{2.6.9}$$

Eq. (2.6.9) imposes the following conditions on the separation constants

$$k_{1,l} = \frac{l\pi}{L_x}, \quad l = 1, 2, 3, \ldots$$

$$k_{2,m} = \frac{m\pi}{L_y}, \quad m = 1, 2, 3, \ldots$$

$$k_{3,n} = \frac{n\pi}{L_z}, \quad n = 1, 2, 3, \ldots \tag{2.6.10}$$

The numbers $k_{1,l}$, $k_{2,m}$, and $k_{3,n}$ can be interpreted as the components of the wave number k in the x, y, and z direction, respectively.

The eigenfunction solution, or a mode, is

$$p_{lmn} = \sin(k_{1,l} x)\sin(k_{2,m} y)\sin(k_{3,n} z) \tag{2.6.11}$$

From Eqs. (2.6.5) and (2.6.10), we see that all wave numbers k (or frequencies $\omega = kc$) are not allowed. A solution of the wave equation exists only for those wave numbers k given by

$$k_{lmn} = \left[\frac{l^2 \pi^2}{L_x^2} + \frac{m^2 \pi^2}{L_y^2} + \frac{n^2 \pi^2}{L_z^2} \right]^{1/2} \tag{2.6.12}$$

Now consider the x component of the wave number $k_{1,l}$. Similar results will hold for the y and z components. Let us compute the difference between two adjacent eigenvalues of k_1, namely, $k_{1,l+1}$ and $k_{1,l}$

$$\Delta k_{1,l+1,l} = \frac{(l+1)\pi}{L_x} - \frac{l\pi}{L_x}$$

$$\Delta k_{1,l+1,l} = \frac{\pi}{L_x}$$

Similarly,

$$\Delta k_{2,m+1,m} = \frac{\pi}{L_y}$$

$$\Delta k_{3,n+1,n} = \frac{\pi}{L_z}$$

Now as the walls of the box recede to infinity, that is, as L_x, L_y, and L_z become infinitely large, we see that $\Delta k_{1,l+1,l}$, $\Delta k_{2,m+1,m}$, and $\Delta k_{3,n+1,n}$ approach zero. That is, the spacing between eigenvalues becomes smaller and smaller as the region enclosed by the bounding surfaces becomes larger and larger. In the limit, the eigenvalue spectrum becomes continuous, since all values of k_1, k_2, and k_3 become allowed.

These two extremes, that is, either a purely discrete spectrum or a purely continuous spectrum, are not the only possibility. We will encounter cases in oceanic propagation where we will have a mixture of discrete and continuous spectra. The theory of the continuous spectrum is appreciably more difficult to handle than the theory of discrete spectra. Hence, it is important to note that for many purposes some or all of the difficulties can be sidestepped by making use of the fact that, by a slight modification of many physical problems of no practical importance as regards the final outcome, one can substitute a purely discrete spectrum for the discrete, continuous one. For this purpose one has only to place the system under discussion in a large imaginary "box," requiring that the field vanish on the surface of the box. It is evident that, if the box is large enough, it will be legitimate to assume that results computed for the modified problem are as good as those computed for the original one in which the coordinate space extends to infinity. The "box" eliminates the continuous spectrum completely.

Let us summarize the results of the theory of continuous spectra. We are concerned with the differential equation

$$-\frac{d}{dz}\left[p(z)\frac{d\psi}{dz}\right] + [q(z) - \lambda w(z)]\psi = 0 \qquad (2.6.13)$$

Let the operator L be defined as

$$L \equiv \frac{1}{w(z)}\left\{-\frac{d}{dz}\left[p(z)\frac{d}{dz}\right] + q(z)\right\} \qquad (2.6.14)$$

Then Eq. (2.6.13) can be witten in operator form as

$$L\psi = \lambda\psi \qquad (2.6.15)$$

The operator L acts on functions $\psi(z)$ which are defined in an interval I and subject to various admissability conditions.

The coefficients $p(z)$ and $w(z)$ are supposed to be positive in the interior of the interval I. If the interval I is finite, if the functions $w(z)$ and $p(z)$ are continuous and positive at the endpoints, and if also $q(z)$ is continuous there, the operator L is said to be regular or of the Sturm-Liouville type. We have seen that for such regular operators a complete sequence of eigenvalues λ_n and eigenfunctions $\psi_n(z, \lambda_n)$ exists provided that appropriate boundary conditions are imposed at the endpoints. The sequence is called complete if an arbitrary function $f(z)$ of a certain class admits an expansion in the mean of the form

$$f(z) \doteq \sum_{n=1}^{\infty} A_n \psi_n(z, \lambda_n) \qquad (2.6.16)$$

where the A_n are the Fourier coefficients.

The operator L is said to be singular if all regularity conditions mentioned above are not satisfied. Thus the operator L is singular if the interval extends to infinity in one or both directions, or if one of the functions $p(z)$, $p^{-1}(z)$, $w(z)$, $w^{-1}(z)$, $q(z)$ does not approach a finite limit at one endpoint at least. It is known that there are singular operators L that do not possess a complete sequence of eigenvalues and eigenfunctions in the sense of regular operators; that is, an arbitrary function cannot be expanded as an infinite series of these eigenfunctions. Nevertheless, for these singular operators there exists an analog to the expansion of arbitrary functions with respect to eigenfunctions as given in Eq. (2.6.16). In this modified expansion, integration occurs instead of summation, involving "improper" eigenfunctions $\phi(z, \lambda)$, which depend continuously on the eigenvalue λ in a certain set S of values of λ. Thus the expansion formula is of the form

$$f(z) = \sum_{n=1}^{N} A_n \psi_n(z, \lambda_n) + \int_S b(\lambda) \phi(z, \lambda) \, d\lambda \qquad (2.6.17)$$

The set S in which the improper eigenfunctions $\phi(z, \lambda)$ are defined is called the continuous spectrum of the operator L; the values λ_n occurring in the sum form the discrete spectrum. The functions ψ_n are called "proper" eigenfunctions.

The eigenfunctions ψ_n are called proper because they are square integrable, that is, they can be normalized according to Eqs. (2.2.15):

$$\int_I w(z) \psi_n(z) \psi_m(z) \, dz = \delta_{nm} \qquad (2.6.18a)$$

The improper eigenfunctions cannot be normalized in the usual sense; that is, they are not square integrable. They are to be normalized according to the requirement

$$\int_I w(z) \phi(z, \lambda) \phi(z, \lambda') \, dz = \delta(\lambda - \lambda') \qquad (2.6.18b)$$

where $\delta(\lambda - \lambda')$ is the Dirac delta function.

The Dirac delta function $\delta(z)$ is defined to be zero for every value of its argument z except at the origin, where it is infinite in such a way that

$$\int_{-\infty}^{\infty} \delta(z) \, dz = 1 \qquad (2.6.19)$$

It can be shown that for every continuous function $f(z)$, we have

$$\int_{-\infty}^{\infty} f(z)\delta(z)\, dz = f(0) \tag{2.6.20}$$

or

$$\int_{-\infty}^{\infty} f(z)\delta(z - z_0)\, dz = f(z_0) \tag{2.6.21}$$

While the delta function is not really a function in the rigorous mathematical sense, it can be defined rigorously by use of the mathematical theory of distributions.

The following is a list of some of the properties of the delta function.

$$\delta(-z) = \delta(z) \tag{2.6.22}$$

$$z\delta(z) = 0 \tag{2.6.23}$$

$$\delta(az) = \frac{1}{|a|}\delta(z) \tag{2.6.24}$$

$$\delta[f(z)] = \frac{1}{df/dz}\delta(z - z_0), \quad f(z_0) = 0 \tag{2.6.25}$$

$$\int_{-\infty}^{\infty} \delta(z - z'')\delta(z'' - z')\, dz'' = \delta(z - z') \tag{2.6.26}$$

$$\frac{d}{dz}\delta(z) = -\frac{1}{z}\delta(z) \tag{2.6.27}$$

2.7. GREEN'S FUNCTIONS

Consider again the following Sturm-Liouville problem which consists of the differential equation

$$-\frac{d}{dz}\left(p\frac{d\psi}{dz}\right) + q\psi - \lambda' w\psi = 0 \tag{2.7.1}$$

defined on the finite interval $a \le z \le b$ and the boundary conditions

$$\psi(a) = 0 \tag{2.7.2}$$

$$\frac{d\psi(b)}{dz} = 0 \tag{2.7.3}$$

78 MATHEMATICAL REVIEW

We are still assuming that the real functions p, q, and w are continuous, p differentiable, and $p > 0$, $w > 0$ on the finite interval $[a, b]$. In Section 2.2 we saw that there was an infinite number of solutions, the eigenfunctions $\psi_1, \ldots, \psi_n$, which satisfy the differential equation (2.7.1) and the boundary conditions (2.7.2) and (2.7.3). For these eigenfunctions, Eq. (2.7.1) takes the form

$$-\frac{d}{dz}\left(p\frac{d\psi_n}{dz}\right) + q\psi_n - \lambda'_n w \psi_n = 0 \tag{2.7.4}$$

where λ'_n are the corresponding eigenvalues.

Associated with Eq. (2.7.1) is a Green's function which will be denoted $G(z, z')$. The Green's function satisfies the inhomogeneous equation

$$-\frac{d}{dz}\left[p(z)\frac{dG}{dz}(z, z')\right] + q(z)G(z, z') - \lambda' w(z)G(z, z') = \delta(z - z') \tag{2.7.5}$$

where $\delta(z - z')$ is the Dirac delta function. We require that $G(z, z')$ satisfy the boundary conditions, Eqs. (2.7.2) and (2.7.3). It is important to note that in Eq. (2.7.5), λ' is an arbitrary parameter and not an eigenvalue. We will see that the value of λ' is not influenced by the boundary conditions. We further assume that $G(z, z')$ is continuous and has a continuous derivative everywhere in the interval $[a, b]$ except at point $z = z'$. At point $z = z'$, the derivative of G will satisfy a discontinuity condition which we will now derive.

Let us integrate Eq. (2.7.5) over an interval, $z' - \varepsilon \leq z \leq z' + \varepsilon$ containing the point z'. Here ε is an arbitrary positive number.

$$-\int_{z'-\varepsilon}^{z'+\varepsilon} \frac{d}{dz}\left[p(z)\frac{dG}{dz}(z, z')\right] dz + \int_{z'-\varepsilon}^{z'+\varepsilon} \{q(z) - \lambda' w(z)\} G(z, z') \, dz$$

$$= \int_{z'-\varepsilon}^{z'+\varepsilon} \delta(z - z') \, dz \tag{2.7.6}$$

Because of the assumed continuity of the functions involved, the second integral in Eq. (2.7.6) will vanish as $\varepsilon \to 0$. Also because of the properties of the delta function, Eq. (2.6.19),

$$\int_{z'-\varepsilon}^{z'+\varepsilon} \delta(z - z') \, dz = 1$$

Finally we need to evaluate the first integral. Integrating yields

$$-\int_{z'-\varepsilon}^{z'+\varepsilon} \frac{d}{dz}\left[p(z)\frac{dG}{dz}(z, z')\right] dz = -\left[p(z)\frac{dG}{dz}(z, z')\right]_{z'-\varepsilon}^{z'+\varepsilon}$$

Combining these results, Eq. (2.7.6) becomes

$$-\left[p(z)\frac{dG}{dz}(z, z')\right]_{z'-\varepsilon}^{z'+\varepsilon} = 1 \qquad (2.7.7a)$$

Since $p(z)$ is continuous, we obtain the discontinuity condition

$$\left[\frac{dG}{dz}(z, z')\right]_{z'-\varepsilon}^{z'+\varepsilon} = \frac{-1}{p(z')} \qquad (2.7.7b)$$

Now let $f(z)$ be an arbitrary continuous function with a piecewise continuous derivative. Then we proved in Section 2.5 that f could be expanded in terms of the complete orthonormal set of eigenfunctions that satisfy the Sturm-Liouville problem, Eqs. (2.7.2), (2.7.3), and (2.7.4). Let

$$f(z) = \sum_{n=1}^{\infty} A_n \psi_n(z) \qquad (2.7.8)$$

where the coefficients are given by

$$A_n = \int_a^b w(z) f(z) \psi_n(z)\, dz \qquad (2.7.9)$$

Substituting Eq. (2.7.9) into Eq. (2.7.8) yields

$$f(z) = \sum_{n=1}^{\infty} \int_a^b w(z') f(z') \psi_n(z') \psi_n(z)\, dz'$$

Now we also saw in Section 2.5 that the series, Eqs. (2.7.8), was uniformly convergent and so, by Theorem 2.5.2, we can interchange the operations of summation and integration to obtain

$$f(z) = \int_a^b \left\{ \sum_{n=1}^{\infty} w(z') \psi_n(z') \psi_n(z) \right\} f(z')\, dz'$$

The only way this last equality can hold is if

$$\sum_{n=1}^{\infty} w(z) \psi_n(z') \psi_n(z) = \delta(z - z') \qquad (2.7.10)$$

where we have used Eq. (2.6.22). Eq. (2.7.10) is called the completeness relation. It represents an expansion of the delta function in terms of the complete, orthogonal set of eigenfunctions ψ_n.

Next let us obtain an expansion for the Green's function in terms of the eigenfunctions. Since $G(z, z')$ has the required continuity properties, we can write

$$G(z, z') = \sum_{n=1}^{\infty} A_n \psi_n(z) \qquad (2.7.11)$$

Substituting Eqs. (2.7.11) and (2.7.10) into Eq. (2.7.5) gives

$$-\frac{d}{dz}p\left[\sum_n A_n \frac{d\psi_n(z)}{dz}\right] + q\sum_n A_n \psi_n(z) - \lambda' w \sum_n A_n \psi_n(z)$$

$$= \sum_n w(z)\psi_n(z')\psi_n(z)$$

Using Eq. (2.7.4), this last equation becomes

$$\sum_n A_n \left\{ -\frac{d}{dz}\left(p\frac{d\psi_n}{dz}\right) + q\psi_n \right\} - \lambda' w(z) \sum_n A_n \psi_n(z) = \sum_n w(z)\psi_n(z')\psi_n(z)$$

or

$$\sum_n A_n \lambda'_n w(z)\psi_n(z) - \sum_n A_n \lambda' w(z)\psi_n(z) = \sum_n w(z)\psi_n(z')\psi_n(z)$$

or

$$\sum_n A_n (\lambda'_n - \lambda') w(z)\psi_n(z) = \sum_n w(z)\psi_n(z')\psi_n(z)$$

Multiplying this last equation by $\psi_m(z)$, integrating over the interval $[a, b]$, and making use of the orthonormality of the eigenfunctions gives

$$\sum_n A_n(\lambda'_n - \lambda')\int_a^b w(z)\psi_n(z)\psi_m(z)\,dz = \sum_n \psi_n(z')\int_a^b w(z)\psi_n(z)\psi_m(z)\,dz$$

$$\sum_n A_n(\lambda'_n - \lambda')\delta_{mn} = \sum_n \psi_n(z')\delta_{nm}$$

$$A_m(\lambda'_m - \lambda') = \psi_m(z')$$

$$A_m = \frac{\psi_m(z')}{\lambda'_m - \lambda'}$$

Consequently, the desired expansion Eq. (2.7.11) becomes

$$G(z, z', \lambda') = \sum_{n=1}^{\infty} \frac{\psi_n(z')\psi_n(z)}{\lambda'_n - \lambda'} \qquad (2.7.12)$$

GREEN'S FUNCTIONS

Now let us treat the arbitrary parameter λ' as a complex variable and integrate $G(z, z', \lambda')$ around a contour $C_{\lambda'}$ in the complex λ' plane which encloses all the singularities of $G(z, z', \lambda')$. From Eq. (2.7.12) we get

$$\int_{C_{\lambda'}} G(z, z', \lambda') \, d\lambda' = \sum_{n=1}^{\infty} \psi_n(z') \psi_n(z) \int_{C_{\lambda'}} \frac{d\lambda'}{\lambda'_n - \lambda'}$$

By the Cauchy residue theorem we get

$$\int_{C_{\lambda'}} \frac{d\lambda'}{\lambda'_n - \lambda'} = -2\pi i$$

so that we obtain the very important result

$$\int_{C_{\lambda'}} G(z, z', \lambda') \, d\lambda' = -2\pi i \sum_{n=1}^{\infty} \psi_n(z) \psi_n(z')$$

or, on using the completeness relation Eq. (2.7.10),

$$-\frac{1}{2\pi i} \int_{C_{\lambda'}} G(z, z', \lambda') \, d\lambda' = \frac{\delta(z-z')}{w(z)} \qquad (2.7.13)$$

There is one final, important property of the Green's function that we need to prove. This is that the Green's function is symmetric in its arguments.

Let the operator L be defined by

$$L = -\frac{d}{dz}\left(p\frac{d}{dz}\right) + q - \lambda' w \qquad (2.7.14)$$

Then the defining equation for the Green's function, namely, Eq. (2.7.5), can be written compactly as

$$LG(z, z') = \delta(z - z') \qquad (2.7.15)$$

For a different point z'' we can also write

$$LG(z, z'') = \delta(z - z'') \qquad (2.7.16)$$

Now multiply Eq. (2.7.15) by $G(z, z'')$ and integrate over z to obtain

$$\int_a^b G(z, z'') LG(z, z') \, dz$$

Using integration by parts twice and the fact that the Green's function satisfies the boundary conditions Eqs. (2.7.2) and (2.7.3), we can show that

$$\int_a^b G(z, z'') LG(z, z') \, dz = \int_a^b G(z, z') LG(z, z'') \, dz \qquad (2.7.17)$$

Now using Eqs. (2.7.15) and (2.7.16), Eq. (2.7.17) becomes

$$\int_a^b G(z, z'') \delta(z - z') \, dz = \int_a^b G(z, z') \delta(z - z'') \, dz$$

or

$$G(z', z'') = G(z'', z') \qquad (2.7.18)$$

which is the desired result.

2.8. METHOD OF STEEPEST DESCENT

Let us first list some definitions and theorems from the theory of complex variables that we will need to develop the method of steepest descent.

Definition 2.8.1. A complex valued function $f(z)$ is said to be analytic at a point z_0 if $f'(z)$ exists in some neighborhood of z_0.

Theorem 2.8.1. Let f be a complex-valued function defined on an open set S and write $f = u + iv$. If f has a derivative $f'(z_0)$ at the point $z_0 = x_0 + iy_0$ of S, then u and v must have finite partial derivatives $\partial u/\partial x$, $\partial u/\partial y$, $\partial v/\partial x$, $\partial v/\partial y$ at (x_0, y_0), and they are related to $f'(z_0)$ by the two equations

$$f'(z_0) = \frac{\partial u}{\partial x}(x_0, y_0) + i\frac{\partial v}{\partial x}(x_0, y_0)$$

and

$$f'(z_0) = \frac{\partial v}{\partial y}(x_0, y_0) - i\frac{\partial u}{\partial y}(x_0, y_0)$$

This implies, in particular, that

$$\frac{\partial u}{\partial x} = \frac{\partial v}{\partial y} \quad \text{and} \quad \frac{\partial v}{\partial x} = -\frac{\partial u}{\partial y}$$

These are called the Cauchy-Riemann conditions.

Theorem 2.8.2. If $f = u + iv$ is an analytic function, then its real and imaginary parts satisfy Laplace's equation, that is, $\nabla^2 u = 0$ and $\nabla^2 v = 0$.

Definition 2.8.2. Functions satisfying Laplace's equation are said to be harmonic.

Theorem 2.8.3. A nonconstant harmonic function has neither a maximum nor a minimum in its region of definition

The method of steepest descent is used to estimate the value of integrals of the form

$$I = \int_C F(z) e^{\rho f(z)} \, dz \qquad (2.8.1)$$

for large values of the real and positive parameter ρ. The functions $F(z)$ and $f(z)$ are arbitrary analytic functions of the complex variable z, and C is the path of integration in the complex z plane. Integrals like Eq. (2.8.1) arise in the solution of the wave equation.

Let the contour $\Gamma = C + C_1$ as shown in Figure 2.8.1 and let $g(z)$ be an analytic function on Γ and inside Γ. Then by Cauchy's theorem

$$\int_\Gamma g \, dz = \int_C g \, dz + \int_{C_1} g \, dz = 0$$

Therefore,

$$\int_C g \, dz = -\int_{C_1} g \, dz$$

Thus under the conditions stated on g we can deform the path C into a path C_1; that is, the path C can be continuously deformed into a path C_1, as long as it does not pass through a singularity of the integrand during the deformation. In Eq. (2.8.1) then, $F(z)$ and $f(z)$ must be analytic only when $g(z)$ is analytic, and C and C_1 must have the same endpoints. Thus the integral I given in Eq. (2.8.1) taken over some arbitrary path C can always be replaced by the integral of the same integrand taken over the new path of integration C_1, provided the integrand is analytic on and between the two curves C and C_1. If the integrand is not analytic between the two curves C and C_1, it may be necessary to add some terms obtained in going around singular points. In particular, if we go around a pole, we must add the residue at the pole. If the integrand is

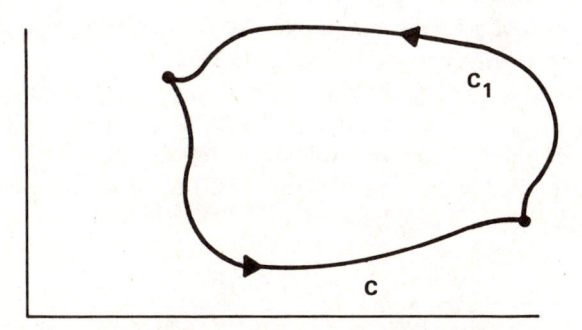

FIGURE 2.8.1. Definition of the contour Γ.

multiple-valued and we encounter a branch point as we deform the contour, we must add the integral over the edges of an appropriate cut made at this point.

The method of steepest descent shows how it may be possible to choose C so that a good approximation to I can be found. Following the method used in Brekhovskikh [2.12], let us now proceed to show how such a contour is chosen. Let

$$f(z) = f_1(x, y) + if_2(x, y) \tag{2.8.2}$$

where f_1, f_2 are real functions of $x, y(z = x + iy)$. The integrand is now of the form

$$F(z)e^{\rho f_1}e^{i\rho f_2} \tag{2.8.3}$$

where the magnitude of this function is determined mainly by $e^{\rho f_1}$ when ρ is large. The object of the method is to deform the contour C in such a way that f_1 for points on C has a maximum and falls away as rapidly as possible on either side of this maximum. In this way the principal contribution to the integral is given over a short interval C surrounding this maximum point.

A necessary condition that $z = z_0$ is a maximum on contour C is that

$$f'(z_0) = 0 \tag{2.8.4}$$

By Theorem 2.8.1, we must have at this point

$$\frac{\partial f_1}{\partial x} = \frac{\partial f_1}{\partial y} = \frac{\partial f_2}{\partial x} = \frac{\partial f_2}{\partial y} = 0$$

By Theorem 2.8.2, f_1 satisfies Laplace's equation

$$\nabla^2 f_1 = 0$$

By Theorem 2.8.3, the point $z = z_0$ is neither a true maximum nor a true minimum of the surface $f_1(x, y)$. In a neighborhood of this point, the surface will have the form of a saddle and, hence, the point is called a saddle point. An example of such a surface is shown in Figure 2.8.2.

It is still possible to vary the path C by changing the direction in which it passes through z_0 in such a way that f_1 decreases as rapidly as possible as we proceed from z_0 along C on either side of the saddle point. This means that we have to choose $|\partial f_1/\partial s|$ (s is arc length on C) to be as large as possible. So differentiating f_1 along C, we have

$$\frac{\partial f_1}{\partial s} = \frac{\partial f_1}{\partial x}\frac{dx}{ds} + \frac{\partial f_1}{\partial y}\frac{dy}{ds} \tag{2.8.5}$$

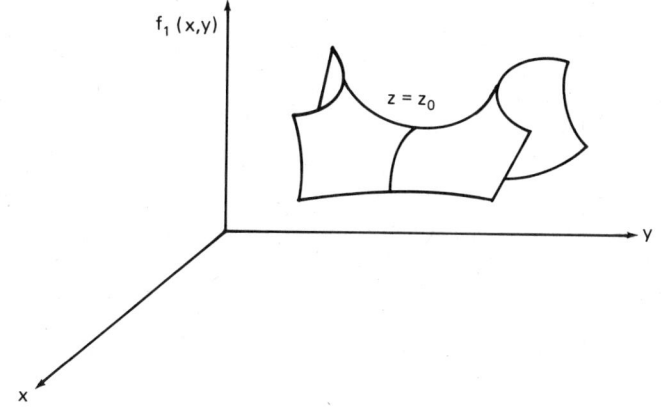

FIGURE 2.8.2. A saddle surface that arises in the application of the method of steepest descent.

But dx/ds and dy/ds are simply the direction cosines, and if we let ψ be the angle between the curve C and the x-axis we can write

$$\frac{dx}{ds} = \cos \psi, \qquad \frac{dy}{ds} = \sin \psi$$

so that Eq. (2.8.5) becomes

$$\frac{\partial f_1}{\partial s} = \frac{\partial f_1}{\partial x} \cos \psi + \frac{\partial f_1}{\partial y} \sin \psi \qquad (2.8.6)$$

Now we can look at all the curves C passing through point $z = z_0$ as a one-parameter family of curves described by the parameter ψ. So the condition for $\partial f_1/\partial s$ to be a maximum is given by

$$\frac{d}{d\psi}\left(\frac{\partial f_1}{\partial s}\right) = 0$$

or, using Eq. (2.8.6),

$$-\frac{\partial f_1}{\partial x} \sin \psi + \frac{\partial f_1}{\partial y} \cos \psi = 0 \qquad (2.8.7)$$

Using the Cauchy-Rieman conditions, Eq. (2.8.7) becomes

$$0 = -\frac{\partial f_2}{\partial y} \sin \psi - \frac{\partial f_2}{\partial x} \cos \psi = -\frac{\partial f_2}{\partial s} \qquad (2.8.8)$$

Hence, $f_2 = $ constant along the optimum path C. This optimum path is called the line of steepest descent.

Thus, the most advantageous path of integration must go through the saddle point determined by Eq. (2.8.4) and must leave this point along the line of the most rapid decrease of f_1, which coincides with the line $f_2 = $ constant.

On such a path C we have

$$f(z) = f_1(x, y) + ik \qquad (2.8.9)$$

where we have put $f_2 = k$ (k constant).

If we now expand both sides of Eq. (2.8.9) in a Taylor series about $z = z_0$ and neglect terms of higher order than the second in the small parameter s, we have

$$f(z_0) + \tfrac{1}{2}(z - z_0)^2 f''(z_0) = f_1(x_0, y_0)$$
$$+ \frac{1}{2}s^2 \left\{ \cos\psi \frac{\partial}{\partial x} + \sin\psi \frac{\partial}{\partial y} \right\}^2 f_1 + ik \qquad (2.8.10)$$

where we have used Eq. (2.8.4). Since

$$f(z_0) = f_1(x_0, y_0) + ik$$

we see, from Eq. (2.8.10),

$$\frac{1}{2}(z - z_0)^2 f''(z_0) = \frac{1}{2}s^2 \left\{ \cos\psi \frac{\partial}{\partial x} + \sin\psi \frac{\partial}{\partial y} \right\}^2 f_1 \qquad (2.8.11)$$

Thus, the left-hand side of Eq. (2.8.11) is real. Moreover, since $f_1(x_0, y_0)$ is a local maximum, the left side of Eq. (2.8.11) is also negative.

Thus, we can write

$$f(z) - f(z_0) = -\sigma^2 \qquad (2.8.12)$$

where σ is real and varies from $-\infty$ to $+\infty$.

Now let us see why the surface $f_1(x, y)$ has the form of a saddle near the point $z = z_0$. Let us consider Eq. (2.8.12) in the complex Σ plane, where $\Sigma = \sigma + i\sigma'$, so that we can write

$$f(z) - f(z_0) = -\Sigma^2 \qquad (2.8.13)$$

We now transform from the z plane to the Σ plane. As we have seen in Eq. (2.8.12), the path of integration coincides with the real axis in the Σ plane. Separating Eq. (2.8.13) into real and imaginary parts, we obtain

$$f_1 = f_1(z_0) - (\sigma^2 - \sigma'^2) \qquad (2.8.14)$$

$$f_2 = f_2(z_0) - 2\sigma\sigma' \qquad (2.8.15)$$

Hence, in terms of the coordinates σ and σ', the lines $f_1 = $ constant and $f_2 = $ constant form two mutually orthogonal systems of hyperbolas.

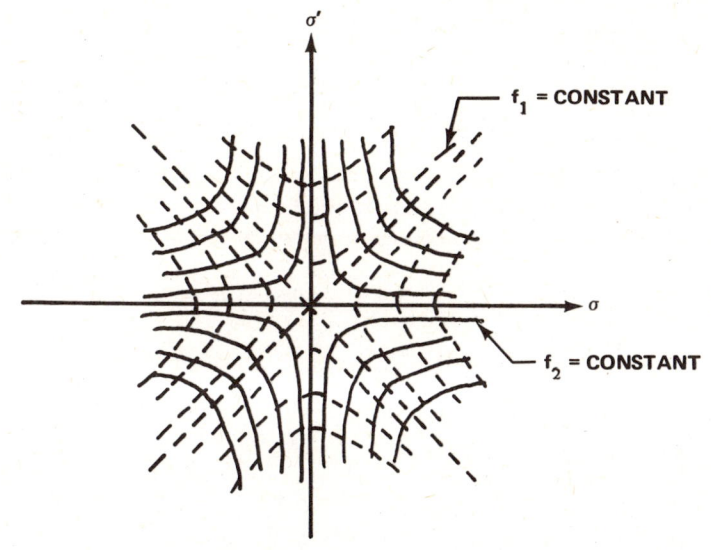

FIGURE 2.8.3. The neighborhood of a saddle point.

As already indicated, the saddle point is $\Sigma = 0$, and the path of integration is the real axis ($\sigma' = 0$). We see from Figure 2.8.3 that the real axis is one of the lines f_2 = constant and is perpendicular to the lines f_1 = constant. We note that one other line of f_2 = constant passes through the saddle point, namely, the imaginary axis ($\sigma = 0$). However, along this line the function f_1 does not decrease with distance from the point $\Sigma = 0$, but rather increases, as can be seen from Eq. (2.8.14). Thus, the path $\sigma = 0$ is not a path of most rapid decrease of f_1, but rather a path of most rapid increase of f_1.

Let us represent the relief of the function f_1 on the Σ plane as follows: at each point (σ, σ') we erect perpendicular to the Σ plane a segment whose length is proportional to the value of the function $f_1(\sigma, \sigma')$, and then pass a surface through the ends of the segments. In the neighborhood of the point $\Sigma = 0$, this relief will have the form of a saddle because on both sides of $\Sigma = 0$, along the real axis, it decreases, and in the perpendicular direction, along the imaginary axis, it increases.

We are now ready to evaluate the integral of Eq. (2.8.1). Let us write

$$I = \int_c F(z) e^{\rho f(z)} \, dz$$

$$I = \int_{-\infty}^{\infty} e^{\rho f(z_0) - \rho \sigma^2} F(z) \frac{dz}{d\sigma} \, d\sigma$$

$$= e^{\rho f(z_0)} \int_{-\infty}^{\infty} \Phi(\sigma) e^{-\rho \sigma^2} \, d\sigma \qquad (2.8.16)$$

88 MATHEMATICAL REVIEW

where we have put

$$\Phi(\sigma) = F(z)\frac{dz}{d\sigma} \qquad (2.8.17)$$

Since ρ is assumed to be large, only small values of σ will contribute to the integral. It is, therefore, convenient to expand $\Phi(\sigma)$ in a power series in powers of σ:

$$\Phi(\sigma) = \Phi(0) + \Phi'(0)\sigma + \tfrac{1}{2}\Phi''(0)\sigma^2 + \cdots \qquad (2.8.18)$$

Substituting this into the integral and using the values of the definite integrals

$$\int_{-\infty}^{\infty} e^{-\rho\sigma^2} d\sigma = \sqrt{\pi/\rho}$$

$$\int_{-\infty}^{\infty} e^{-\rho\sigma^2}\sigma^2 d\sigma = \frac{1}{2}\sqrt{\pi/\rho^3}$$

we obtain

$$I = e^{\rho f(z_0)}\sqrt{\pi/\rho}\left\{\Phi(0) + \frac{1}{4\rho}\Phi''(0) + \cdots\right\} \qquad (2.8.19)$$

Thus, the method of steepest descent allows us to represent the integral in the form of a series of inverse powers of the large parameter ρ.

If the function $\Phi(\sigma)$ varies sufficiently slowly compared with $e^{-\rho\sigma^2}$, that is, if its derivatives are sufficiently small, we can limit ourselves to the first term in Eq. (2.8.19),

Let us now write the series in Eq. (2.8.19) in powers of σ. Differentiating Eq. (2.8.12), we have

$$f'(z)\frac{dz}{d\sigma} = -2\sigma$$

Solving for $dz/d\sigma$ and substituting into Eq. (2.8.17) yields

$$\Phi(\sigma) = -2\sigma\frac{F(z)}{f'(z)} \qquad (2.8.20)$$

We now represent $f(z)$ and $F(z)$ as a power series in $(z - z_0)$:

$$f(z) = f(z_0) - A(z - z_0)^2 + B(z - z_0)^3 + C(z - z_0)^4 + \cdots$$

$$(2.8.21)$$

METHOD OF STEEPEST DESCENT

where $A = -\frac{1}{2}f''(z_0)$, $B = \frac{1}{6}f'''(z_0)$, $C = \frac{1}{24}f^{(4)}(z_0)$
and

$$F(z) = F(z_0)\left[1 + P(z - z_0) + Q(z - z_0)^2 + \cdots\right]$$

where
$$P = \frac{F'(z_0)}{F(z_0)}, \qquad Q = \frac{1}{2}\frac{F''(z_0)}{F(z_0)} \qquad (2.8.22)$$

Substituting $f(z)$ from Eq. (2.8.21) into Eq. (2.8.12), we obtain

$$-A(z - z_0)^2 + B(z - z_0)^3 + C(z - z_0)^4 + \cdots = -\sigma^2 \qquad (2.8.23)$$

We now want to invert this series. We set

$$(z - z_0) = \frac{\sigma}{\sqrt{A}}\left(1 + a_1\sigma + a_2\sigma^2 + \cdots\right) \qquad (2.8.24)$$

Substituting this expression into Eq. (2.8.23) and equating coefficients of like powers of σ, we obtain

$$a_1 = \frac{B}{2A^{3/2}}, \qquad a_2 = \frac{C}{2A^2} + \frac{5}{8}\frac{B^2}{A^3} \qquad (2.8.25)$$

From Eq. (2.8.20), taking into account Eqs. (2.8.21) and (2.8.22), we have

$$\Phi(\sigma) = -2\sigma F(z_0)\frac{1 + P(z - z_0) + Q(z - z_0)^2}{-2A(z - z_0) + 3B(z - z_0)^2 + 4C(z - z_0)^3}$$

Substituting for $(z - z_0)$ and using Eq. (2.8.24), we obtain

$$\Phi(\sigma) = \frac{F(z_0)}{\sqrt{A}}\left[1 + \left(\frac{P}{A^{1/2}} + \frac{B}{A^{3/2}}\right)\sigma\right.$$
$$\left. + \left(\frac{Q}{A} + \frac{15}{8}\frac{B^2}{A^3} + \frac{3}{2}\frac{C}{A^2} + \frac{3}{2}\frac{BP}{A^2}\right)\sigma^2 + \cdots\right] \qquad (2.8.26)$$

Using the values of A, B, C, P, and Q given by Eqs. (2.8.21) and (2.8.22), we obtain

$$\Phi(0) = \left[-\frac{2}{f''(z_0)}\right]^{1/2} F(z_0)$$

$$\frac{1}{2}\Phi''(0) = \Phi(0)\left[\frac{f'''}{(f'')^2}\frac{F'}{F} + \frac{1}{4}\frac{f^{(4)}}{(f'')^2} - \frac{5}{12}\frac{(f''')^2}{(f'')^3} - \frac{F''}{Ff''}\right] \qquad (2.8.27)$$

Suppose $f(z)$ has more than one saddle point, say saddle points $z_s (= x_s + iy_s)$ and $0 \leq s \leq (n-1)$ where n is finite. Then there is only one point, say z_0, for which

$$f_1(x_0, y_0) \geq f_1(x_s, y_s), 1 \leq s < (n-1)$$

If $f_1(x_0, y_0) > f_1(x_s, y_s)$, then $z = z_0$ is the dominant saddle point, and only the contribution to the integral from this one saddle point is considered. If $f_1(x_0, y_0) = f_1(x_s, y_s)$ for some other values of s, then the approximate value of the integral I is simply the sum of terms of the form of Eq. (2.8.19) over each saddle point.

A method closely related to the method of steepest descent is Kelvin's method of stationary phase. This is discussed in the works of Kelvin [2.13], Wilf [2.14], and Watson [2.15].

In Kelvin's method of stationary phase, integrals of the type

$$\int_C F(z) e^{pf(z)} dz$$

are evaluated by using paths through saddle points such that $f_1 = $ constant instead of f_2. For this path the modulus of e^{pf} is constant while the phase varies.

The method of steepest descent and the method of stationary phase are nearly equivalent since the paths pass through the saddle point at an angle $\pi/4$ to each other.

Just which method yields the better approximation depends on the particular problem, and for each particular problem one must obtain an estimate of the error incurred for the approximation used. However, a general observation can be made concerning the usefulness of the two methods. The self-cancelling of the phase oscillations in the method of stationary phase is a weaker decay mechanism than the possible exponential decay of the exponential factor in the method of steepest descent. So the method of stationary phase can only be used when the exponential factor has a constant absolute magnitude over the integration path.

2.9. BESSEL'S DIFFERENTIAL EQUATION OF ARBITRARY ORDER

Since Bessel functions are used extensively in the solution of the wave equation for oceanic propagation, we will spend some time in studying their properties.

Bessel's equation of order ν, where ν is any real number, is given by

$$x^2 y'' + xy' + (x^2 - \nu^2) y = 0 \tag{2.9.1}$$

We will use the method of Frobenius to solve this equation. We will assume the solution has the form of an infinite series. In particular, we choose the following form for that series:

$$y = x^k \sum_{n=0}^{\infty} A_n x^{jn} \qquad (2.9.2)$$

For the moment k and j are arbitrary numbers. We could have written the series simply as

$$y = \sum_{n=0}^{\infty} A_n x^n$$

but the choice of Eq. (2.9.2) will facilitate the solution.

Let us define the operator L as

$$L = x^2 \frac{d^2}{dx^2} + x \frac{d}{dx} + (x^2 - \nu^2) \qquad (2.9.3)$$

Then Bessel's equation (2.9.1) can be written compactly as

$$L(y) = 0 \qquad (2.9.4)$$

Now a general term of the series (2.9.2) is

$$x^r = x^{k+jn}$$

Let us determine how L operates on a single term x^r.

$$\begin{aligned} L(x^r) &= \left\{ x^2 \frac{d^2}{dx^2} + x \frac{d}{dx} + (x^2 - \nu^2) \right\} x^r \\ &= x^2 r(r-1) x^{r-2} + xrx^{r-1} + (x^2 - \nu^2) x^r \\ &= r(r-1) x^r + rx^r + x^{r+2} - \nu^2 x^r \\ &= (r^2 - \nu^2) x^r + x^{r+2} \\ &= (r + \nu)(r - \nu) x^r + x^{r+2} \qquad (2.9.5) \end{aligned}$$

First we note that there is a "jump" of 2 between terms, that is, the power of x increases by 2. Therefore, we can set

$$j = 2$$

The series solution Eq. (2.9.2) can then be written

$$y = \sum_{n=0}^{\infty} A_n x^{k+2n} \qquad (2.9.6)$$

We will assume A_0 does not vanish. Now we substitute the series Eq. (2.9.6) into Bessel's equation (2.9.4):

$$L(y) = \sum_{n=0}^{\infty} A_n L(x^{k+2n}) = 0$$

Using Eq. (2.9.5), we get, on noting $r = k + 2n$,

$$L(y) = \sum_{n=0}^{\infty} A_n \{(k + 2n + \nu)(k + 2n - \nu)x^{k+2n} + x^{k+2n+2}\}$$

$$= \sum_{n=0}^{\infty} A_n (k + 2n + \nu)(k + 2n - \nu)x^{k+2n} + \sum_{n'=0}^{\infty} A_{n'} x^{k+2n'+2}$$

(2.9.7)

where we have denoted the summation index n' instead of n in the second summation because it is a dummy index and can be called anything we choose.

We wish to combine the two series. For this to be possible, the exponent of x must be the same. Since n and n' are dummy indices, let us set

$$k + 2n = k + 2n' + 2$$

or

$$n = n' + 1 \qquad (2.9.8)$$

Using Eq. (2.9.8) in the second summation, we can write Eq. (2.9.7) as

$$L(y) = \sum_{n=0}^{\infty} A_n (k + 2n + \nu)(k + 2n - \nu)x^{k+2n} + \sum_{n=1}^{\infty} A_{n-1} x^{k+2n}$$

$$= A_0 (k + \nu)(k - \nu)x^k$$

$$+ \sum_{n=1}^{\infty} \{A_n (k + 2n + \nu)(k + 2n - \nu) + A_{n-1}\} x^{k+2n} = 0$$

(2.9.9)

In order for the series Eq. (2.9.6) to be a solution of Bessel's equation, each coefficient must vanish. Since we have assumed $A_0 \neq 0$, we require that

$$(k + \nu)(k - \nu) = 0 \qquad (2.9.10)$$

and

$$A_n (k + 2n + \nu)(k + 2n - \nu) + A_{n-1} = 0 \qquad (2.9.11)$$

BESSEL'S DIFFERENTIAL EQUATION OF ARBITRARY ORDER   93

Eq. (2.9.10) is called the indicial equation. This will be satisfied if we choose

$$k = \pm \nu \tag{2.9.12}$$

Eq. (2.9.11) is known as a recursion relation. That is, given A_0, all other coefficients are determined from Eq. (2.9.11).

The recursion formula Eq. (2.9.11) for the coefficients A_n becomes

$$A_n = \frac{-A_{n-1}}{(k + 2n + \nu)(k + 2n - \nu)} \tag{2.9.13}$$

First, consider the case when $k = \nu$. For this case we have

$$A_n = \frac{-A_{n-1}}{2^2 n(n + \nu)} \tag{2.9.14}$$

So that

$$A_1 = \frac{-A_0}{1 \cdot (\nu + 1)2^2}$$

$$A_2 = \frac{A_0}{2(\nu + 1)(\nu + 2)2^4}$$

$$\vdots$$

$$A_n = \frac{(-1)^n A_0}{n!(\nu + 1)(\nu + 2) \cdots (\nu + n)2^{2n}}$$

Thus the solution Eq. (2.9.2) corresponding to $k = \nu$ can be written

$$A_0 y_1(x) = A_0 x^\nu \left\{ 1 - \frac{1}{(\nu + 1)} \frac{x^2}{2} + \frac{1}{1 \cdot 2 \cdot (\nu + 1)(\nu + 2)} \left(\frac{x}{2}\right)^4 \right.$$

$$\left. - \cdots + \frac{(-1)^n}{n!(\nu + 1)(\nu + 2) \cdots (\nu + n)} \left(\frac{x}{2}\right)^{2n} + \cdots \right\}$$

$$\tag{2.9.15}$$

For $k = -\nu$, Eq. (2.9.13) becomes

$$A_n = \frac{-A_{n-1}}{4n(n - \nu)} \tag{2.9.16}$$

94 MATHEMATICAL REVIEW

and the series Eq. (2.9.6) becomes

$$A_0 y_2(x) = A_0 x^{-\nu} \left\{ 1 - \frac{1}{(1-\nu)} \frac{x^2}{2} + \frac{1}{1 \cdot 2(1-\nu)(2-\nu)} \left(\frac{x}{2}\right)^4 \right.$$
$$\left. - \cdots + \frac{(-1)^n}{n!(1-\nu)(2-\nu)\cdots(n-\nu)} \left(\frac{x}{2}\right)^{2n} + \cdots \right\}$$

(2.9.17)

In order to define the Bessel functions, we need to choose a normalization factor A_0. Since the normalization factor is defined in terms of the gamma function $\Gamma(\xi)$, we need to review some of its properties.

The gamma function, denoted $\Gamma(\xi)$, for any positive real ξ is defined as

$$\Gamma(\xi) = \int_0^\infty e^{-t} t^{\xi-1} \, dt, \quad \xi > 0 \qquad (2.9.18)$$

This integral only converges for $\xi > 0$. Now,

$$\Gamma(1) = \int_0^\infty e^{-t} \, dt = 1$$

$$\Gamma(2) = \int_0^\infty t e^{-t} \, dt = 1 \qquad (2.9.19)$$

For $\xi > 1$, using integration by parts,

$$\Gamma(\xi) = \int_0^\infty e^{-t} t^{\xi-1} \, dt$$
$$= [-t^{\xi-1} e^{-t}]_0^\infty + \int_0^\infty (\xi-1) t^{\xi-2} e^{-t} \, dt$$
$$= (\xi-1) \int_0^\infty e^{-t} t^{\xi-2} \, dt$$
$$= (\xi-1)\Gamma(\xi-1)$$

So, for $\xi > 0$,

$$\Gamma(\xi+1) = \xi \Gamma(\xi) \qquad (2.9.20)$$

We want to show that when $\xi = m$, m a positive integer,

$$\Gamma(m+1) = m! \qquad (2.9.21)$$

The proof will be by induction on m:

1. For $m = 1$ we have shown $\Gamma(2) = 1$
2. Assume proposition is true for $m - 1$, that is, assume $\Gamma(m) = (m - 1)!$ is true. Then

$$\Gamma(m + 1) = m\Gamma(m)$$
$$= m(m - 1)! = m!$$

Now we want to extend the definition of the gamma function to include negative, nonintegral values of ξ. We cannot use the definition Eq. (2.9.18) because it does not converge for $\xi \leq 0$. So we use the property given by Eq. (2.9.20).

For definition, if $\xi < 0$, but $\xi \neq -1, -2, -3, \ldots$, then we define $\Gamma(\xi)$ by Eq. (2.9.20), namely,

$$\Gamma(\xi) = \frac{\Gamma(\xi + 1)}{\xi}$$

We are now ready to define the Bessel function.

2.9.1. Bessel Functions of the First Kind of Nonintegral Order

For all nonintegral values of ν, we choose the normalization factor for the solution $k = \nu$ to be

$$A_0 = \frac{1}{2^\nu \Gamma(\nu + 1)}$$

and define the Bessel function of the first kind of order ν

$$J_\nu(x) = \frac{1}{2^\nu \Gamma(\nu + 1)} y_1(x)$$

or

$$J_\nu(x) = \sum_{n=0}^{\infty} \frac{(-1)^n}{n! \Gamma(\nu + n + 1)} \left(\frac{x}{2}\right)^{2n+\nu} \qquad (2.9.22a)$$

and for the solution $k = -\nu$, we choose the normalization factor

$$A_0 = \frac{1}{2^{-\nu} \Gamma(1 - \nu)}$$

and define the Bessel functions of the first kind of order $-\nu$ as

$$J_{-\nu}(x) = \frac{1}{2^{-\nu} \Gamma(1 - \nu)} y_2(x)$$

or

$$J_{-\nu}(x) = \sum_{n=0}^{\infty} \frac{(-1)^n}{n!\Gamma(1+n-\nu)} \left(\frac{x}{2}\right)^{2n-\nu} \quad (2.9.22b)$$

It can be shown that these series converge uniformly for all finite values of x.

Now we can show that $J_\nu(x)$ and $J_{-\nu}(x)$ are two independent solutions of Bessel's equation. Theorem 2.2.1 states that, if we can show that the Wronskian of $J_\nu(x)$ and $J_{-\nu}(x)$ does not vanish, then the two solutions are independent.

Putting Bessel's equation in the form of the Sturm-Liouville equation (2.2.1) yields

$$\frac{d}{dx}(xy') + \left(x - \frac{\nu^2}{x}\right)y = 0 \quad (2.9.23)$$

Hence, $p = x$ and the Wronskian given by Eq. (2.2.8) becomes

$$W[J_\nu(x), J_{-\nu}(x)] = \frac{\text{constant}}{x} = \frac{c}{x} \quad (2.9.24)$$

We have to show that the constant is not zero. We consider for this purpose the series expansion of the Bessel functions involved and their derivatives

$$J_\nu(x) = \sum_{n=0}^{\infty} \frac{(-1)^n (x/2)^{2n+\nu}}{n!\Gamma(n+\nu+1)} \quad (2.9.25)$$

$$J_{-\nu}(x) = \sum_{n=0}^{\infty} \frac{(-1)^n (x/2)^{2n-\nu}}{n!\Gamma(n-\nu+1)} \quad (2.9.26)$$

$$J'_\nu(x) = \sum_{n=0}^{\infty} \frac{(-1)^n (2n+\nu)(x/2)^{2n+\nu-1}}{2n!\Gamma(n+\nu+1)} \quad (2.9.27)$$

$$J'_{-\nu}(x) = \sum_{n=0}^{\infty} \frac{(-1)^n (2n-\nu)(x/2)^{2n-\nu-1}}{2n!\Gamma(n-\nu+1)} \quad (2.9.28)$$

In evaluating the Wronskian we multiply Eq. (2.9.25) by Eq. (2.9.28) and Eq. (2.9.26) by Eq. (2.9.27), compute the difference, and have to obtain according to Eq. (2.9.24) a constant, divided by x. Carrying out these operations, we find that only multiplication of the first term of Eq. (2.9.25) with the first term of Eq. (2.9.28) yields a term in $1/x$. The same holds true for the product of Eqs. (2.9.26) and (2.9.27). All other terms cancel and there remains

$$\frac{c}{x} = \frac{-\nu(x/2)^{-1}}{2\Gamma(\nu+1)\Gamma(1-\nu)} - \frac{\nu(x/2)^{-1}}{2\Gamma(1-\nu)\Gamma(\nu+1)}$$

Upon using Eq. (2.9.20), we get the final result

$$W[J_\nu(x), J_{-\nu}(x)] = \frac{-2}{x\Gamma(1-\nu)\Gamma(\nu)} \tag{2.9.29}$$

Since ν is not an integer, the Wronskian does not vanish, and so J_ν and $J_{-\nu}$ are two independent solutions when ν is not an integer.

2.9.2. Bessel Functions of the First Kind of Integral Order

When ν is a nonnegative integer m, we see from Eq. (2.9.17) that the solution $y_2(x)$ does not exist, since the denominator of A_n becomes zero. Consequently, there is only one solution and this is defined by Eq. (2.9.25) when use is made of Eq. (2.9.21). Thus we define the Bessel function of the first kind of integral order m as

$$J_m(x) = \sum_{n=0}^{\infty} \frac{(-1)^n}{n!(n+m)!} \left(\frac{x}{2}\right)^{2n+m} \tag{2.9.30}$$

2.9.3. Bessel Functions of the Second Kind — The Neumann Functions

We need to find a second independent solution of Bessel's equation when ν is an integer. We have already shown that when ν is a nonnegative real number, $J_\nu(x)$ is a solution and when ν is any nonintegral real number, $J_{-\nu}(x)$ is a solution. We will show that

$$\left(\frac{\partial J_\nu}{\partial \nu}\right)_{\nu=m} \quad \text{and} \quad \left(\frac{\partial J_{-\nu}}{\partial \nu}\right)_{\nu=m}$$

are also solutions of Bessel's equation. We will then define a Bessel function of the second kind as a particular combination of these solutions, and show that it is indeed an independent solution for $\nu = m$, an integer.

First, let us show that $(\partial J_\nu/\partial \nu)_{\nu=m}$ is a solution.

Using Eq. (2.9.3), we get

$$L\left\{\left(\frac{\partial J_\nu}{\partial \nu}\right)_{\nu=m}\right\} = \frac{\partial}{\partial \nu}\{L(J_\nu)\}_{\nu=m} = 0$$

Similarly, we can show $(\partial J_{-\nu}/\partial \nu)_{\nu=m}$ is a solution.

We now define a Bessel function of the second kind, or a Neumann function, by

$$N_m(x) \equiv \frac{1}{\pi}\left\{\left(\frac{\partial J_\nu}{\partial \nu}\right)_{\nu=m} - (-1)^m \left(\frac{\partial J_{-\nu}}{\partial \nu}\right)_{\nu=m}\right\} \tag{2.9.31}$$

MATHEMATICAL REVIEW

We will now indicate, without showing all the details, how a series representation for N_m is arrived at. We begin with

$$\frac{\partial J_\nu}{\partial \nu} = \sum_{n=0}^{\infty} \frac{(-1)^n}{n!}$$

$$\times \left\{ \frac{\ln(x/2) \cdot (x/2)^{2n+\nu} \Gamma(n+\nu+1) - (x/2)^{2n+\nu} \partial \Gamma(n+\nu+1)/\partial \nu}{\Gamma^2(n+\nu+1)} \right\}$$

$$= \ln\frac{x}{2} \sum_{n=0}^{\infty} \frac{(-1)^n (x/2)^{2n+\nu}}{n! \Gamma(n+\nu+1)} - \sum_{n=0}^{\infty} \frac{(-1)^n (x/2)^{2n+\nu}}{n! \Gamma(n+\nu+1)} \frac{\partial}{\partial \nu} \ln \Gamma(n+\nu+1)$$

This expression has to be considered for $\nu = m$. That the first term does not cause any difficulties is clear enough. The second factor in the second term requires a little more attention. In view of Eq. (2.9.20) we have

$$\Gamma(\nu + n + 1) = (n + \nu)(n + \nu - 1) \cdots (1 + \varepsilon)\Gamma(1 + \varepsilon), \qquad 0 < \varepsilon < 1$$

and therefore

$$\frac{\partial \ln \Gamma(n+\nu+1)}{\partial v} = \frac{1}{n+\nu} + \frac{1}{n+\nu-1} + \cdots + \frac{1}{2+\varepsilon}$$

$$+ \frac{1}{1+\varepsilon} + \frac{1}{\Gamma(1+\varepsilon)} \frac{\partial \Gamma(1+\varepsilon)}{\partial \varepsilon}$$

Hence,

$$\left[\frac{\partial \ln \Gamma(n+\nu+1)}{\partial \nu} \right]_{\nu=m} = \frac{1}{\nu+m} + \frac{1}{n+m-1} + \cdots + \frac{1}{2} + 1 + \Gamma' \quad (1)$$

It follows from Eq. (2.9.18) that

$$\Gamma'(1) = \int_0^\infty e^{-t} \ln t \, dt$$

The number

$$\gamma = -\int_0^\infty e^{-t} \ln t \, dt = 0.57721566 \cdots$$

is called Euler's constant.

We therefore conclude that

$$\left(\frac{\partial J_\nu}{\partial \nu} \right)_{\nu=m} = \left[\ln\left(\frac{x}{2}\right) + \gamma \right] J_m(x)$$

$$- \sum_{n=0}^{\infty} \frac{(-1)^n (x/2)^{2n+m}}{n! \Gamma(n+m+1)} \left[\frac{1}{n+m} + \cdots + \frac{1}{2} + 1 \right]$$

(2.9.32)

In case $m = 0$, the first term in the latter sum ($n = 0$) has to be set equal to zero.

There is a problem when we try to differentiate $J_{-\nu}$ with respect to ν. Now

$$J_{-\nu} = \sum_{n=0}^{\infty} \frac{(-1)^n (x/2)^{-\nu+2n}}{n!\Gamma(n - \nu + 1)}$$

The denominator $\Gamma(n - \nu + 1)$ has vertical asymptotes for $\nu = m$ and $n = 0, 1, 2, \ldots, (m - 1)$, thus making a differentiation with respect to ν at $\nu = m$ impossible. We therefore split the summation at $n = m - 1$,

$$J_{-\nu}(x) = \sum_{n=0}^{m-1} \frac{(-1)^n (x/2)^{-\nu+2n}}{n!\Gamma(n - \nu + 1)} + \sum_{n=m}^{\infty} \frac{(-1)^n (x/2)^{-\nu+2n}}{n!\Gamma(n - \nu + 1)}$$

and transform the terms of the first sum, using the relation from Whittaker and Watson [2.16],

$$\Gamma(\xi)\Gamma(1 - \xi) = \frac{\pi}{\sin \pi \xi} \quad (2.9.33)$$

If we subsequently differentiate with respect to ν, set $\nu = m$, make the substitution $-m + 2n = m + 2n'$, and write n again instead of n', we see that

$$\left(\frac{\partial J_{-\nu}}{\partial \nu}\right)_{\nu=m} = (-1)^m \sum_{n=1}^{m-1} \frac{(m - n - 1)!(x/2)^{-m+2n}}{n!}$$

$$+ \sum_{n=0}^{\infty} \frac{(-1)^{m+n}(x/2)^{m+2n}}{\Gamma(n + 1)(n + m)!}\left(\ln(x/2) - \frac{1}{n} - \cdots - \frac{1}{2} - 1 + \gamma\right)$$

$$= (-1)^m \sum_{n=0}^{m-1} \frac{(m - n - 1)!(x/2)^{-m+2n}}{n!}$$

$$+ (-1)^{m-1}[\ln(x/2) + \gamma] J_m(x)$$

$$+ (-1)^m \sum_{n=0}^{\infty} \frac{(-1)^n (x/2)^{m+2n}}{n!(n + m)!}\left(\frac{1}{n} + \cdots + \frac{1}{2} + 1\right) \quad (2.9.34)$$

Combining Eqs. (2.9.32) and (2.9.34), according to Eq. (2.9.31), yields

$$N_m(x) = \frac{2}{\pi}[\ln(x/2) + \gamma] J_m(x)$$

$$- \sum_{n=0}^{m-1} \frac{(m - n - 1)!(x/2)^{2n-m}}{\pi n!} - \sum_{n=0}^{\infty} \frac{(-1)^n (x/2)^{m+2n}}{\pi n!(m + n)!}$$

$$\times \left(\frac{1}{n + m} + \cdots + \frac{1}{n + 1} + \frac{2}{n} + \cdots + \frac{2}{2} + \frac{2}{1}\right) \quad (2.9.35)$$

This particular Bessel function of the second kind is called a Neumann function. Eq. (2.9.35) holds only for integral values of m. However a Neumann function N_ν can be defined for any real value of ν. This is accomplished by the definition

$$N_\nu(x) \equiv \frac{\cos(\pi\nu)J_\nu(x) - J_{-\nu}(x)}{\sin(\pi\nu)} \qquad (2.9.36)$$

Since J_ν and $J_{-\nu}$ are independent solutions of Bessel's equation when ν is not an integer, N_ν is clearly a second solution independent of J_ν. The series expansion for $N_\nu(x)$ when ν is not an integer can be obtained by simply inserting the series expansion for J_ν and $J_{-\nu}$ as given by Eqs. (2.9.25) and (2.9.26), respectively, into Eq. (2.9.36).

Furthermore, it can be shown by the use of L'Hospital's rule that

$$N_m(x) = \lim_{\nu \to m} \left\{ \frac{\cos(\pi\nu)J_\nu(x) - J_{-\nu}(x)}{\sin(\pi\nu)} \right\}$$

$$= \frac{1}{\pi} \left\{ \frac{\partial J_\nu}{\partial \nu} - (-1)^m \left(\frac{\partial J_{-\nu}}{\partial \nu} \right) \right\}_{\nu=m}$$

which is identical with Eq. (2.9.31).

We can show by the same method that was employed to derive Eq. (2.9.29) that

$$W[J_\nu(x), N_\nu(x)] = \frac{2}{\pi x} \qquad (2.9.37)$$

2.9.4. Bessel Functions of the Third Kind — Hankel Functions

To define the Hankel functions, we will seek a solution of Bessel's equation (2.9.1) in the form of a contour integral. For the complex variable z, Bessel's equation takes the form

$$z^2 w'' + zw' + (z^2 - \nu^2)w = 0 \qquad (2.9.38)$$

In place of the unknown function $w(z)$, we introduce a new unknown function $V(\zeta)$ of the complex variable ζ by means of the contour integral

$$w(z) = \int_C K(z, \zeta) V(\zeta) \, d\zeta \qquad (2.9.39)$$

where the kernel of the transform $K(z, \zeta)$, assumed to be analytic, and the contour C are to be determined so that the integral representative given by Eq. (2.9.39) is a solution of Bessel's equation.

BESSEL'S DIFFERENTIAL EQUATION OF ARBITRARY ORDER 101

Substituting the expression for $w(z)$ into Bessel's equation (2.9.38) gives

$$\int_c \left\{ z^2 \frac{\partial^2 K}{\partial z^2} + z \frac{\partial K}{\partial z} + z^2 K - \nu^2 K \right\} V(\zeta) \, d\zeta = 0 \qquad (2.9.40)$$

At this point K is a completely arbitrary function, and we can impose any condition on it that we please. So we will require K to be a solution of the differential equation

$$z^2 \frac{\partial^2 K}{\partial z^2} + z \frac{\partial K}{\partial z} + z^2 K = -\frac{\partial^2 K}{\partial \zeta^2} \qquad (2.9.41)$$

This equation has the simple solution

$$K(z, \zeta) = a e^{iz \sin \zeta} + b e^{-iz \sin \zeta}$$

We will choose $a = 0$ and $b = 1/\pi$, so that

$$K(z, \zeta) = \frac{1}{\pi} e^{-iz \sin \zeta} \qquad (2.9.42)$$

Substituting Eq. (2.9.41) into Eq. (2.9.40) results in

$$-\int_c \left\{ \frac{\partial^2 K}{\partial \zeta^2} + \nu^2 K \right\} V(\zeta) \, d\zeta = 0$$

Integrating the first term twice by parts yields

$$-\int_c \left\{ \frac{\partial^2 K}{\partial \zeta^2} + \nu^2 K \right\} V(\zeta) \, d\zeta = -\left\{ V \frac{\partial K}{\partial \zeta} - K \frac{dV}{d\zeta} \right\}_A^B$$

$$-\int_c K \left(\frac{d^2 V}{d\zeta^2} + \nu^2 V \right) d\zeta \qquad (2.9.43)$$

We now choose V to satisfy the differential equation

$$\frac{d^2 V}{d\zeta^2} + \nu^2 V = 0 \qquad (2.9.44)$$

Of the two independent solutions of Eq. (2.9.44), we will choose the two arbitrary constants so that

$$V(\zeta) = e^{i\nu \zeta} \qquad (2.9.45)$$

102 MATHEMATICAL REVIEW

At this point the contour is still arbitrary. Thus, if we choose C with end points A and B so that the integrated term in Eq. (2.9.43), namely,

$$\left[V \frac{\partial K}{\partial \zeta} - K \frac{dV}{d\zeta} \right]_A^B \tag{2.9.46}$$

vanishes, then w given by

$$w(z) = \frac{1}{\pi} \int_C e^{-iz \sin \zeta + i\nu \zeta} \, d\zeta \tag{2.9.47}$$

will be a solution of Bessel's equation (2.9.38).

Consider the two contours C_1 and C_2 in the $\zeta = \alpha + i\beta$ plane shown in Figures 2.9.1 and 2.9.2, respectively. For the contour C_1, $A = -\pi + i\infty$ and $B = -i\infty$. For the contour C_2, $A = -i\infty$ and $B = \pi + i\infty$. If we use the expression for K given by Eq. (2.9.42) and the expression for V given by Eq. (2.9.45), and substitute these into Eq. (2.9.46), it is relatively easy to see that this integrated expression vanishes at end points A and B for both contours C_1 and C_2, provided Re $z > 0$.

We can now use the solution Eq. (2.9.47) evaluated along these two contours to define the Hankel functions. The Hankel function of the first kind of order ν is defined by

$$H_\nu^{(1)}(z) = \frac{1}{\pi} \int_{C_1} e^{-iz \sin \zeta + i\nu \zeta} \, d\zeta \tag{2.9.48}$$

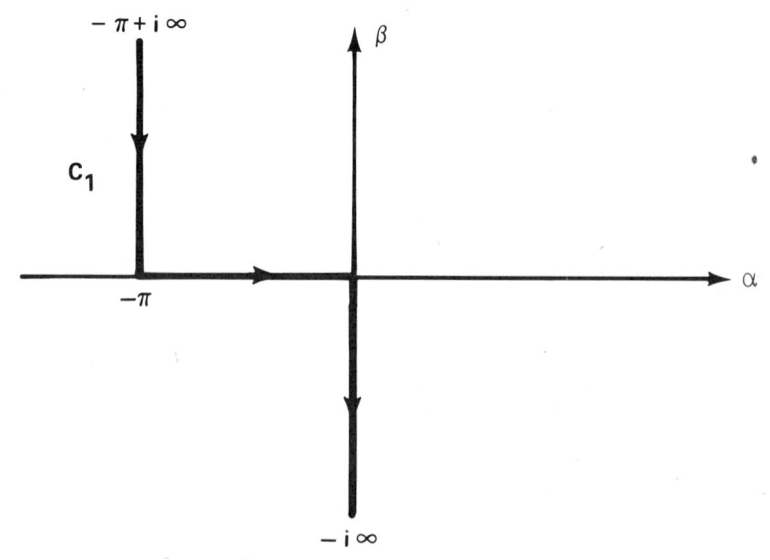

FIGURE 2.9.1. Contour for the Hankel function of the first kind.

BESSEL'S DIFFERENTIAL EQUATION OF ARBITRARY ORDER   103

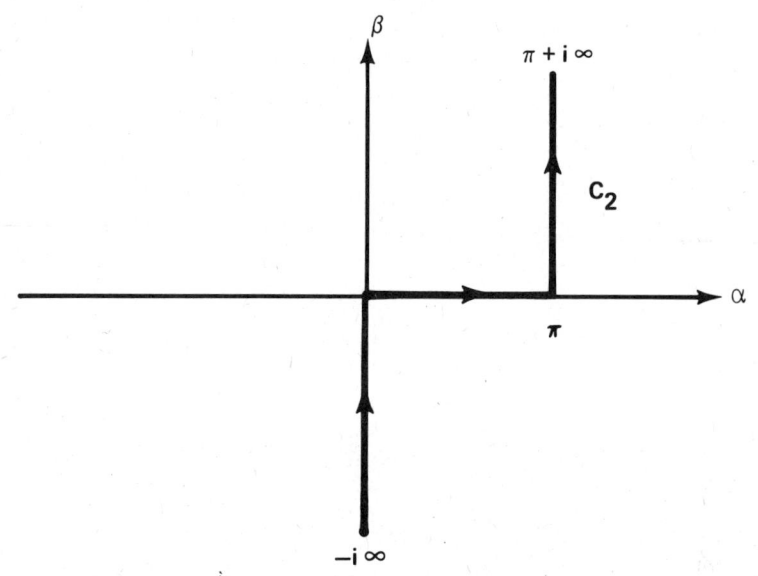

FIGURE 2.9.2. Contour for the Hankel function of the second kind.

and the Hankel function of the second kind of order ν is defined by

$$H_\nu^{(2)}(z) = \frac{1}{\pi} \int_{C_2} e^{-iz\sin\zeta + i\nu\zeta}\, d\zeta \qquad (2.9.49)$$

Now let $z = \rho$ where ρ is a positive real number. We want to find the asymptotic form for the Hankel functions for large values of ρ. Consider the Hankel function of the first kind given by Eq. (2.9.48). This integral has the same form as the integral given by Eq. (2.8.1):

$$H_\nu^{(1)}(\rho) = \frac{1}{\pi} \int_{C_1} e^{-i\rho\sin\zeta + i\nu\zeta}\, d\zeta \qquad (2.9.50)$$

Comparing Eq. (2.9.50) with Eq. (2.8.1), we see that

$$F = \frac{1}{\pi} e^{i\nu\zeta}$$

$$\rho = \rho$$

$$f = -i\sin\zeta$$

We want to use the expression given by Eq. (2.8.19) and we will retain only the first term.

$$\frac{1}{\pi} \int_{C_1} e^{-i\rho\sin\zeta + i\nu\zeta}\, d\zeta \approx e^{\rho f(\zeta_0)} \left(\frac{\pi}{\rho}\right)^{1/2} \Phi(0)$$

where $\Phi(0)$ is given by Eq. (2.8.27).

104 MATHEMATICAL REVIEW

Taking the derivative of f, we find

$$\frac{df}{d\zeta} = -i\cos\zeta = 0$$

The solution of this equation along contour C_1 is

$$\zeta_0 = -\frac{\pi}{2}$$

Thus,

$$f(\zeta_0) = i$$

$$f''(\zeta_0) = -i$$

$$\Phi(0) = \left[\frac{-2}{-i}\right]^{1/2}\frac{1}{\pi}e^{-i\nu\pi/2} = \frac{\sqrt{2}}{\pi}e^{-i\pi/4}e^{-i\nu\pi/2}$$

Putting all of these results together, we get

$$H_\nu^{(1)}(\rho) \approx \left(\frac{2}{\pi\rho}\right)^{1/2}e^{i(\rho-\nu\pi/2-(\pi/4))} \qquad (2.9.51)$$

Similarly, we get

$$H_\nu^{(2)}(\rho) = \left(\frac{2}{\pi\rho}\right)^{1/2}e^{i(\rho-\nu\pi/2-(\pi/4))} \qquad (2.9.52)$$

It can be shown that the Hankel functions are analytic for $\text{Re}\, z > 0$. We now need to establish a connection between the Hankel functions $H_\nu^{(1)}(z)$ and $H_\nu^{(2)}(z)$ and the Bessel functions $J_\nu(z)$ and $N_\nu(z)$. If we form the combination

$$\tfrac{1}{2}\left[H_\nu^{(1)}(z) + H_\nu^{(2)}(z)\right]$$

use the integral representations Eqs. (2.9.48) and (2.9.49) for $H_\nu^{(1)}$ and $H_\nu^{(2)}$, perform a suitable change of variables, expand the transformed integrand in a Laurent series, and integrate term by term, we get the result

$$J_\nu(z) = \frac{H_\nu^{(1)}(z) + H_\nu^{(2)}(z)}{2}$$

Similarly, it can be shown that

$$N_\nu(z) = \frac{H_\nu^{(1)}(z) - H_\nu^{(2)}(z)}{2i}$$

Since $J_\nu(z)$ and $N_\nu(z)$ are independent, it follows that $H_\nu^{(1)}(z)$ and $H_\nu^{(2)}(z)$ are independent.

Inverting these two expressions, we can express the Hankel functions of the first and second kind in terms of the Bessel functions J_ν and Neumann functions N_ν by the following relations:

$$H_\nu^{(1)}(z) = J_\nu(z) + iN_\nu(z) \tag{2.9.53}$$

$$H_\nu^{(2)}(z) = J_\nu(z) - iN_\nu(z) \tag{2.9.54}$$

By the techniques used in Section 2.9.1 above, we can show that the Wronskian of the Hankel function is given by

$$W\left[H_\nu^{(1)}(z), H_\nu^{(2)}(z)\right] = \frac{4}{\pi i z} \tag{2.9.55}$$

2.9.5. Some Recursion Formulas

Let $f_\nu(x)$ represent any of the Bessel functions. Then $f_\nu(x)$ satisfies the following recursion formulas.

$$f_{\nu-1}(x) + f_{\nu+1}(x) = \frac{2\nu}{x} f_\nu(x) \tag{2.9.56}$$

$$f_{\nu-1}(x) - f_{\nu+1}(x) = \frac{2\, df_\nu(x)}{dx} \tag{2.9.57}$$

$$\frac{\nu}{x} f_\nu(x) - f_{\nu+1}(x) = \frac{df_\nu(x)}{dx} \tag{2.9.58}$$

$$f_{\nu-1}(x) - \frac{\nu}{x} f_\nu(x) = \frac{df_\nu(x)}{dx} \tag{2.9.59}$$

These relations are proved simply by inserting the appropriate series representation in the recursive formula.

2.9.6. Some Limiting Forms for Bessel Functions

For reference purposes, some limiting forms of the various kinds of Bessel functions will be given for small and large values of their arguments. Only the leading terms will be given for simplicity.

$$x \ll 1, \quad J_\nu(x) \to \frac{1}{\Gamma(\nu+1)} \left(\frac{x}{2}\right)^\nu \tag{2.9.60}$$

$$N_\nu(x) \to \begin{cases} \dfrac{2}{\pi}\left[\ln\left(\dfrac{x}{2}\right) + \gamma\right], & \nu = 0 \tag{2.9.61} \\ -\dfrac{\Gamma(\nu)}{\pi}\left(\dfrac{2}{x}\right)^\nu, & \nu \neq 0 \tag{2.9.62} \end{cases}$$

106 MATHEMATICAL REVIEW

In these formulas, ν is real and nonnegative.

$$x \gg 1, \quad J_\nu(x) \to \sqrt{\frac{2}{\pi x}} \cos\left(x - \frac{\nu\pi}{2} - \frac{\pi}{4}\right) \tag{2.9.63}$$

$$N_\nu(x) \to \sqrt{\frac{2}{\pi x}} \sin\left(x - \frac{\nu\pi}{2} - \frac{\pi}{4}\right) \tag{2.9.64}$$

The transition from the small x behavior to the large x behavior occurs in the region of $x \sim \nu$. The asymptotic results in Eqs. (2.9.63) and (2.9.64) follow directly from Eqs. (2.9.51) and (2.9.52).

2.9.7. Modified Bessel Functions

A form of Bessel's equation (2.9.1) that is encountered quite frequently arises when the argument is purely imaginary. If we let $x = iz$, then Bessel's equation (2.9.1) becomes

$$z^2 \frac{d^2 y}{dz^2} + z \frac{dy}{dz} - (z^2 + \nu^2) y = 0 \tag{2.9.65}$$

The solutions of this equation are called modified Bessel functions. It is evident that they are just Bessel functions of pure imaginary argument. The usual choices of linearly independent solutions are denoted by $I_\nu(z)$ and $K_\nu(z)$. They are defined by

$$I_\nu(z) = i^{-\nu} J_\nu(iz) \tag{2.9.66}$$

$$K_\nu(z) = \frac{\pi}{2} i^{\nu+1} H_\nu^{(1)}(iz) \tag{2.9.67}$$

and are real functions for real z. Their limiting forms for small and large z are assuming real $\nu \geq 0$:

$$z \ll 1, \quad I_\nu(z) \to \frac{1}{\Gamma(\nu+1)} \left(\frac{z}{2}\right)^\nu \tag{2.9.68}$$

$$K_\nu(z) \begin{cases} -\left[\ln\left(\frac{z}{2}\right) + 0.5772\right], & \nu = 0 \\ \frac{\Gamma(\nu)}{2} \left(\frac{2}{z}\right)^\nu, & \nu \neq 0 \end{cases} \tag{2.9.69}$$

$$z \gg 1, \nu$$

$$I_\nu(z) \to \frac{1}{\sqrt{2\pi z}} e^z \tag{2.9.70}$$

$$K_\nu(z) \to \sqrt{\frac{\pi}{2z}} e^{-z} \tag{2.9.71}$$

2.10. AIRY'S DIFFERENTIAL EQUATION

A differential equation that will arise in the solution of the inhomogeneous, layered oceanic wave guide problem is the Airy differential equation:

$$\frac{d^2y}{dx^2} - xy = 0 \qquad (2.10.1)$$

By a transformation of variables, we change this into Bessel's equation. Substituting

$$y(x) = (x)^{1/2} z(x) \quad \text{for} \quad x \geq 0$$
$$= (-x)^{1/2} z(x) \quad \text{for} \quad x < 0$$

we obtain the following differential equation:

$$z'' + \frac{1}{x} z' - \left(\frac{1}{4x^2} + x\right) z = 0 \qquad (2.10.2)$$

To derive the general solution of this equation, let us make the further substitution:

$$t = \tfrac{2}{3} x^{3/2} \quad \text{for} \quad x \geq 0$$
$$= \tfrac{2}{3}(-x)^{3/2} = \tfrac{2}{3} x^{3/2} \quad \text{for} \quad x < 0$$

Eq. (2.10.2) will then become

$$\frac{d^2z}{dt^2} + \frac{1}{t}\frac{dz}{dt} - \left[\frac{1/9}{t^2} + 1\right] z = 0 \quad \text{for } x > 0 \qquad (2.10.3)$$

$$\frac{d^2z}{dt^2} + \frac{1}{t}\frac{dz}{dt} + \left[1 - \frac{1/9}{t^2}\right] z = 0 \quad \text{for } x < 0 \qquad (2.10.4)$$

Eq. (2.10.3) is Bessel's equation (2.9.65) for modified Bessel functions and Eq. (2.10.4) is Bessel's equation (2.9.1) for the ordinary Bessel functions. The general solutions of Eqs. (2.10.3) and (2.10.4) can be written in the form

$$z(t) = C_1 I_{-1/3}(t) + C_2 I_{1/3}(t) \quad \text{for } x \geq 0$$
$$z(t) = D_1 J_{-1/3}(t) + D_2 J_{1/3}(t) \quad \text{for } x < 0$$

If we set

$$C_1 = -C_2 = D_1 = D_3 = \tfrac{1}{3}$$

and

$$C_1 = C_2 = D_1 = -D_2 = \tfrac{1}{3}$$

we obtain two independent solutions denoted by $Ai(x)$ and $Bi(x)$:

$$Ai(x) = \frac{\sqrt{x}}{3}\left\{ I_{-1/3}\left(\tfrac{2}{3}x^{3/2}\right) - I_{1/3}\left(\tfrac{2}{3}x^{3/2}\right) \right\} \quad \text{for } x \geq 0$$

$$= \frac{\sqrt{|x|}}{3}\left\{ J_{-1/3}\left(\tfrac{2}{3}|x|^{3/2}\right) + J_{1/3}\left(\tfrac{2}{3}|x|^{3/2}\right) \right\} \quad \text{for } x < 0 \quad (2.10.5)$$

and

$$Bi(x) = \frac{\sqrt{x}}{3}\left\{ I_{-1/3}\left(\tfrac{2}{3}x^{3/2}\right) + I_{1/3}\left(\tfrac{2}{3}x^{3/2}\right) \right\} \quad \text{for } x \geq 0$$

$$= \frac{\sqrt{|x|}}{3}\left\{ J_{-1/3}\left(\tfrac{2}{3}|x|^{3/2}\right) - J_{1/3}\left(\tfrac{2}{3}|x|^{3/2}\right) \right\} \quad \text{for } x < 0 \quad (2.10.6)$$

From the power series definition of I_ν and J_ν, it follows that for

$$x \to 0, \; Ai(x) \to \frac{1}{\sqrt[3]{18}\,\Gamma(2/3)}$$

$$Bi(x) \to \frac{1}{\sqrt[3]{18}\,\Gamma(2/3)} \quad (2.10.7)$$

For large absolute values of x, we have the limiting forms for Ai and Bi:

$$Ai(x) = \frac{1}{2\sqrt{\pi}} x^{-1/4} e^{-(2/3)x^{3/2}}, \quad \text{for } x \to \infty$$

$$= \frac{1}{\sqrt{\pi}} x^{-1/4} \sin\left(\tfrac{2}{3}|x|^{3/2} + \tfrac{\pi}{4}\right), \quad \text{for } x \to -\infty \quad (2.10.8)$$

$$Bi(x) = \frac{\sqrt{3}}{\sqrt{\pi}} x^{-1/4} e^{(2/3)x^{3/2}}, \quad \text{for } x \to \infty$$

$$= \frac{\sqrt{3}}{\sqrt{\pi}} x^{-1/4} \sin\left(\tfrac{2}{3}|x|^{3/2}\right) + \tfrac{\pi}{4}, \quad \text{for } x \to -\infty \quad (2.10.9)$$

Graphs of the Airy functions are shown in Figures 2.10.1 and 2.10.2.

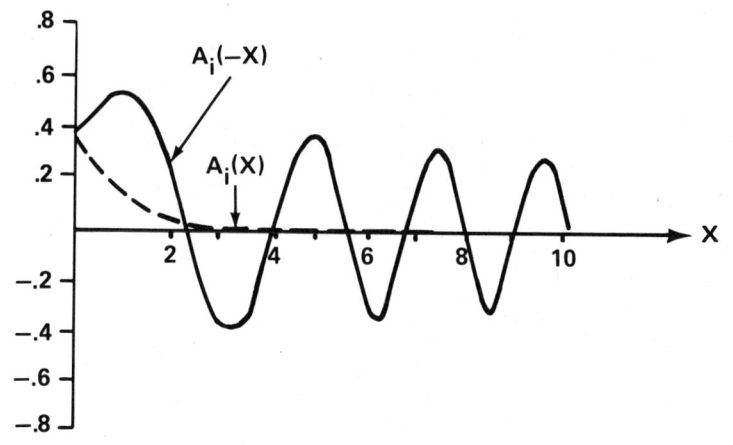

FIGURE 2.10.1. The Airy function $Ai(x)$.

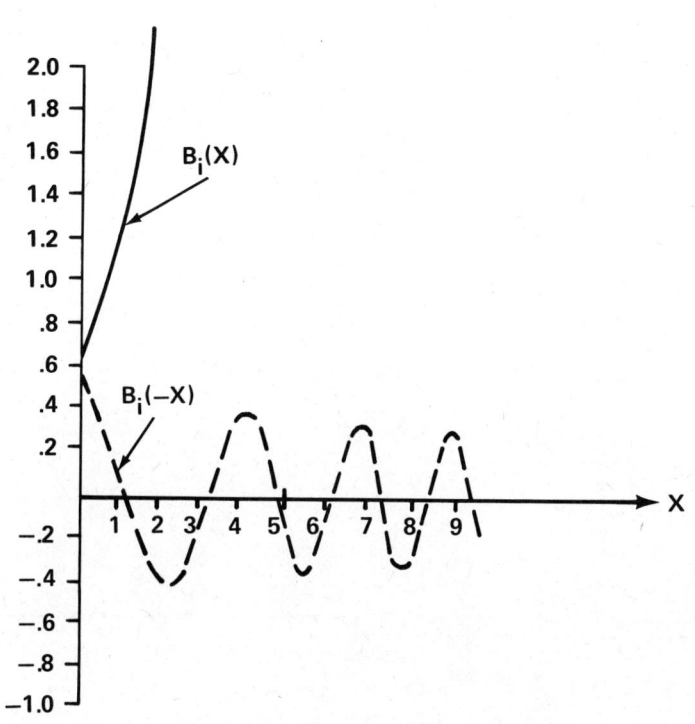

FIGURE 2.10.2. The Airy function $Bi(x)$.

REFERENCES

2.1. H. Sagan, *Boundary and Eigenvalue Problems in Mathematical Physics* (Wiley, New York, 1961), pp. 88–91.

2.2. P. Morse and H. Feshbach, *Methods of Theoretical Physics*, Part I, (McGraw-Hill, New York, 1953), Chapter 5, Section 1.

2.3 E. L. Ince, *Ordinary Differential Equations*, (Dover Publications, New York, 1956), Chapter 10.

2.4. See Sagan, Chapter 5.

2.5. R. Courant and D. Hilbert, *Methods of Mathematical Physics*, Vol. 1, (Interscience, New York, 1953).

2.6. H. S. Wilf, *Mathematics for the Physical Sciences*, (Wiley, New York, 1962), p. 222.

2.7. See Morse and Feshbach, pp. 278–279.

2.8. See Wilf, pp. 226–228.

2.9. See Morse and Feshbach, pp. 736–737.

2.10. W. E. Boyce and R. C. DiPrima, *Elementary Differential Equations and Boundary Value Problems*, (Wiley, New York, 1977), Chapter 11, Section 7.

2.11. See Morse and Feshbach, pp. 738–739.

2.12. L. M. Brekhovskikh, *Waves in Layered Media*, (Academic Press, New York, 1960), pp. 245–250.

2.13. Lord Kelvin, *Proc. Royal Soc.* **42** 80 (1887).

2.14. See Wilf, pp. 131–136.

2.15. G. N. Watson, "The Limits of Applicability of the Principle of Stationary Phase," *Proc. Cambridge Phil. Soc.* **19**, 49–55.

2.16. E. T. Whittaker and G. N. Watson, *Modern Analysis*, American Ed. (Macmillan, New York, 1948).

3 PROPAGATION IN THE OCEAN AS A BOUNDARY VALUE PROBLEM: THE HOMOGENEOUS LAYERED MODEL

3.0. INTRODUCTION

In this chapter the basic techniques for solving the wave equation are introduced by treating fairly simply models of the ocean, namely, models that consist of one and two homogeneous layers. In Section 3.1 the problem of a single homogeneous layer is solved by the complex λ-plane representation for the Green's function. In Section 3.2 the same problem is solved using an eigenfunction expansion technique for the Green's function. The same two techniques are then applied to a model that consists of two homogeneous layers. This is done in Sections 3.3 and 3.4. In both models the ocean is assumed to be bounded by a pressure-release air-water interface and a rigid bottom. These boundary conditions give rise to a discrete spectrum of eigenvalues. In Section 3.5 the rigid bottom of the second layer is allowed to recede to infinity. This gives rise to an eigenvalue spectrum that is part discrete and part continuous.

3.1. THE COMPLEX λ-PLANE REPRESENTATION OF THE GREEN'S FUNCTION FOR A HOMOGENEOUS LAYER BOUNDED BY A PRESSURE-RELEASE SURFACE AND A RIGID BOTTOM

We are now ready to solve the wave equation for the simplest oceanic waveguide model. The model consists of a single layer of constant density ρ_0 and constant sound speed c, bounded by a pressure release surface, the ocean-air interface, and a rigid bottom. A layer of constant density and sound speed is said to be homogeneous; that is, it has the same properties at every

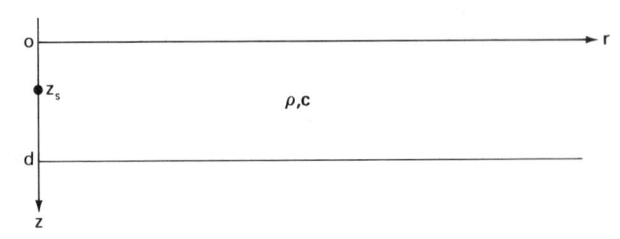

FIGURE 3.1.1. Coordinate geometry for a single homogeneous layer.

point. We will take the depth of the water column to be d. The geometry is shown in Figure 3.1.1.

A pressure-release surface is characterized by the condition that the total acoustic pressure vanishes everywhere on the surface. While this is only an approximate condition, it is an excellent approximation. This boundary condition implies that all the energy incident on the surface is reflected and that an incident plane wave experiences a 180-degree phase shift after reflection. A rigid surface is characterized by the condition that the total acoustic particle velocity vanishes everywhere on the surface. This boundary condition implies that all the energy incident on the surface is reflected and that an incident plane wave experiences no phase shift after reflection. Both of these boundary conditions are discussed in Kinsler and Frey [3.1] for the case of a plane wave.

In this chapter we want to examine solutions of the wave equation given by Eq. (1.7.30). Because we will be dealing with a homogeneous medium in this chapter, the $\nabla \rho_0$ will vanish. If, for convenience, we let $\sigma = \partial Q_1/\partial t$, then Eq. (1.7.30) becomes for our specific case

$$\nabla^2 \mathcal{P} - \frac{1}{c^2}\frac{\partial^2 \mathcal{P}}{\partial t^2} = -\sigma \qquad (3.1.1)$$

where we are now denoting the solution to this equation by the symbol $\mathcal{P}$ instead of p_1.

In this book we will only consider a point source radiating at a single frequency. Physically, a point source can be viewed as a pulsating sphere, that is, a sphere whose radius varies sinusoidally with time and whose radius is much less than a wavelength. The characteristics of such a source are described in Kinsler and Frey [3.2]. From symmetry considerations it is clear that such a source will radiate a spherical wave in a homogeneous medium with no boundaries.

A point source can be described mathematically by a source function σ of the form

$$\sigma = A\delta(\mathbf{r} - \mathbf{r}_s)e^{-i\omega t} \qquad (3.1.2)$$

Here $\delta(\mathbf{r} - \mathbf{r}_s)$ is the Dirac delta function, $\mathbf{r}_s$ is the position vector of the source, A is an amplitude factor, and $\omega/2\pi$ is the frequency of the source.

Since the source is radiating at a single frequency, we would expect the solution of Eq. (3.1.1) to also have the same time variation. Thus we let

$$\mathscr{P}(\mathbf{r}, t) = p(\mathbf{r})e^{-i\omega t} \tag{3.1.3}$$

where the acoustic pressure amplitude $p(\mathbf{r})$ is not a function of time. If we substitute this expression for $\mathscr{P}$ into Eq. (3.1.1), we see that the quantity $p(\mathbf{r})$ satisfies the Helmholtz equation

$$\nabla^2 p + k^2 p = -\delta(\mathbf{r} - \mathbf{r}_s) \tag{3.1.4}$$

To obtain Eq. (3.1.4), we have set A equal to 1 and defined the wave number k to be

$$k = \frac{\omega}{c} \tag{3.1.5}$$

The quantity A is set equal to 1 only for convenience because we will only be interested in a ratio of pressures and, hence, A would cancel out. For other applications it is trivial to reinsert A in the final solution.

With the exception of Chapter 7, Sections 5 and 6, we will work in cylindrical coordinates and assume that the waveguide is cylindrically symmetric. The relation between cylindrical coordinates (r, z, θ) and Cartesian coordinates (x, y, z) is given mathematically by

$$x = r\cos\theta$$
$$y = r\sin\theta$$
$$z = z \tag{3.1.6}$$

and is shown geometrically in Figure 3.1.2. The condition of cylindrical symmetry means that the solution is independent of the angle θ.

The Jacobian of the transformation is

$$J = \begin{vmatrix} \frac{\partial x}{\partial r} & \frac{\partial x}{\partial \theta} & \frac{\partial x}{\partial z} \\ \frac{\partial y}{\partial r} & \frac{\partial y}{\partial \theta} & \frac{\partial y}{\partial z} \\ \frac{\partial z}{\partial r} & \frac{\partial z}{\partial \theta} & \frac{\partial z}{\partial z} \end{vmatrix} = \begin{vmatrix} \cos\theta & -r\sin\theta & 0 \\ \sin\theta & r\cos\theta & 0 \\ 0 & 0 & 1 \end{vmatrix} = r \tag{3.1.7}$$

Let ξ_1, ξ_2, ξ_3 be any arbitrary coordinates. Is this arbitrary coordinate system, the delta function only has meaning when it is expressed as an integral in the form

$$\iiint \delta(\xi_1 - \xi_1')\delta(\xi_2 - \xi_2')\delta(\xi_3 - \xi_3') \, d\xi_1 \, d\xi_2 \, d\xi_3 \tag{3.1.8}$$

or, in vector notation,

$$\iiint \delta(\mathbf{r} - \mathbf{r}') \, d\mathbf{r}$$

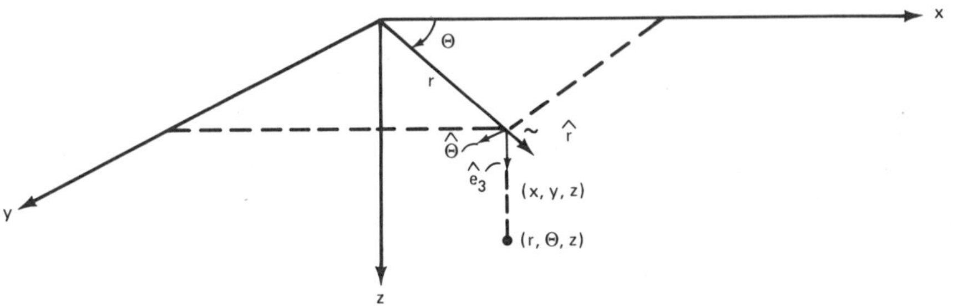

FIGURE 3.1.2. Relationship between Cartesian coordinates and cylindrical coordinates.

where (ξ_1, ξ_2, ξ_3) are the components of the vector $\mathbf{r}$ in the arbitrary coordinate system and $d\mathbf{r} = d\xi_1 \, d\xi_2 \, d\xi_3$. In Cartesian coordinates this integral becomes

$$\iiint \delta(x - x')\delta(y - y')\delta(z - z') \, dx \, dy \, dz \tag{3.1.9}$$

To transform a volume element for Cartesian coordinates to an arbitrary coordinate system, we saw in Chapter 1, Eq. (1.2.3), that

$$dx \, dy \, dz = |J| \, d\xi_1 \, d\xi_2 \, d\xi_3 \tag{3.1.10}$$

where J is the Jacobian of the transformation.

Thus if wish to write the delta function in terms of the ξ coordinates, we must set

$$\delta(x - x')\delta(y - y')\delta(z - z') = \frac{1}{|J|}\delta(\xi_1 - \xi_1')\delta(\xi_2 - \xi_2')\delta(\xi_3 - \xi_3') \tag{3.1.11}$$

Then Eq. (3.1.9) becomes Eq. (3.1.8) on using Eqs. (3.1.10) and (3.1.11), that is,

$$\iiint \delta(x - x')\delta(y - y')\delta(z - z') \, dx \, dy \, dz$$

$$= \iiint \frac{1}{|J|}\delta(\xi_1 - \xi_1')\delta(\xi_2 - \xi_2')\delta(\xi_3 - \xi_3')|J| \, d\xi_1 \, d\xi_2 \, d\xi_3$$

$$= \iiint \delta(\xi_1 - \xi_1')\delta(\xi_2 - \xi_2')\delta(\xi_3 - \xi_3') \, d\xi_1 \, d\xi_2 \, d\xi_3$$

We can now express the delta function in terms of cylindrical coordinates by letting $\xi_1 = r$, $\xi_2 = \theta$, and $\xi_3 = z$. Then using Eq. (3.1.7), Eq. (3.1.11) becomes

$$\delta(x - x')\delta(y - y')\delta(z - z') = \frac{1}{r}\delta(r - r')\delta(\theta - \theta')\delta(z - z') \tag{3.1.12}$$

We have already expressed the Helmholtz equation in cylindrical coordinates in Section 2.1. Consequently, Eq. (3.1.4) becomes in cylindrical coordinates

$$\frac{\partial^2 p}{\partial r^2} + \frac{1}{r}\frac{\partial p}{\partial r} + \frac{\partial^2 p}{\partial z^2} + k^2 p = -\frac{1}{r}\delta(r)\delta(\theta)\delta(z - z_s) \quad (3.1.13)$$

The assumption of cylindrical symmetry, of course, means that the acoustic pressure field p is independent of angle θ and, consequently, no derivatives of θ appear on the left side of Eq. (3.1.13). However, a function of θ does appear on the right side. We can easily remove the θ dependence by simply integrating over all values of θ from 0 to 2π. This integration yields

$$\int_0^{2\pi} d\theta \left(\frac{\partial^2 p}{\partial r^2} + \frac{1}{r}\frac{\partial p}{\partial r} + \frac{\partial^2 p}{\partial z^2} + k^2 p \right) = -\frac{1}{r}\delta(r)\delta(z - z_s)\int_0^{2\pi}\delta(\theta)\,d\theta$$

$$2\pi\left(\frac{\partial^2 p}{\partial r^2} + \frac{1}{r}\frac{\partial p}{\partial r} + \frac{\partial^2 p}{\partial z^2} + k^2 p \right) = -\frac{1}{r}\delta(r)\delta(z - z_s)$$

or, finally, we get the desired wave equation:

$$\frac{\partial^2 p}{\partial r^2} + \frac{1}{r}\frac{\partial p}{\partial r} + \frac{\partial^2 p}{\partial z^2} + k^2 p = -\frac{1}{2\pi r}\delta(r)\delta(z - z_s) \quad (3.1.14)$$

We saw in Section 2.1 that if we use separation of variables on the homogeneous Helmholtz equation (no source term), we obtained the separated differential equations

$$\frac{d^2 R}{dr^2} + \frac{1}{r}\frac{dR}{dr} + \lambda R = 0 \quad (3.1.15)$$

$$\frac{d^2 Z}{dz^2} + (k^2 + \lambda)Z = 0 \quad (3.1.16)$$

where we have assumed for the separated solution that

$$p(r, z) = R(r)Z(z) \quad (3.1.17)$$

and, in the notation of Section 2.1, we have here put $\lambda = k_0^2 \xi^2$ and $k = k_0 n$. We are now ready to define the Green's function for this problem. We begin by defining two Green's functions which are separate functions of r and z. We used Eqs. (3.1.15) and (3.1.16) to motivate this definition. We denote the radial Green's function by $G_1(r, r_s, \lambda)$ and require that it satisfy the following inhomogeneous differential equation:

$$\frac{d^2 G_1}{dr^2} + \frac{1}{r}\frac{dG_1}{dr} + \lambda G_1 = -\frac{1}{2\pi r}\delta(r) \quad (3.1.18)$$

If we put Eq. (3.1.18) into the standard Sturm-Liouville form exhibited by Eq. (2.7.5), we obtain

$$-\frac{d}{dr}\left[2\pi r \frac{dG_1}{dr}\right] - 2\pi r \lambda G_1 = \delta(r) \tag{3.1.19}$$

and identify $q = 0$ and

$$w = 2\pi r \tag{3.1.20}$$

$$\lambda' = \lambda \tag{3.1.21}$$

$$p = 2\pi r \tag{3.1.22}$$

Further we require that G_1 satisfy the boundary conditions

(R.1) $$\lim_{r \to \infty} \sqrt{r}\left[\frac{\partial G_1}{\partial r} - i\sqrt{\lambda}\, G_1\right] = 0$$

This is the Sommerfeld radiation condition for the behavior of the field at infinity. It is easier to understand the need for this condition after we have discussed the solution of Eq. (3.1.18), so we will postpone that discussion until later in this section. There we will see that it is not sufficient to require that only the field $G_1(r, r_s, \lambda)$ vanish at infinity, but that this stronger condition is needed.

(R.2) $$\lim_{\varepsilon \to 0}\left\{\varepsilon \frac{dG_1}{dr}(\varepsilon)\right\} = -\frac{1}{2\pi}$$

This is the discontinuity condition given by Eq. (2.7.7a) where in that notation, $z' = r_s = 0$, $p = 2\pi r$, and ε is a small positive quantity that is allowed to approach zero. The physical interpretation of this boundary condition is identical with the same condition for $G_2(z, z_s)$ and will be discussed under boundary condition Z.3.

We denote the Green's function for the z-direction by $G_2(z, z_s, \lambda)$ and require that it satisfy the differential equation

$$\left(\frac{d^2 G_2}{dz^2}\right) + (k^2 - \lambda)G_2 = -\delta(z - z_s) \tag{3.1.23}$$

If we put Eq. (3.1.23) into the standard form of the inhomogeneous Sturm-Liouville equation (2.7.5), we obtain

$$-\frac{d}{dz}\left(\frac{dG_2}{dz}\right) - k^2 G_2 + \lambda G_2 = \delta(z - z_s) \tag{3.1.24}$$

and identify $q = -k^2$ and

$$w = 1 \tag{3.1.25}$$

$$\lambda' = -\lambda \tag{3.1.26}$$

$$p = 1 \tag{3.1.27}$$

Further, we require that G_2 satisfy the boundary conditions:

(Z.1) $$G_2(o, z_s, \lambda) = 0$$

This is the pressure-release condition at the air-water interface. This condition states that the total acoustic pressure (incident wave plus reflected wave) must vanish at the sea surface $z = 0$.

(Z.2) $$\lim_{\varepsilon \to 0} \{G_2(z_s + \varepsilon, z_s, \lambda) - G_2(z_s - \varepsilon, z_s, \lambda)\} = 0$$

This is the continuity requirement on G_2.

(Z.3) $$\lim_{\varepsilon \to 0} \left\{ \frac{dG_2}{dz}(z_s + \varepsilon, z_s, \lambda) - \frac{dG_2}{dz}(z_s - \varepsilon, z_s, \lambda) \right\} = -1$$

This is the discontinuity condition given by Eq. (2.7.7b). Physically dG_2/dz is proportional to the z component of the acoustic particle velocity. This is easily seen by expressing Eq. (1.7.11) in cylindrical coordinates after neglecting ∇p_0. Since a point source can be assumed to be a small, radially vibrating sphere, and since the particle velocity is a vector quantity, the fluid particles are moving in opposite directions on the two sides of the vibrating sphere which are separated at $z = z_s$. Consequently, there is a real, physical discontinuity in the particle velocity.

(Z.4) $$\frac{dG_2}{dz}(d, z_s, \lambda) = 0$$

This is the condition for a rigid bottom. Since dG_2/dz is proportional to the z component of the particle velocity, this condition states that the normal component of the particle velocity must vanish at the bottom.

It is important to note that while Eqs. (3.1.18) and (3.1.23) are similar in form to the separated equations (3.1.15) and (3.1.16) which actually motivated the choice for Eqs. (3.1.18) and (3.1.23), the latter two inhomogeneous differential equations cannot be obtained from the wave equation (3.1.14) by separation of variables because of the inhomogeneous terms involving the delta function. So at this point we are defining $G_1(r, r_s, \lambda)$ and $G_2(z, z_s, \lambda)$ by the above conditions.

We are now ready to obtain a solution $p(r, z)$ for the wave equation. Note that $p(r, z)$ is really the Green's function for the partial differential equation. Assume that the variable λ', which appears in Eq. (2.7.5) and which is given by Eqs. (3.1.21) and (3.1.26) for our particular case, is a complex variable. Then we shall prove that the desired solution of the wave equation can be represented in the form

$$p(r, r_s, z, z_s) = -\frac{1}{2\pi i} \int_{C_\lambda} G_1(r, r_s, \lambda) G_2(z, z_s, \lambda) \, d\lambda \quad (3.1.28)$$

where C_λ is a contour in the complex λ plane taken in the positive sense and to be more fully specified later.

118 PROPAGATION IN THE OCEAN: THE HOMOGENEOUS LAYERED MODEL

For the radial Green's function G_1, condition (2.7.13) becomes

$$\frac{1}{2\pi i}\int_{C_\lambda} G_1(r, r_s, \lambda)\, d\lambda = -\frac{\delta(r - r_s)}{2\pi r} \qquad (3.1.29)$$

where we have used Eqs. (3.1.20) and (3.1.21), and for G_2

$$\frac{1}{2\pi i}\int_{C_\lambda} G_2(z, z_s, \lambda)\, d\lambda = \delta(z - z_s) \qquad (3.1.30)$$

where we have used Eqs. (3.1.25) and (3.1.26).

Now let us substitute Eq. (3.1.28) into the left side of Eq. (3.1.14):

$$\frac{\partial^2 p}{\partial r^2} + \frac{1}{r}\frac{\partial p}{\partial r} + \frac{\partial^2 p}{\partial z^2} + k^2 p$$

$$= -\frac{1}{2\pi i}\int \frac{d^2 G_1}{dr^2} G_2\, d\lambda - \frac{1}{2\pi i r}\int \frac{dG_1}{dr} G_2\, d\lambda - \frac{1}{2\pi i}\int G_1 \frac{d^2 G_2}{dz^2}\, d\lambda$$

$$- \frac{k^2}{2\pi i}\int G_1 G_2\, d\lambda$$

$$\frac{\partial^2 p}{\partial r^2} + \frac{1}{r}\frac{\partial p}{\partial r} + \frac{\partial^2 p}{\partial z^2} + k^2 p$$

$$= -\frac{1}{2\pi i}\int \left\{\frac{d^2 G_1}{dr^2} + \frac{1}{r}\frac{dG_1}{dr}\right\} G_2\, d\lambda$$

$$- \frac{1}{2\pi i}\int G_1 \frac{d^2 G_2}{dz^2}\, d\lambda - \frac{k^2}{2\pi i}\int G_1 G_2\, d\lambda$$

$$= -\frac{1}{2\pi i}\int \left\{-\lambda G_1 - \frac{1}{2\pi r}\delta(r - r_s)\right\} G_2\, d\lambda$$

$$- \frac{1}{2\pi i}\int G_1\{-k^2 G_2 + \lambda G_2 - \delta(z - z_s)\}\, d\lambda - \frac{k^2}{2\pi i}\int G_1 G_2\, d\lambda$$

$$= \frac{1}{2\pi i}\int \lambda G_1 G_2\, d\lambda + \frac{1}{4\pi^2 i r}\delta(r - r_s)\int G_2\, d\lambda$$

$$+ \frac{1}{2\pi i}\int k^2 G_1 G_2\, d\lambda - \frac{1}{2\pi i}\int \lambda G_1 G_2\, d\lambda$$

$$+ \frac{1}{2\pi i}\delta(z - z_s)\int G_1\, d\lambda - \frac{k^2}{2\pi i}\int G_1 G_2\, d\lambda$$

$$= \frac{1}{4\pi^2 i r}\delta(r - r_s)\int G_2\, d\lambda + \frac{1}{2\pi i}\delta(z - z_s)\int G_1\, d\lambda$$

COMPLEX GREEN'S FUNCTION FOR A HOMOGENEOUS LAYER 119

We have used the differential equations (3.1.18) and (3.1.23) to derive this result.

To summarize, we have shown

$$\frac{\partial^2 p}{\partial r^2} + \frac{1}{r}\frac{\partial p}{\partial r} + \frac{\partial^2 p}{\partial z^2} + k^2 p = \frac{1}{4\pi^2 ir}\delta(r - r_s)\int_{C_\lambda} G_2 \, d\lambda$$

$$+ \frac{1}{2\pi i}\delta(z - z_s)\int_{C_\lambda} G_1 \, d\lambda \quad (3.1.31)$$

Now there are two choices for C_λ.

Choice 1. Choose C_λ to be the contour C_r^+ enclosing the singularities of G_1 but not of G_2 and taken in the positive sense (counterclockwise). Then using Eq. (3.1.29) and noting that C_λ defined for that integral is already taken in the positive sense, we get

$$\frac{\delta(r - r_s)}{4\pi^2 ir}\int_{C_z^-} G_2 \, d\lambda + \frac{1}{2\pi i}\delta(z - z_s)\int_{C_z^-} G_1 \, d\lambda$$

$$= -\frac{1}{2\pi r}\delta(z - z_s)\delta(r - r_s)$$

and so the right side of Eq. (3.1.31) equals the right side of Eq. (3.1.14) and p as given by Eq. (3.1.28) is the Green's function solution for the wave equation.

Choice 2. Choose C_λ to be the contour C_z^- enclosing the singularities of G_2 but not of G_1 and taken in the negative sense (clockwise). Then using Eq. (3.1.30), and noting that C_λ defined for that integral is taken in the positive sense, we get

$$\frac{1}{4\pi^2 ir}\delta(r - r_s)\int_{C_z^-} G_2 \, d\lambda + \frac{1}{2\pi i}\delta(z - z_s)\int_{C_z^-} G_1 \, d\lambda$$

$$= -\frac{1}{2\pi r}\delta(r - r_s)\delta(z - z_s)$$

and so the right side of Eq. (3.1.31) again equals the right side of Eq. (3.1.14), and p as given by Eq. (3.1.28) is the Green's function for the wave equation.

For any particular problem, either contour C_r^+ or C_z^- can be chosen. The one chosen is the one that is most convenient.

We now obtain explicit solutions for G_1 and G_2. First let us determine $G_1(r, r_s, \lambda)$. We note that Eq. (3.1.18) reduces to Eq. (3.1.15) everywhere except for the single point $r = 0$. Therefore, we can solve Eq. (3.1.15) for the region $r > 0$ and apply the discontinuity condition at $r = 0$ given by boundary condition R.2, so that the Green's function G_1 has the correct singularity at the source location. The general solution of Eq. (3.1.15) which is Bessel's equation

120 PROPAGATION IN THE OCEAN: THE HOMOGENEOUS LAYERED MODEL

of order zero is

$$G_1 = AH_0^{(1)}(\sqrt{\lambda}\,r) + BH_0^{(2)}(\sqrt{\lambda}\,r)$$

The asymptotic expansions for the Hankel functions were derived in Section 2.9 and are given by Eqs. (2.9.51) and (2.9.52). These are, for our case,

$$H_0^{(1)}(\sqrt{\lambda}\,r) \approx \left[\frac{2}{\pi r \sqrt{\lambda}}\right]^{1/2} e^{i(\sqrt{\lambda}\,r - \pi/4)} \qquad (3.1.32)$$

$$H_0^{(2)}(\sqrt{\lambda}\,r) \approx \left[\frac{2}{\pi r \sqrt{\lambda}}\right]^{1/2} e^{-i(\sqrt{\lambda}\,r - \pi/4)} \qquad (3.1.33)$$

For the time factor $e^{-i\omega t}$, $H_0^{(1)}(\sqrt{\lambda}\,r)$ asymptotically represents outgoing waves and $H_0^{(2)}(\sqrt{\lambda}\,r)$ asymptotically represents incoming waves which are absorbed at the origin.

The boundary condition R.1 requires $B = 0$. So for $r > 0$ the solution of Bessel's equation that satisfies the boundary condition R.1 is

$$G_1(r, \lambda) = AH_0^{(1)}(\sqrt{\lambda}\,r), \qquad r > 0 \qquad (3.1.34)$$

We will now use boundary condition R.2 to evaluate the remaining constant A. Application of the boundary conditions yields

$$\varepsilon A k H_0^{(1)'}(k\varepsilon) = -\frac{1}{2\pi}$$

Using Eq. (2.9.53) yields

$$\varepsilon A k \{ J_0'(k\varepsilon) + i N_0'(k\varepsilon) \} = -\frac{1}{2\pi}$$

and application of Eq. (2.9.58) results in

$$-\varepsilon A k \{ J_1(k\varepsilon) + i N_1(k\varepsilon) \} = -\frac{1}{2\pi} \qquad (3.1.35)$$

Since ε is very small, we can use the limiting forms for J_1 and N_1 given by Eqs. (2.9.60) and (2.9.62), that is,

$$J_1(k\varepsilon) \approx \frac{k\varepsilon}{2} \qquad \begin{matrix} \to 0 \\ \varepsilon \to 0 \end{matrix}$$

$$N_1(k\varepsilon) \approx -\frac{2}{\pi k \varepsilon} \qquad (3.1.36)$$

Combining Eq. (3.1.36) with Eq. (3.1.35) yields

$$\varepsilon A k i \left(\frac{-2}{\pi k \varepsilon} \right) = \frac{1}{2\pi}$$

or

$$A = \frac{i}{4}$$

Thus, the radial Green's function is

$$G_1(r, \lambda) = \frac{i}{4} H_0^{(1)}(\sqrt{\lambda}\, r) \qquad (3.1.37)$$

At this point we can say a few words on why the expression given by boundary condition R.1 is chosen rather than the simpler condition of the field vanishing at infinity. The Green's function $G_1(r, r_s, \lambda)$ is defined to be a particular solution of the inhomogeneous differential equation (3.1.18). In general, this solution is not unique because any constant multiple of a solution of the homogeneous equation (3.1.15) can be added to it and it will still be a particular solution.

For example, the particular solution of the inhomogeneous equation (3.1.18) is

$$G_i = \frac{i}{4} H_0^{(1)}(\sqrt{\lambda}\, r)$$

A solution of the homogeneous equation (3.1.15), regular at the origin, is

$$R = J_0(\sqrt{\lambda}\, r)$$

Thus, a second particular solution of Eq. (3.1.18) is

$$g = H_0^{(1)}(\sqrt{\lambda}\, r) + a J_0(\sqrt{\lambda}\, r)$$

If the boundary condition at infinity is only for the field to vanish, then both G_1 and g satisfy it because of Eqs. (2.9.63) and (2.9.51). Thus, the solution is not unique. Hence, radiation problems would not be determined by their prescribed sources in the finite domain. This paradoxical result shows that the condition of vanishing at infinity is not sufficient, and that we have to replace it by a stronger condition at infinity. We call it the radiation condition: the sources must be sources, not sinks of energy. The energy that is radiated from the sources must scatter to infinity; no energy may be radiated from infinity into the prescribed singularities of the field. The boundary condition at infinity given by R.1, namely,

$$\lim_{r \to \infty} r^{1/2} \left(\frac{\partial G_1}{\partial r} - i\sqrt{\lambda}\, G_1 \right) = 0$$

prevents solutions of the homogeneous equation such as $J_0(\sqrt{\lambda}\, r)$ from being

122 PROPAGATION IN THE OCEAN: THE HOMOGENEOUS LAYERED MODEL

added to the correct particular solution; that is, it guarantees a unique solution.

The proof of this uniqueness can be found in Sommerfeld [3.3].

While we will not prove that the Sommerfeld radiation condition leads to a unique solution, we can easily show how such a form arises. The details are shown in Appendix A at the end of this chapter.

Next let us determine $G_2(z, z_s, \lambda)$. Again we note that the inhomogeneous differential equation for G_2, namely, Eq. (3.1.24), reduces to the homogeneous equation (3.1.16) everywhere except at the single point $z = z_s$. So we can solve Eq. (3.1.16) separately for the two regions $0 \leq z \leq z_s$ and $z_s \leq z \leq d$ and then match the two solutions at $z = z_s$ so they have the correct singularity. They will have the correct singularity if they satisfy boundary condition Z.3. The general solutions of Eq. (3.1.16) in the two regions are

$$G_2(z, z_s, \lambda) = A_2(z_s, \lambda)e^{i\gamma z} + B_2(z_s, \lambda)e^{-i\gamma z}, \quad 0 \leq z \leq z_s \tag{3.1.38}$$

$$= A_3(z_s, \lambda)e^{i\gamma z} + B_3(z_s, \lambda)e^{-i\gamma z}, \quad z_s \leq z \leq d \tag{3.1.39}$$

where for convenience we have put

$$\gamma^2 = k^2 - \lambda \tag{3.1.40}$$

Applying the boundary conditions Z.1 through Z.4 results in

$$A_2 + B_2 = 0 \tag{3.1.41}$$

$$A_2 e^{i\gamma z_s} + B_2 e^{-i\gamma z_s} = A_3 e^{i\gamma z_s} + B_3 e^{-i\gamma z_s} \tag{3.1.42}$$

$$i\gamma A_3 e^{i\gamma z_s} - i\gamma B_3 e^{-i\gamma z_s} - i\gamma A_2 e^{i\gamma z_s} + i\gamma B_2 e^{-i\gamma z_s} = -1 \tag{3.1.43}$$

$$i\gamma A_3 e^{i\gamma d} - i\gamma B_3 e^{-i\gamma d} = 0 \tag{3.1.44}$$

It is trivial to solve these four equations, we will just state the final result.

$$A_2 = -\frac{i}{2\gamma} \frac{\cos \gamma(z_s - d)}{\cos \gamma d} \tag{3.1.45}$$

$$B_2 = \frac{i}{2\gamma} \frac{\cos \gamma(z_s - d)}{\cos \gamma d} \tag{3.1.46}$$

$$A_3 = \frac{e^{-i\gamma d}}{2\gamma} \frac{\sin \gamma z_s}{\cos \gamma d} \tag{3.1.47}$$

$$B_3 = \frac{e^{i\gamma d}}{2\gamma} \frac{\sin \gamma z_s}{\cos \gamma d} \tag{3.1.48}$$

COMPLEX GREEN'S FUNCTION FOR A HOMOGENEOUS LAYER

Substituting Eqs. (3.1.45) through (3.1.48) into Eqs. (3.1.38) and (3.1.39) gives the depth-dependent Green's function:

$$G_2(z, z_s, \lambda) = \frac{\cos\gamma(z_s - d)}{\gamma \cos \gamma d} \sin \gamma z, \qquad 0 \le z \le z_s \qquad (3.1.49)$$

$$= \frac{\sin \gamma z_s}{\gamma \cos \gamma d} \cos \gamma(z - d), \qquad z_s \le z \le d \qquad (3.1.50)$$

If we now substitute Eqs. (3.1.37), (3.1.49), and (3.1.50) into Eq. (3.1.28), we obtain the solution for the wave equation $p(r, z, r_s, z_s)$ which is actually the Green's function for the wave equation.

For $0 \le z \le z_s$,

$$p(r, z, z_s) = -\frac{1}{8\pi} \int_{C_\lambda} \frac{1}{\gamma} \left\{ \frac{\cos\gamma(z_s - d)}{\cos \gamma d} \right\} \sin \gamma z H_0^{(1)}(\sqrt{\lambda}\, r)\, d\lambda$$

$$(3.1.51)$$

and, for $z_s \le z \le d$,

$$p(r, z, z_s) = -\frac{1}{8\pi} \int_{C_\lambda} \frac{1}{\gamma} \left\{ \frac{\sin \gamma z_s}{\cos \gamma d} \right\} \cos \gamma(z - d) H_0^{(1)}(\sqrt{\lambda}\, r)\, d\lambda$$

$$(3.1.52)$$

We will choose the contour $C_\lambda = C_z^-$. In Appendix 3.B we show the consequences of choosing the contour C_r^+.

Now let us investigate the λ-plane singularities of the integrands of Eqs. (3.1.51) and (3.1.52). Since G_2 is an even function of γ, we do not have to worry about the branch point for G_2. The only other singularities of G_2 are poles and these are given by

$$\cos \gamma d = 0 \qquad (3.1.53)$$

Note the factor γ in the denominator of the integrals; $\gamma = 0$ cannot be a pole because it appears in combination with $\sin \gamma z/\gamma$ and $\sin \gamma z_s/\gamma$ which does not have a singularity.

Let

$$\gamma = \gamma' + i\gamma''$$

We make use of a well-known result of complex variable theory, namely,

$$\cos \gamma d = \cos \gamma' d \cosh \gamma'' d - i \sin \gamma' d \sinh \gamma'' d$$

124 PROPAGATION IN THE OCEAN: THE HOMOGENEOUS LAYERED MODEL

Suppose $\gamma'' \neq 0$. Then we must have

$$\sin \gamma' d = 0$$

Since $\cosh \gamma'' d$ does not vanish for any value of $\gamma'' d$, we must then have

$$\cos \gamma' d = 0$$

But this is impossible since we already assumed that γ' satisfies $\sin \gamma' d = 0$. Hence, there are no complex solutions of Eq. (3.1.53) and $\gamma'' = 0$.

Thus the solutions of Eq. (3.1.53) are real and are given by

$$\gamma_n = \left(n - \frac{1}{2}\right)\frac{\pi}{d}, \qquad n = 1, 2, 3 \ldots \qquad (3.1.54)$$

or, using Eq. (3.1.40),

$$\lambda_n = k^2 - \left[\left(n - \frac{1}{2}\right)\frac{\pi}{d}\right]^2, \qquad n = 1, 2, 3 \ldots \qquad (3.1.55)$$

Hence, the poles in the λ plane are real.

Now the residue theorem requires the integral to be analytic on and within the closed contour C_z^- except for a finite number of singular points interior to the contour. However, the radial Green's function $H_0^{(1)}(\sqrt{\lambda}\, r)$ has a branch point at $\lambda = 0$ due to the presence of the radical $\sqrt{\lambda}$ in the argument. Consequently, the integrands of Eqs. (3.1.51) and (3.1.52) are not analytic on the entire λ plane. In order to make the integrands analytic, we need to make a branch cut. Mathematically, any branch cut will do; however, physical considerations dictate a particular cut.

Let

$$\kappa = \sqrt{\lambda} = \kappa' + i\kappa'' \qquad (3.1.56)$$

Substituting this expression for κ into the asymptotic expression for the Hankel function as given by Eq. (3.1.32), we get

$$e^{i\kappa r} = e^{i\kappa' r} e^{-\kappa'' r}$$

In order for this expression to vanish as $r \to \infty$, we must require

$$\kappa'' = \mathrm{Im}\sqrt{\lambda} > 0 \qquad (3.1.57)$$

Thus we must choose the two branches of $\sqrt{\lambda}$ so that on one branch $\mathrm{Im}\sqrt{\lambda} > 0$ and on the second branch $\mathrm{Im}\sqrt{\lambda} < 0$. Therefore, the branch cut must be taken to be

$$\mathrm{Im}\,\kappa = \mathrm{Im}\sqrt{\lambda} = 0 \qquad (3.1.58)$$

We note that if we let

$$\kappa^2 = \delta^2$$

so that
$$\kappa = \delta$$
then the condition $\mathrm{Im}\,\kappa = 0$ is satisfied on the branch cut. Thus κ^2 must be real and positive. Now let
$$\kappa^2 = \lambda \equiv \alpha + i\beta \tag{3.1.59}$$

So the equations describing the branch cut are
$$\beta = 0 \tag{3.1.60}$$
$$\alpha > 0 \tag{3.1.61}$$

Thus, the appropriate branch cut runs from the origin along the positive real axis.

We now have the problem that the poles of $G_2(z)$ also lie in the real axis. To circumvent this difficulty, we allow k to have a small, positive imaginary part. We will then allow this to approach zero in the end.

So let
$$k = k' + ik'', \qquad k'' > 0 \tag{3.1.62}$$

Then the poles as given by Eq. (3.1.55) are complex and are now
$$\lambda_n = (k'^2 - k''^2) - \left[\left(n - \frac{1}{2}\right)\frac{\pi}{d}\right]^2 + 2ik'k'' \tag{3.1.63}$$

The poles and branch cuts are shown in Figure 3.1.3. Also shown in the figure is the choice for the contour C_z^-. We can now apply the residue theorem to Eq. (3.1.51), and then let $k'' \to 0$. Thus we get

$$\begin{aligned} p(r, z, z_s) &= \frac{1}{8\pi}\int_{C_z^-}\left[\frac{\cos\gamma(z_s - d)}{\gamma\cos\gamma d}\right]\sin\gamma z H_0^{(1)}(\sqrt{\lambda}\,r)\,d\lambda \\ &= \frac{2\pi i}{8\pi}\sum_{n=1}^{\infty}\frac{\cos\gamma_n(z_s - d)\sin\gamma_n z H_0^{(1)}(\sqrt{\lambda_n}\,r)}{\gamma_n\dfrac{\partial}{\partial\lambda}(\cos\gamma d)_{\lambda=\lambda_n}} \end{aligned} \tag{3.1.64}$$

Now, using Eq. (3.1.40),
$$\begin{aligned} \left[\frac{\partial}{\partial\lambda}\cos\gamma d\right]_{\lambda=\lambda_n} &= -\left[d(\sin\gamma d)\frac{\partial\gamma}{\partial\lambda}\right]_{\lambda=\lambda_n} \\ &= -\left[d(\sin\gamma d)\frac{\partial}{\partial\lambda}(k^2 - \lambda)^{1/2}\right]_{\lambda=\lambda_n} \\ &= d\frac{\sin\gamma_n d}{2\gamma_n} = \frac{d(-1)^{n+1}}{2\gamma_n}, \qquad n = 1, 2, \ldots \tag{3.1.65} \end{aligned}$$

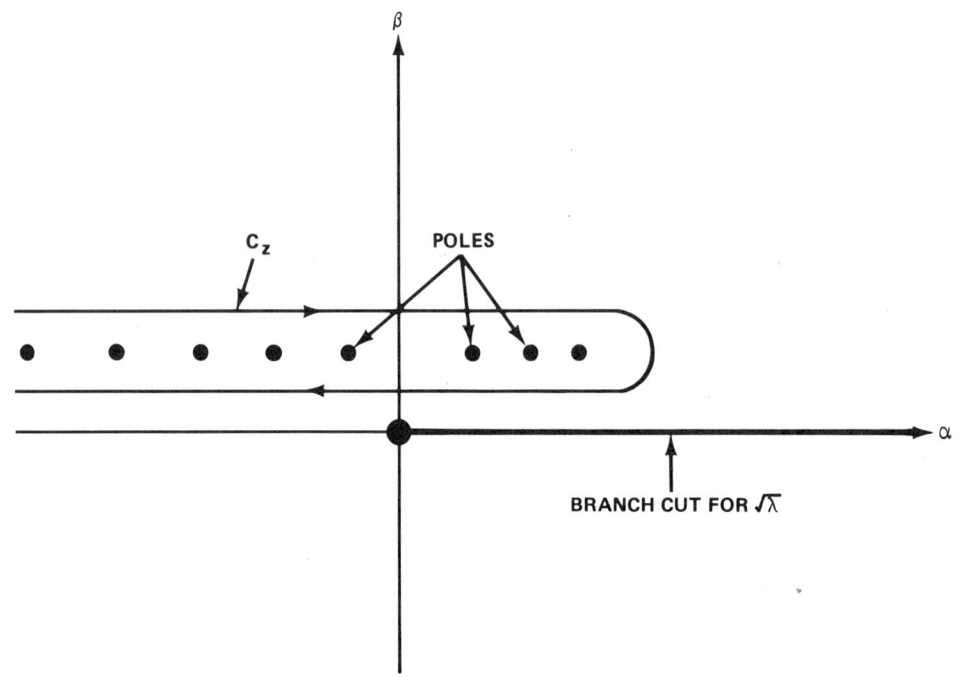

126 PROPAGATION IN THE OCEAN: THE HOMOGENEOUS LAYERED MODEL

FIGURE 3.1.3. Choice for the contour C_z^- for a single, homogeneous layer.

Also, for the one factor in the numerator of the series Eq. (3.1.64),

$$\cos \gamma_n (z_s - d) = \cos \gamma_n z_s \cos \gamma_n d + \sin \gamma_n z_s \sin \gamma_n d$$

$$= (-1)^{n+1} \sin \gamma_n z_s \qquad (3.1.66)$$

where we have used Eq. (3.1.53); that is, we have used the fact that

$$\cos \gamma_n d = 0$$

and the implication from this is that

$$\sin \gamma_n d = (-1)^{n+1}, \qquad n = 1, 2, \ldots$$

This last result was also used in Eq. (3.1.65).

Substituting Eqs. (3.1.65) and (3.1.66) into Eq. (3.1.64), we obtain

$$p(r, z, z_s) = \frac{i}{2d} \sum_{n=1}^{\infty} \sin(\gamma_n z_s) \sin(\gamma_n z) H_0^{(1)}\left(\sqrt{\lambda_n}\, r\right) \qquad (3.1.67)$$

EIGENFUNCTION EXPANSION FOR A HOMOGENEOUS LAYER

If we start with Eq. (3.1.52) and proceed in the same manner to evaluate the complex integral, we end up with Eq. (3.1.67). Consequently, Eq. (3.1.67) holds for the entire range $0 \leq z \leq d$.

For convenience, let

$$\kappa_n = \sqrt{\lambda_n} \tag{3.1.68}$$

Then the solution of the wave equation for our problem becomes

$$p(r, z, z_s) = \frac{i}{2d} \sum_{n=1}^{\infty} \sin(\gamma_n z_s)\sin(\gamma_n z) H_0^{(1)}(\kappa_n r) \tag{3.1.69}$$

The most useful form of this solution is the asymptotic form for large values of $\kappa_n r$. Using Eq.(3.1.32), this solution can be written as

$$p(r, z, z_s) = \frac{1}{\sqrt{2\pi d}} \sum_{n=1}^{\infty} \sin(\gamma_n z_s)\sin(\gamma_n z) \frac{e^{i(\kappa_n r + \pi/4)}}{\sqrt{\kappa_n r}} \tag{3.1.70}$$

It is important to note that in deriving this solution we never talked about eigenfunctions or eigenvalues. The series Eq. (3.1.70) arose as a sum of the residues of our complex integral. In the next section we will rederive this result using the eigenfunctions of the wave equation.

3.2. EIGENFUNCTION EXPANSION OF THE GREEN'S FUNCTION FOR A HOMOGENEOUS LAYER BOUNDED BY A PRESSURE-RELEASE SURFACE AND A RIGID BOTTOM

As we showed in Section 3.1, Eq. (3.1.14), the Green's function for the Helmholtz equation, $p(r, z, z_s)$ must satisfy

$$\frac{\partial^2 p}{\partial r^2} + \frac{1}{r}\frac{\partial p}{\partial r} + \frac{\partial^2 p}{\partial z^2} + k^2 p = -\frac{1}{2\pi r}\delta(r)\delta(z - z_s) \tag{3.2.1}$$

Consider the homogeneous Helmholtz equation

$$\frac{\partial^2 p}{\partial r^2} + \frac{1}{r}\frac{\partial p}{\partial r} + \frac{\partial^2 p}{\partial z^2} + k^2 p = 0 \tag{3.2.2}$$

Letting

$$p = \phi(r)\psi(z) \tag{3.2.3}$$

and using the technique of separation of variables, we arrive, as before, at the

128 PROPAGATION IN THE OCEAN: THE HOMOGENEOUS LAYERED MODEL

separated equations

$$\frac{d^2\phi}{dr^2} + \frac{1}{r}\frac{d\phi}{dr} + \lambda\phi = 0 \tag{3.2.4}$$

$$\frac{d^2\psi}{dz^2} + \gamma^2\psi = 0 \tag{3.2.5}$$

where $-\lambda$ is the separation constant and, for convenience, we have set

$$\gamma^2 = k^2 - \lambda \tag{3.2.6}$$

The boundary conditions on the depth function $\psi(z)$ are

(Z.1) $\qquad\qquad\qquad\qquad \psi(0) = 0$

This is the pressure-release condition at the sea surface.

(Z.2) $\qquad\qquad\qquad\qquad \frac{d\psi}{dz}(d) = 0$

This is the condition for a rigid bottom. The general solution of Eq. (3.2.5) is

$$\psi = A \sin \gamma z + B \cos \gamma z \tag{3.2.7}$$

Applying boundary condition Z.1, we get

$$B = 0$$

Applying boundary condition Z.2, we get

$$\frac{d\psi}{dz}(d) = \gamma A \cos \gamma d = 0$$

or

$$\cos \gamma d = 0 \tag{3.2.8}$$

This is the characteristic equation for the eigenvalues given in general by Eq. (2.2.7). This is also identical to Eq. (3.1.53) which was the equation for the singularities of the depth-dependent Green's function $G_2(z, z_s, \lambda)$. Using Eq. (3.2.6), the eigenvalue solutions of Eq. (3.2.8) are

$$\lambda_n = k^2 - \left[\frac{(2n-1)}{d}\frac{\pi}{2}\right]^2 \tag{3.2.9a}$$

Thus the eigenvalue spectrum is discrete. The depth-dependent eigenfunctions are then

$$\psi_n(z) = A \sin \gamma_n z$$

EIGENFUNCTION EXPANSION FOR A HOMOGENEOUS LAYER

We now normalize ψ_n according to Eq. (2.2.14); that is, we require

$$\int_0^d \psi_n^2(z)\, dz = 1$$

This requires that

$$A = \sqrt{\frac{2}{d}}$$

Therefore, the normalized eigenfunctions are

$$\psi_n(z) = \sqrt{\frac{2}{d}}\, \sin\left[(k^2 - \lambda_n)^{1/2} z\right] \qquad (3.2.9b)$$

We have seen in Section 2.5 that the set of eigenfunctions $\psi_n(z)$ is complete. Thus we can expand the Green's function solution p of Eq. (3.2.1) in terms of these eigenfunctions. Let

$$p(r, z) = \sum_{n=1}^{\infty} \phi_n(r) \psi_n(z) \qquad (3.2.10)$$

where the coefficients ϕ_n are to be determined. Since we are trying to expand a function of r and z in terms of a complete set of eigenfunctions which are functions of z only, the coefficients ϕ_n must be functions of r.

Substituting Eq. (3.2.10) into Eq. (3.2.1) yields

$$\sum_n \psi_n \phi_n'' + \frac{1}{r}\sum_n \psi_n \phi_n' + \sum_n \phi_n \psi_n'' + k^2 \sum_n \phi_n \psi_n = -\frac{1}{2\pi r}\delta(r)\delta(z - z_s)$$

Using Eqs. (3.2.5) and (3.2.6) gives

$$\sum_n \psi_n \phi_n'' + \frac{1}{r}\sum_n \psi_n' \phi_n' + \sum_n \phi_n(\lambda_n - k^2)\psi_n + k^2 \sum_n \phi_n \psi_n$$

$$= -\frac{1}{2\pi r}\delta(r)\delta(z - z_s)$$

and finally, on collecting terms,

$$\sum_n \psi_n \left\{ \phi_n'' + \frac{1}{r}\phi_n' + \lambda_n \phi_n \right\} = -\frac{1}{2\pi r}\delta(r)\delta(z - z_s)$$

Multiplying this last equation by $\psi_m(z)$ and using the orthonormality of the eigenfunctions, we obtain

$$\sum_n \int_0^d \psi_m \psi_n \, dz \left\{ \phi_n'' + \frac{1}{r} \phi_n' + \lambda_n \phi_n \right\} = -\frac{1}{2\pi r} \delta(r) \int_0^d \psi_m \delta(z - z_s) \, dz$$

$$\sum_n \delta_{mn} \left\{ \phi_n'' + \frac{1}{r} \phi_n' + \lambda_n \phi_n \right\} = -\frac{1}{2\pi r} \delta(r) \psi_m(z_s)$$

$$\phi_m'' + \frac{1}{r} \phi_m' + \lambda_m \phi_m = -\frac{1}{2\pi r} \delta(r) \psi_m(z_s)$$

$$\frac{d^2}{dr^2}\left[\frac{\phi_m}{\psi_m(z_s)}\right] + \frac{1}{r}\frac{d}{dr}\left[\frac{\phi_m}{\psi_m(z_s)}\right] + \lambda_n \left[\frac{\phi_m}{\psi_m(z_s)}\right] = -\frac{1}{2\pi r} \delta(r)$$

This is exactly the same equation and the same boundary conditions that we solved in Section 3.1. There in Eq. (3.1.37) we saw that the solution was

$$\frac{\phi_m}{\psi_m(z_s)} = \frac{i}{4} H_0^{(1)}(\sqrt{\lambda} \, r)$$

or

$$\phi_m(r) = \frac{i}{4} \psi_m(z_s) H_0^{(1)}(\sqrt{\lambda} \, r) \qquad (3.2.11)$$

Thus using Eqs. (3.2.9) and (3.2.11), the solution of Eq. (3.2.10) becomes

$$p(r, z) = \frac{i}{2d} \sum_{n=1}^{\infty} \sin(\gamma_n z_s) \sin(\gamma_n z) H_0^{(1)}(\kappa_n r) \qquad (3.2.12)$$

This is identical to the Green's function solution we found in Section 3.1 using the complex λ-plane method. The results of this section allow us to interpret the complex λ-plane technique. We see that the singularities of the Green's function in the complex λ plane are the eigenvalues of the associated homogeneous problem. The expansion of the complex integral in λ space, namely, Eq. (3.1.28) into a series of residues, is equivalent to the expansion of the Green's function into a series of eigenfunctions. Since in this case the singularities of the Green's function $G_2(z, z_s, \lambda)$ are a discrete set of poles, the eigenvalue spectrum consists of a discrete but infinite set of values. Thus the eigenvalue spectrum can be completely determined by examining the singularities of the Green's function. We will see in Section 3.5 that when the branch point cannot be ignored, its contribution leads to the continuous spectrum. Moreover, the limits of integration over the continuous spectrum will be automatically determined by the nature of the branch point. To attempt to solve a problem that has both a discrete and continuous spectrum contribution by the method

EIGENFUNCTION EXPANSION FOR A HOMOGENEOUS LAYER 131

of this section can sometimes be difficult. It means assuming an expansion of the form Eq. (2.6.17) and then substituting it into the wave equation following the procedure of this section. Before leaving this single-layer model, there are two interesting items to note. Consider Eq. (3.2.9a). When

$$k < \frac{\pi}{2d} \qquad (3.2.13)$$

then the eigenvalue κ_n is imaginary, say $\kappa_n = i\alpha$ where α is a real positive quantity. Then the asymptotic form of the solution of the wave equation given by Eq. (3.1.70) becomes

$$p = \frac{1}{d\sqrt{2\pi}} \sum_{n=1}^{\infty} \sin(\gamma_n z_s)\sin(\gamma_n z) \frac{e^{-\alpha r + i\pi/4}}{\sqrt{\kappa_n r}}$$

Consequently, for values of the wave number $k < \pi/2d$, there is no radiation field in the waveguide, but only a heavily attenuated acoustic field. In terms of the wavelength, Eq. (3.2.13) becomes

$$\lambda > 4d$$

that is, when the wavelength is greater than four times the water depth, energy ceases to radiate down the waveguide. The frequency associated with $\lambda_c = 4d$, namely $f_c = c/4d$, is called the cutoff frequency of the waveguide.

The second item of interest is associated with the symmetry of the Green's function that was discussed in Section 2.7 and is expressed by Eq. (2.7.18). Let $\mathbf{r}_s = r_s\hat{r} + z_s\hat{e}_3$ be the position vector for the source point. Here $\hat{r}$ and $\hat{e}_3$ are the unit base vectors in our cylindrical coordinate system (see Figure 3.1.2). Let $\mathbf{r} = r\hat{r} + z\hat{e}_3$ be the position vector for the field point or the point where a receiver would be located. Examination of the Green's function solution Eq. (3.2.12) shows that

$$p(\mathbf{r}, \mathbf{r}_s) = p(\mathbf{r}_s, \mathbf{r}) \qquad (3.2.14)$$

As we would expect, symmetry of the individual one-dimensional Green's function implies symmetry of the two-dimensional Green's function $p(\mathbf{r}, \mathbf{r}_s)$.

Eq. (3.2.14) has an interesting physical interpretation. It says that if we interchange the source and receiver locations, holding all other quantities fixed, the pressure will remain unchanged. That is, $p(\mathbf{r}, \mathbf{r}_s)$ is the pressure at position $\mathbf{r}$ due to a source at position $\mathbf{r}_s$, and $p(\mathbf{r}_s, \mathbf{r})$ is the pressure at position $\mathbf{r}_s$ due to a source at position $\mathbf{r}$. This is called the acoustic reciprocity principle. It is seen that it is a consequence of the symmetry of the Green's function.

3.3. THE COMPLEX λ-PLANE REPRESENTATION OF THE GREEN'S FUNCTION FOR AN OCEAN CONSISTING OF TWO HOMOGENEOUS LAYERS BOUNDED BY A PRESSURE-RELEASE SURFACE AND A RIGID BOTTOM

We are now in a position to solve a more interesting problem than the single homogeneous layer model of the last two sections. We will consider the problem of an ocean consisting of two homogeneous layers bounded by a pressure-release surface above and a rigid bottom below. Figure 3.3.1 shows the geometry for this case.

Let ρ_1, c_1 and ρ_2, c_2 be the constant density and sound velocity in layers 1 and 2, respectively. Assume there is a point source at $r = r_s = 0$ and $z = z_s$. We will assume the source lies in layer 1, but, as we will see later, it is easy to do the problem with the source in any layer.

The Helmholtz equation in layer 1 is

$$\frac{\partial^2 p^{(1)}}{\partial r^2} + \frac{1}{r}\frac{\partial p^{(1)}}{\partial r} + \frac{\partial^2 p^{(1)}}{\partial z^2} + k_1^2 p^{(1)} = -\frac{1}{2\pi r}\delta(r)\delta(z - z_s) \quad (3.3.1)$$

where
$$k_1 = \frac{\omega}{c_1} \quad (3.3.2)$$

Again we are assuming a time factor of $e^{-i\omega t}$. The differential equations for the range-dependent Green's function $G_r^{(1)}(r, \lambda)$ and the depth-dependent Green's function $G_z^{(1)}(z, z_s, \lambda)$ are

$$\frac{d^2 G_r^{(1)}}{dr^2} + \frac{1}{r}\frac{dG_r^{(1)}}{dr} + \lambda G_r^{(1)} = -\frac{1}{2\pi r}\delta(r) \quad (3.3.3)$$

$$\frac{d^2 G_z^{(1)}}{dz^2} + (k_1^2 - \lambda) G_z^{(1)} = -\delta(z - z_s) \quad (3.3.4)$$

The Helmholtz equation in layer 2 is

$$\frac{\partial^2 p^{(2)}}{\partial r^2} + \frac{1}{r}\frac{\partial p^{(2)}}{\partial r} + \frac{\partial^2 p^{(2)}}{\partial z^2} + k_2^2 p^{(2)} = 0 \quad (3.3.5)$$

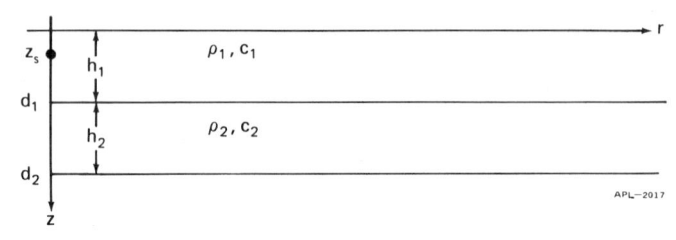

FIGURE 3.3.1. Coordinate geometry for two homogeneous layers.

where
$$k_2 = \frac{\omega}{c_2} \quad (3.3.6)$$

The differential equations for the range-dependent Green's function $G_r^{(2)}(r, \lambda)$ and the depth-dependent Green's function $G_z^{(2)}(z, z_s, \lambda)$ are

$$\frac{d^2 G_r^{(2)}}{dr^2} + \frac{1}{r} \frac{dG_r^{(2)}}{dr} + \lambda G_r^{(2)} = -\frac{1}{2\pi r} \delta(r) \quad (3.3.7)$$

$$\frac{d^2 G_z^{(2)}}{dz^2} + (k_2^2 - \lambda) G_z^{(2)} = 0 \quad (3.3.8)$$

We require that $G_r^{(1)}$ and $G_r^{(2)}$ satisfy the boundary conditions

(R.1) $$\lim_{r \to \infty} \sqrt{r} \left[\frac{\partial G_r^{(\alpha)}}{\partial r} - i\sqrt{\lambda}\, G_r^{(\alpha)} \right] = 0$$

(R.2) $$\lim_{\varepsilon \to 0} \left[\varepsilon \frac{dG_r^{(\alpha)}}{dr}(\varepsilon) \right] = -\frac{1}{2\pi}$$

where $\alpha = 1, 2$.

As we saw in Section 3.1, the solution of Eqs. (3.3.3) and (3.3.7), subject to the boundary conditions R.1 and R.2, is

$$G_r^{(\alpha)}(r, \lambda) = \frac{i}{4} H_0^{(1)}(\sqrt{\lambda}\, r) \quad (3.3.9)$$

We require that $G_z^{(\alpha)}$ satisfy the boundary conditions

(Z.1) $$G_z^{(1)}(0, z_s, \lambda) = 0$$

(Z.2) $$\lim_{\varepsilon \to 0} \left\{ G_z^{(1)}(z_s + \varepsilon, z_s, \lambda) - G_z^{(1)}(z_s - \varepsilon, z_s, \lambda) \right\} = 0$$

(Z.3) $$\lim_{\varepsilon \to 0} \left\{ \frac{dG_z^{(1)}}{dz}(z_s + \varepsilon, z_s, \lambda) - \frac{dG_z^{(1)}}{dz}(z_s - \varepsilon, z_s, \lambda) \right\} = -1$$

(Z.4) $$G_z^{(1)}(d_1, z_s, \lambda) = G_z^{(2)}(d_1, z_s, \lambda)$$

(Z.5) $$\frac{1}{\rho_1} \frac{dG_z^{(1)}}{dz}(d_1, z_s, \lambda) = \frac{1}{\rho_2} \frac{dG_z^{(2)}}{dz}(d_1, z_s, \lambda)$$

This is the requirement of continuity of the normal derivative of particle velocity.

(Z.6) $$\frac{dG_z^{(2)}}{dz}(d_2, z_s, \lambda) = 0$$

134 PROPAGATION IN THE OCEAN: THE HOMOGENEOUS LAYERED MODEL

For convenience, let

$$\gamma_1^2 = k_1^2 - \lambda \tag{3.3.10}$$

$$\gamma_2^2 = k_2^2 - \lambda \tag{3.3.11}$$

The general solutions of Eqs. (3.3.4) and (3.3.8) are

$$G_z^{(1)} = A \sin \gamma_1 z + B \cos \gamma_1 z, \quad 0 \le z \le z_s \tag{3.3.12}$$

$$= C \sin \gamma_1 z + D \cos \gamma_1 z, \quad z_s \le z \le d_1 \tag{3.3.13}$$

$$G_z^{(2)} = E \sin \gamma_2 z + F \cos \gamma_2 z, \quad d_1 \le z \le d_2 \tag{3.3.14}$$

Applying the boundary conditions Z.1 through Z.6 yields

$$B = 0 \tag{3.3.15}$$

$$A \sin \gamma_1 z_s = C \sin \gamma_1 z_s + D \cos \gamma_1 z_s \tag{3.3.16}$$

$$-\gamma_1 C \cos \gamma_1 z_s + \gamma_1 D \sin \gamma_1 z_s + \gamma_1 A \cos \gamma_1 z_s = 1 \tag{3.3.17}$$

$$C \sin \gamma_1 d_1 + D \cos \gamma_1 d_1 = E \sin \gamma_2 d_1 + F \cos \gamma_2 d_1 \tag{3.3.18}$$

$$\rho_2 \gamma_1 C \cos \gamma_1 d_1 - \rho_2 \gamma_1 D \sin \gamma_1 d_1 = \rho_1 \gamma_2 E \cos \gamma_2 d_1 - \rho_1 \gamma_2 F \sin \gamma_2 d_1 \tag{3.3.19}$$

$$E \cos \gamma_2 d_2 - F \sin \gamma_2 d_2 = 0 \tag{3.3.20}$$

From Eqs. (3.3.16) and (3.3.17) we get

$$(A - C)\sin \gamma_1 z_s = D \cos \gamma_1 z_s$$

$$(A - C)\gamma_1 \cos \gamma_1 z_s = 1 - \gamma_1 D \sin \gamma_1 z_s \tag{3.3.21}$$

Eliminating $(A - C)$ between these two equations gives

$$D \frac{\cos \gamma_1 z_s}{\sin \gamma_1 z_s} \gamma_1 \cos \gamma_1 z_s + \gamma_1 z_s = 1$$

$$D(\cos^2 \gamma_1 z_s + \sin^2 \gamma_1 z_s) = \frac{\sin \gamma_1 z_s}{\gamma_1}$$

$$D = \frac{\sin \gamma_1 z_s}{\gamma_1} \tag{3.3.22}$$

COMPLEX GREEN'S FUNCTION FOR TWO HOMOGENEOUS LAYERS

Consequently, using Eq. (3.3.22) in Eq. (3.3.21) results in

$$A - C = \frac{\cos \gamma_1 z_s}{\gamma_1} \quad (3.3.23)$$

Now Eqs. (3.3.18), (3.3.19), (3.3.20), and (3.3.23) are four equations in the four unknowns A, C, E, and F. Substituting the expression for D given by Eq. (3.3.22) into Eqs. (3.3.18) and (3.3.19), we arrive at three equations for the three unknowns C, E, and F.

$$C \sin \gamma_1 d_1 - E \sin \gamma_2 d_1 - F \cos \gamma_2 d_1 = -\frac{\sin \gamma_1 z_s \cos \gamma_1 d_1}{\gamma_1} \quad (3.3.24)$$

$$\rho_2 \gamma_1 C \cos \gamma_1 d_1 - \rho_1 \gamma_2 E \cos \gamma_2 d_1 + \rho_1 \gamma_2 F \sin \gamma_2 d_1 = \rho_2 \sin \gamma_1 z_s \sin \gamma_1 d_1 \quad (3.3.25)$$

$$E \cos \gamma_2 d_2 - F \sin \gamma_2 d_2 = 0 \quad (3.3.26)$$

Let M denote the determinant of the coefficients

$$M = \begin{vmatrix} \sin \gamma_1 d_1 & -\sin \gamma_2 d_1 & -\cos \gamma_2 d_1 \\ \rho_2 \gamma_1 \cos \gamma_1 d_1 & -\rho_1 \gamma_2 \cos \gamma_2 d_1 & \rho_1 \gamma_2 \sin \gamma_2 d_2 \\ 0 & \cos \gamma_2 d_2 & -\sin \gamma_2 d_2 \end{vmatrix}$$

or

$$M = \rho_1 \gamma_2 \sin \gamma_1 h_1 \sin \gamma_2 h_2 - \rho_2 \gamma_1 \cos \gamma_1 h_1 \cos \gamma_2 h_2 \quad (3.3.27)$$

where we have used $h_1 = d_1$ and $h_2 = d_2 - d_1$.

Using Cramer's rule to solve the system of Eqs. (3.3.24) through (3.3.26), we get for C

$$C = \frac{1}{M} \begin{vmatrix} -(\sin \gamma_1 z_s \cos \gamma_1 d_1)/\gamma_1 & -\sin \gamma_2 d_1 & -\cos \gamma_2 d_1 \\ \rho_2 \sin \gamma_1 z_s \sin \gamma_1 d_1 & -\rho_1 \gamma_2 \cos \gamma_2 d_1 & \rho_1 \gamma_2 \sin \gamma_2 d_1 \\ 0 & \cos \gamma_2 d_2 & -\sin \gamma_2 d_2 \end{vmatrix}$$

and, consequently,

$$C = \frac{-\sin \gamma_1 z_s}{\gamma_1 M} \{ \rho_1 \gamma_2 \cos \gamma_1 h_1 \sin \gamma_2 h_2 + \gamma_1 \rho_2 \sin \gamma_1 h_1 \cos \gamma_2 h_2 \} \quad (3.3.28)$$

Using this expression for C, we get from Eq. (3.3.23)

$$A = \frac{1}{\gamma_1 M} \{ \rho_1 \gamma_2 \sin \gamma_2 h_2 \sin \gamma_1 (h_1 - z_s) - \gamma_1 \rho_2 \cos \gamma_2 h_2 \cos \gamma_1 (d_1 - z_s) \}$$

$$(3.3.29)$$

136 PROPAGATION IN THE OCEAN: THE HOMOGENEOUS LAYERED MODEL

Again, using Cramer's rule to solve the system of Eqs. (3.3.24) through (3.3.26), we get for E

$$E = \frac{1}{M} \begin{vmatrix} \sin \gamma_1 d_1 & -(\sin \gamma_1 z_s \cos \gamma_1 d_1)/\gamma_1 & -\cos \gamma_2 d_1 \\ \rho_2 \gamma_1 \cos \gamma_1 d_1 & \rho_2 \sin \gamma_1 z_s \sin \gamma_1 d_1 & \rho_1 \gamma_2 \sin \gamma_2 d_1 \\ 0 & 0 & -\sin \gamma_2 d_2 \end{vmatrix}$$

and, consequently,

$$E = \frac{-1}{M}(\rho_2 \sin \gamma_2 d_2 \sin \gamma_1 z_s) \qquad (3.3.30)$$

Substituting the expression for E into Eq. (3.3.26), we get for F the expression

$$F = -\frac{1}{M}(\rho_2 \cos \gamma_2 d_2 \sin \gamma_1 z_s) \qquad (3.3.31)$$

Using these expressions for A, B, C, D, E, and F, the depth-dependent Green's function given by Eqs. (3.3.12) through (3.3.14) becomes for $0 \le z \le z_s$

$$G_z^{(1)} = \{\rho_1 \gamma_2 \sin \gamma_2 h_2 \sin \gamma_1 (h_1 - z_s)$$

$$- \gamma_1 \rho_2 \cos \gamma_2 h_2 \cos \gamma_1 (h_1 - z_s)\} \frac{\sin \gamma_1 z}{\gamma_1 M} \qquad (3.3.32)$$

and, for $z_s \le z \le d_1$,

$$G_z^{(1)} = \{\rho_1 \gamma_2 \sin \gamma_2 h_2 \sin \gamma_1 (h_1 - z)$$

$$- \rho_2 \gamma_1 \cos \gamma_2 h_2 \cos \gamma_1 (h_1 - z)\} \frac{\sin \gamma_1 z_s}{\gamma_1 M} \qquad (3.3.33)$$

and, for $d_1 \le z \le d_2$,

$$G_z^{(2)} = -\frac{1}{M} \rho_2 \sin \gamma_1 z_s \cos \gamma_2 (d_2 - z) \qquad (3.3.34)$$

First, let us determine the solution of the Helmholtz equation in the range $0 \le z \le z_s$. The general solution is given by Eq. (3.1.28) which in this particular case becomes

$$P^{(1)}(r, r_s, z, z_s) = \frac{-1}{2\pi i} \int_{C_\lambda} G_r^{(1)}(r, r_s, \lambda) G_z^{(1)}(z, z_s, \lambda) \, d\lambda \qquad (3.3.35)$$

Using Eqs. (3.3.9) and (3.3.32), this last equation becomes

$$P^{(1)} = -\frac{1}{8\pi} \int_{C_\lambda} \{\rho_1 \gamma_2 \sin \gamma_2 h_2 \sin \gamma_1 (d_1 - z_s)$$

$$- \rho_2 \gamma_1 \cos \gamma_2 h_2 \cos \gamma_1 (d_1 - z_s)\} \frac{\sin \gamma_1 z}{\gamma_1 M} H_0^{(1)}(\sqrt{\lambda} r) \, d\lambda \qquad (3.3.36)$$

where we choose $C_\lambda = C_z^-$ to enclose the λ-plane singularities of $G_z^{(\alpha)}$ in the negative sense (clockwise), but no singularities of $G_r^{(\alpha)}$.

The only radicals and hence possible branch points of $G_z^{(\alpha)}$ are associated with γ_1 and γ_2. However, inspection of Eq. (3.3.32) shows that $G_z^{(\alpha)}$ is an even function of γ_1 and γ_2. Consequently, we do not have to worry about any branch cuts. The only singularities of $G_z^{(\alpha)}$ are poles given by the condition that the denominator of Eq. (3.3.36) vanish; that is,

$$M = 0$$

or, on using Eq. (3.3.27),

$$\rho_1\gamma_2 \sin\gamma_1 h_1 \sin\gamma_2 h_2 = \rho_2\gamma_1 \cos\gamma_1 h_1 \cos\gamma_2 h_2 \qquad (3.3.37a)$$

or

$$\rho_1\gamma_2 \tan\gamma_2 h_2 = \rho_2\gamma_1 \operatorname{ctn}\gamma_1 h_1 \qquad (3.3.37b)$$

As in Section 3.1, the Hankel function has a branch point at $\lambda = 0$ and we must use the same branch, namely $\operatorname{Im}\sqrt{\lambda} > 0$, as we used there. Also, since the poles that are the solution of Eq. (3.3.37a) are real numbers, we must use the same technique used in Section 3.1 to move the poles off the real axis. So again we must let k have a small imaginary part and then allow it to vanish after we have done the integration. We will not go into the details in this section. However, we will go into the details in Section 3.5 and they are similar here.

So applying the residue theorem to Eq. (3.3.36) gives

$$P^{(1)} = \frac{i}{4} \sum_{n=1}^{\infty} \left\{ \rho_1 \gamma_{2_n} \sin\gamma_{2_n} h_2 \sin\gamma_{1_n}(h_1 - z_s) \right.$$

$$\left. - \gamma_{1_n} \rho_2 \cos\gamma_{2_n} h_2 \cos\gamma_{1_n}(h_1 - z_s) \right\}$$

$$\cdot \frac{\sin\gamma_{1_n} z}{\gamma_{1_n}[\partial M/\partial \lambda]_{\lambda=\lambda_n}} H_0^{(1)}(\sqrt{\lambda}_n r)$$

Using Eq. (3.3.37a), this last expression can be simplified to

$$P^{(1)} = \frac{-i\rho_2\rho_1}{4\rho_1} \sum_{n=1}^{\infty} \frac{\cos\gamma_{2_n} h_2 \sin\gamma_{1_n} z_s}{\sin\gamma_{1_n} h_1 [\partial M/\partial \lambda]_{\lambda=\lambda_n}} \sin\gamma_{1_n} z H_0^{(1)}(\sqrt{\lambda}_n r) \qquad (3.3.38)$$

Now we need to evaluate $[\partial M/\partial \lambda]_{\lambda=\lambda_n}$.
From Eqs. (3.3.10) and (3.3.11),

$$\gamma = (k_\alpha^2 - \lambda)^{1/2}, \alpha = 1, 2$$

138 PROPAGATION IN THE OCEAN: THE HOMOGENEOUS LAYERED MODEL

and so

$$\frac{\partial \gamma_\alpha}{\partial \lambda} = -\frac{1}{2\gamma_\alpha}$$

Consequently, using Eq. (3.3.27),

$$\begin{aligned}\frac{\partial M}{\partial \lambda} &= \frac{\partial}{\partial \lambda}\{\rho_1\gamma_2\sin\gamma_1 h_1\sin\gamma_2 h_2 - \rho_2\gamma_1\cos\gamma_1 h_1\cos\gamma_2 h_2\}\\ &= -\frac{\rho_1}{2\gamma_2}\sin\gamma_1 h_1\sin\gamma_2 h_2 - \frac{h_1\rho_1\gamma_2}{2\gamma_1}\sin\gamma_2 h_2\cos\gamma_1 h_1\\ &\quad -\frac{h_2\rho_1}{2}\sin\gamma_1 h_1\cos\gamma_2 h_2 + \frac{\rho_2}{2\gamma_1}\cos\gamma_1 h_1\cos\gamma_2 h_2\\ &\quad -\frac{\rho_2 h_1}{2}\sin\gamma_1 h_1\cos\gamma_2 h_2 - \frac{\rho_2\gamma_1 h_2}{2\gamma_2}\cos\gamma_1 h_1\sin\gamma_2 h_2\end{aligned}$$

We see from inspection of the series solution Eq. (3.3.38) that the quantity of interest to us is

$$\frac{-\sin\gamma_1 h_1}{\rho_1\rho_2\cos\gamma_2 h_2}\frac{\partial M}{\partial \lambda}$$

Using this expression for $\partial M/\partial \lambda$, this quantity becomes, after employing Eq. (3.3.37a) for simplification,

$$\begin{aligned}\frac{-\sin\gamma_{1_n} h_1}{\rho_1\rho_2\cos\gamma_{2_n} h_2}\left[\frac{\partial M}{\partial \lambda}\right]_{\lambda=\lambda_n} &= \frac{h_1}{2\rho_1} - \frac{\sin\gamma_{1_n} h_1\cos\gamma_{1_n} h_1}{2\gamma_{1_n}\rho_1}\\ &\quad + \frac{1}{2\rho_2\gamma_{2_n}}\sin^2\gamma_{1_n} h_1\frac{\sin\gamma_{2_n} h_2}{\cos\gamma_{2_n} h_2} + \frac{h_2}{2\rho_2}\frac{\sin^2\gamma_{1_n} h_1}{\cos^2\gamma_{2_n} h_2}\\ &\equiv \frac{1}{N_n^2}\end{aligned} \quad (3.3.39)$$

The solution of Eq. (3.3.38) can then be written in the form

$$p^{(1)} = \frac{i}{4\rho_1}\sum_{n=1}^{\infty}N_n^2\sin\gamma_{1_n}z_s\sin\gamma_{1_n}zH_0^{(1)}(\sqrt{\lambda_n}\,r) \quad (3.3.40)$$

It is easy to show by the same techniques that were used to derive the solution of Eq. (3.3.40) for the range $0 \le z \le z_s$ that Eq. (3.3.40) also holds for the range $z_s \le z \le d_1$. So $p^{(1)}$ given by Eq. (3.3.40) holds for the entire layer.

Finally, let us derive an expression for $p^{(2)}$. The solution $p^{(2)}$ is given by an expression identical to Eq. (3.3.35), namely,

$$p^{(2)} = -\frac{1}{2\pi i}\int_{C_z^-} G_r^{(2)}(r, r_s, \lambda) G_z^{(2)}(z, z_s, \lambda)\, d\lambda \qquad (3.3.41)$$

where again C_z^- encloses the singularities of $G_z^{(2)}$ with negative sense, but not the singularities of $G_r^{(2)}$.

Using Eqs. (3.3.9) and (3.3.34), we get

$$p^{(2)} = \frac{1}{8\pi}\int_{C_z^-}\{\rho_2\sin\gamma_1 z_s\cos\gamma_2(d_2 - z)\}\left(\frac{1}{M}\right) H_0^{(1)}(\sqrt{\lambda}\,r)\, d\lambda \qquad (3.3.42)$$

Applying the residue theorem to Eq. (3.3.42) yields

$$p^{(2)} = -\frac{i}{4}\sum_{n=1}^{\infty} \frac{\rho_2\sin\gamma_{1_n} z_s\cos\gamma_{2_n}(d_2 - z)}{[\partial M/\partial\lambda]_{\lambda=\lambda_n}} H_0^{(1)}(\sqrt{\lambda_n}\,r) \qquad (3.3.43)$$

Using the expression for $[\partial M/\partial\lambda]_{\lambda=\lambda_n}$ given by Eq. (3.3.39), we can write the solution for $p^{(2)}$ in the form

$$p^{(2)} = \frac{i}{4\rho_1}\sum_{n=1}^{\infty} \frac{N_n^2 \sin\gamma_{1_n} h_1}{\cos\gamma_2 h_2} \sin\gamma_{1_n} z_s\cos\gamma_{2_n}(d_2 - z) H_0(\sqrt{\lambda_n}\,r) \qquad (3.3.44)$$

We see that for this case the spectrum of eigenvalues is discrete. The reason here is the same as the single-layer model, that is, because the ocean is assumed to be bounded in the z direction. As in the case of the single-layer model, we will see in the next section that Eqs. (3.3.40) and (3.3.44) represent the eigenfunction expansion of the Green's function.

3.4. EIGENFUNCTION EXPANSION OF THE GREEN'S FUNCTION FOR AN OCEAN CONSISTING OF TWO HOMOGENEOUS LAYERS BOUNDED BY A PRESSURE-RELEASE SURFACE AND A RIGID BOTTOM

Here we are going to solve the same problem as we did in Section 3.3 except we will use the eigenfunction expansion techniques. The geometry is the same as in Figure 3.3.1. To use this technique we have to take into account the variation of density with depth. Consequently, we must start with Eq. (1.7.30) with the source term equated to zero.

$$\nabla^2\mathcal{P} - \frac{1}{\rho_0}\nabla\rho_0 \cdot \nabla\mathcal{P} - \frac{1}{c^2}\frac{\partial^2\mathcal{P}}{\partial t^2} = 0$$

140 PROPAGATION IN THE OCEAN: THE HOMOGENEOUS LAYERED MODEL

For the time factor $e^{-i\omega t}$, this becomes

$$\nabla^2 p - \frac{1}{\rho_0} \nabla \rho_0 \cdot \nabla p + k^2 p = 0 \qquad (3.4.1)$$

where, as usual, $k = \omega/c$. Here,

$$\rho_0(z) = \rho_1, \, 0 \le z \le d_1$$
$$= \rho_2, \, d_1 < z \le d_2 \qquad (3.4.2)$$

and

$$c(z) = c_1, \, 0 \le z \le d_1$$
$$= c_2, \, d_1 < z \le d_2 \qquad (3.4.3)$$

In cylindrical coordinates, Eq. (3.4.1) becomes

$$\frac{\partial^2 p}{\partial r^2} + \frac{1}{r} \frac{\partial p}{\partial r} + \frac{\partial^2 p}{\partial z^2} - \frac{1}{\rho_0} \frac{d\rho_0}{dz} \frac{\partial p}{\partial z} + k^2 p = 0 \qquad (3.4.4)$$

Let

$$p = p^{(1)} \quad \text{for} \quad 0 \le z \le d_1$$
$$= p^{(2)} \quad \text{for} \quad 0 \le z \le d_2 \qquad (3.4.5)$$

Using separation of variables,

$$p^{(\alpha)} = \phi(r)\psi^{(\alpha)}(z) \qquad (3.4.6)$$

where $\alpha = 1, 2$, we get for the separated depth equations for Layer 1

$$\frac{d}{dz}\left(\frac{1}{\rho_1} \frac{d\psi^{(1)}}{dz}\right) + \left(\frac{k_1^2}{\rho_1} - \frac{\lambda}{\rho_1}\right)\psi^{(1)} = 0 \qquad (3.4.7)$$

and for Layer 2

$$\frac{d}{dz}\left(\frac{1}{\rho_2} \frac{d\psi^{(2)}}{dz}\right) + \left(\frac{k_2^2}{\rho_2} - \frac{\lambda}{\rho_2}\right)\psi^{(2)} = 0 \qquad (3.4.8)$$

The boundary conditions on the depth functions are

(Z.1) $\qquad \psi^{(1)}(0) = 0$

(Z.2) $\qquad \psi^{(1)}(d_1) = \psi^{(2)}(d_1)$

(Z.3) $\qquad \dfrac{1}{\rho_1} \dfrac{d\psi^{(1)}(d_1)}{dz} = \dfrac{1}{\rho_2} \dfrac{d\psi^{(2)}(d_1)}{dz}$

(Z.4) $\qquad \dfrac{d\psi^{(2)}}{dz}(d_2) = 0$

EIGENFUNCTION EXPANSION FOR TWO HOMOGENEOUS LAYERS

The general solutions of Eqs. (3.4.7) and (3.4.8) are

$$\psi^{(1)}(z) = A \sin \gamma_1 z + D \cos \gamma_1 z \qquad (3.4.9)$$

$$\psi^{(2)}(z) = B \cos \gamma_2 z + C \sin \gamma_2 z \qquad (3.4.10)$$

Application of the boundary conditions yields the systems of equations

$$D = 0 \qquad (3.4.11)$$

$$A \sin \gamma_1 d_1 = B \cos \gamma_2 d_1 + C \sin \gamma_2 d_1 \qquad (3.4.12)$$

$$\gamma_1 \rho_2 A \cos \gamma_1 d_1 = -\gamma_2 \rho_1 B \sin \gamma_2 d_1 + \gamma_2 \rho_1 C \cos \gamma_2 d_1 \qquad (3.4.13)$$

$$-B \sin \gamma_2 d_2 + C \cos \gamma_2 d_2 = 0 \qquad (3.4.14)$$

By Cramer's rule, in order for these equations to have a nontrivial solution, the determinant of the coefficients of the unknown quantities A, B, and C must vanish. Denoting this determinant by M, we have

$$M = \begin{vmatrix} \sin \gamma_1 d_1 & -\cos \gamma_2 d_1 & -\sin \gamma_2 d_1 \\ \gamma_1 \rho_2 \cos \gamma_1 d_1 & \gamma_2 \rho_1 \sin \gamma_2 d_1 & -\gamma_2 \rho_1 \cos \gamma_2 d_1 \\ 0 & -\sin \gamma_2 d_2 & \cos \gamma_2 d_2 \end{vmatrix} = 0$$

or, upon expanding the determinant,

$$M = \gamma_1 \rho_2 \cos \gamma_1 h_1 \cos \gamma_2 h_2 - \gamma_2 \rho_1 \sin \gamma_1 h_1 \sin \gamma_2 h_2 = 0$$

This is the characteristic equation discussed in Section 2.2 and given by Eq. (2.2.7). The values of λ that satisfy this equation are the eigenvalues. We also note that this equation is identical with Eq. (3.3.37a), giving the poles of the depth-dependent Green's function. Thus, once again we see that the singularities of the Green's function are the eigenvalues of the problem.

Solving Eq. (3.4.14) for C yields

$$C = B \tan \gamma_2 d_2 \qquad (3.4.15)$$

Substituting this expression for C into Eq. (3.4.12) results in

$$B = A \frac{\sin \gamma_1 d_1 \cos \gamma_2 d_2}{\cos \gamma_2 h_2} \qquad (3.4.16)$$

Consequently, combining Eqs. (3.4.15) and (3.4.16) gives the expression

$$C = A \frac{\sin \gamma_1 d_1 \sin \gamma_2 d_2}{\cos \gamma_2 h_2} \qquad (3.4.17)$$

142 PROPAGATION IN THE OCEAN: THE HOMOGENEOUS LAYERED MODEL

Since there are not enough conditions to specify the coefficient A, it remains arbitrary. It can therefore be chosen as we please. We shall choose it later to normalize the depth eigenfunctions.

The solutions of the depth equations (3.4.9) and (3.4.10) now become

$$\psi^{(1)}(z) = A \sin \gamma_1 z \qquad (3.4.18)$$

$$\psi^{(2)}(z) = A \frac{\sin \gamma_1 d_1}{\cos \gamma_2 h_2} \cos \gamma_2 (d_2 - z) \qquad (3.4.19)$$

We now need to normalize the depth eigenfunctions. If we compare the differential equations for the depth functions, Eqs. (3.4.7) and (3.4.8), with the Sturm-Liouville equation (2.1.1), we see the weight function

$$w = \frac{1}{\rho_0}$$

where ρ_0 is given by Eq. (3.4.2). Thus we require that the depth functions satisfy the condition Eq. (2.2.14):

$$\int_0^{d_2} \frac{1}{\rho_0} \psi_n^2 \, dz = \int_0^{d_1} \frac{1}{\rho_1} \psi_n^{(1)^2} \, dz + \int_{d_1}^{d_2} \frac{1}{\rho_2} \psi_n^{(2)^2} \, dz = 1 \qquad (3.4.20)$$

Using Eqs. (3.4.18) and (3.4.19), Eq. (3.4.20) becomes

$$\frac{A^2}{\rho_1} \int_0^{d_1} \sin^2 \gamma_1 z \, dz + \frac{A^2}{\rho_2} \int_{d_1}^{d_2} \frac{\sin^2 \gamma_1 d_1}{\cos^2 \gamma_2 h_2} \cos^2 \gamma_2 (d_2 - z) \, dz = 1$$

Upon carrying out the trivial integration, we get

$$\frac{1}{\rho_1} \left(\frac{d_1}{2} - \frac{1}{2\gamma_1} \cos \gamma_1 d_1 \sin \gamma_1 d_1 \right)$$

$$+ \frac{1}{\rho_2} \frac{\sin^2 \gamma_1 d_1}{\cos^2 \gamma_2 h_2} \left(\frac{h_2}{2} + \frac{1}{2\gamma_2} \sin \gamma_2 h_2 \cos \gamma_2 h_2 \right) = \frac{1}{N_n^2} \qquad (3.4.21)$$

where we have put $N_n = A$. The depth functions now satisfy the orthonormality condition

$$\int_0^{d_2} \frac{1}{\rho_0} \psi_n \psi_m \, dz = \delta_{nm}$$

given by Eq. (2.2.15).

We now wish to determine the Green's function for the wave equation

$$\frac{\partial^2 p}{\partial r^2} + \frac{1}{r} \frac{\partial p}{\partial r} + \frac{\partial^2 p}{\partial z^2} + k^2 p = -\frac{1}{2\pi r} \delta(r) \delta(z - z_s) \qquad (3.4.22)$$

EIGENFUNCTION EXPANSION FOR TWO HOMOGENEOUS LAYERS

We have omitted the term involving the density gradient, namely,

$$\frac{1}{\rho_0}\frac{d\rho_0}{dz}\frac{\partial p}{\partial z}$$

because the differential equation is a "local" or "point" condition and ρ_0 is constant in both layers, so that $d\rho_0/dz$ vanishes. Consequently we do not need to include it here. It was carried along in Eq. (3.4.4) only to show how the density entered as a weight function in the depth equations (3.4.7) and (3.4.8). The depth eigenfunctions ψ_n are only orthonormal relative to the weight function $w = \rho_0^{-1}$, even though ρ_0 is constant in both layers.

Since the set of depth eigenfunctions ψ_n is complete, we can expand the Green's function in terms of them. Let

$$p = \sum_{n=1}^{\infty} \phi_n(r)\psi_n(z) \qquad (3.4.23)$$

Substituting Eq. (3.4.23) into Eq. (3.4.22) yields

$$\sum_{n=1}^{\infty} \psi_n \frac{d^2\phi_n}{dr^2} + \frac{1}{r}\sum_{n=1}^{\infty} \psi_n \frac{d\phi_n}{dr} + \sum_{n=1}^{\infty} \phi_n \frac{d^2\psi_n}{dz^2}$$

$$+ k^2 \sum_{n=1}^{\infty} \phi_n \psi_n = -\frac{1}{2\pi r}\delta(r)\delta(z - z_s)$$

Using the depth equations (3.4.7) and (3.4.8), this last equation becomes

$$\sum_n \psi_n \frac{d^2\phi_n}{dr^2} + \frac{1}{r}\sum_n \psi_n \frac{d\phi_n}{dr} + \sum_n \lambda \phi_n \psi_n = -\frac{1}{2\pi r}\delta(r)\delta(z - z_s)$$

Multiplying this equation by $\rho_0^{-1}\psi_m$, integrating over the interval 0 to d_2, and using the orthonormality property of the eigenfunctions ψ_n yields

$$\sum_n \left\{ \frac{d^2\phi_n}{dr^2} + \frac{1}{r}\frac{d\phi_n}{dr} + \lambda\phi_n \right\} \int_0^{d_2} \frac{1}{\rho_0}\psi_m\psi_n\, dz$$

$$= -\frac{\delta(r)}{2\pi r}\int_0^{d_2} \frac{1}{\rho_0}\psi_m\delta(z - z_s)\, dz$$

$$\frac{d^2\phi_m}{dr^2} + \frac{1}{r}\frac{d\phi_m}{dr} + \lambda\phi_m = \frac{-\psi_m^{(1)}(z_s)}{2\pi\rho_1 r}\delta(r) \qquad (3.4.24)$$

or

$$\frac{d^2}{dr^2}\left[\frac{\rho_1}{\psi_m^{(1)}(z_s)}\phi_m\right] + \frac{1}{r}\frac{d}{dr}\left[\frac{\rho_1}{\psi_m^{(1)}(z_s)}\phi_m\right] + \frac{\lambda_m \rho_1 \phi_m}{\psi_m^{(1)}(z_s)} = -\frac{\delta(r)}{2\pi r}$$

$$(3.4.25)$$

144 PROPAGATION IN THE OCEAN: THE HOMOGENEOUS LAYERED MODEL

As before, ϕ_m satisfies the boundary conditions R.1 and R.2 of Section 3.3. We saw previously that the function ϕ_n, which is a solution of Eq. (3.4.25) and satisfies the boundary conditions R.1 and R.2, is

$$\frac{\rho_1}{\psi_m^{(1)}(z_s)}\phi_m = \frac{i}{4}H_0^{(1)}(\sqrt{\lambda}\,r)$$

or

$$\phi_m = \frac{i}{4\rho_1}\psi_m^{(1)}(z_s)H_0^{(1)}(\sqrt{\lambda}\,r) \qquad (3.4.26)$$

If we combine Eqs. (3.4.18), (3.4.19), and (3.4.26) with Eq. (3.4.23), we arrive at the solution given by Eqs. (3.3.40) and (3.3.44).

3.5. THE COMPLEX λ-PLANE REPRESENTATION OF THE GREEN'S FUNCTION FOR AN OCEAN CONSISTING OF A HOMOGENEOUS LAYER WITH A PRESSURE-RELEASE SURFACE OVERLYING A HOMOGENEOUS HALFSPACE—THE DISCRETE PLUS CONTINUOUS SPECTRUM

We now wish to consider the problem posed in Section 3.3 where we allow the rigid bottom to recede to infinity. The geometry is shown in Figure 3.5.1.

As we shall see, this problem is far more complicated than the problem discussed in Section 3.3, because this problem introduces a mixed spectrum, part discrete and part continuous. The solution for this case can be obtained from the solution of the two-layer problem solved in Section 3.3 by allowing the thickness of the second layer to go to infinity.

The solution for layer 1 is given by Eq. (3.3.36), namely,

$$p^{(1)} = -\frac{1}{8\pi}\int_{c_z^-}\{\rho_1\gamma_2\sin\gamma_2 h_2\sin\gamma_1(h_1-z_s)$$

$$-\rho_2\gamma_1\cos\gamma_2 h_2\cos\gamma_1(h_1-z_s)\}\frac{\sin\gamma_1 z}{\gamma_1 M}H_0^{(1)}(\sqrt{\lambda}\,r)\,d\lambda \qquad (3.5.1)$$

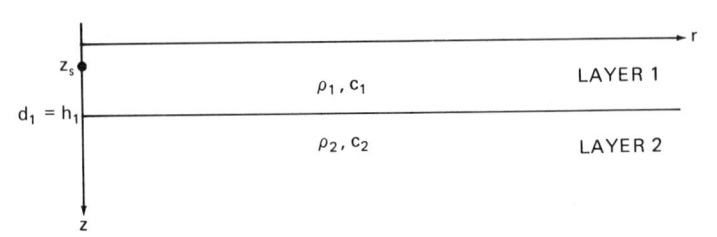

FIGURE 3.5.1. Coordinate geometry for a homogeneous layer over an infinite, homogeneous halfspace.

COMPLEX GREEN'S FUNCTION WITH A DISCRETE PLUS CONTINUOUS SPECTRUM

and M is given by Eq. (3.3.27), namely,

$$M = \rho_1\gamma_2\sin\gamma_1 h_1\sin\gamma_2 h_2 - \rho_2\gamma_1\cos\gamma_1 h_1\cos\gamma_2 h_2 \quad (3.5.2)$$

Let

$$\gamma_2 = i\nu_2 = i\sqrt{\lambda - k_2^2} \quad (3.5.3)$$

and

$$\delta_1 = \frac{\rho_1}{\rho_2} \quad (3.5.4)$$

Then the characteristic equation

$$M = 0$$

becomes

$$\rho_1\gamma_2\sin\gamma_1 h_1\sin\gamma_2 h_2 - \rho_2\gamma_1\cos\gamma_1 h_1\cos\gamma_2 h_2 = 0$$

$$\delta_1\gamma_2\sin\gamma_1 h_1\tan\gamma_2 h_2 - \gamma_1\cos\gamma_1 h_1 = 0$$

$$\delta_1\gamma_2\sin\gamma_1 h_1\tan i\nu_2 h_2 - \gamma_1\cos\gamma_1 h_1 = 0$$

$$i\delta_i\gamma_2\sin\gamma_1 d_1\tanh\nu_2 h_2 - \gamma_1\cos\gamma_1 h_1 = 0$$

where we have used the relation

$$\tan iy = i\tanh y$$

for y real and Eqs. (3.5.3) and (3.5.4).

From the property of the hyperbolic tangent we have

$$\lim_{h_2\to\infty}\tanh(\nu_2 h_2) = 1 \quad (3.5.5)$$

Thus the characteristic equation becomes for the liquid layer over a liquid halfspace

$$M_\infty = i\delta_1\gamma_2\sin\gamma_1 h_1 - \gamma_1\cos\gamma_1 h_1 = 0 \quad (3.5.6)$$

We can similarly evaluate the integral of Eq. (3.5.1). The factor contained within the braces in the integral of Eq. (3.5.1) can be rewritten as

$$\frac{\rho_1\gamma_2\sin\gamma_2 h_2\sin\gamma_1(h_1 - z_s) - \gamma_1\rho_2\cos\gamma_1(h_1 - z_s)\cos\gamma_2 h_2}{\rho_1\gamma_2\sin\gamma_1 h_1\sin\gamma_2 h_2 - \rho_2\gamma_1\cos\gamma_1 h_1\cos\gamma_2 h_2}$$

$$= \frac{\delta_1\gamma_2 i\tanh(\nu_2 h_2)\sin\gamma_1(h_1 - z_s) - \gamma_1\cos\gamma_1(h_1 - z_s)}{\delta_1\gamma_2 i\tanh(\nu_2 h_2)\sin\gamma_1 h_1 - \gamma_1\cos\gamma_1 d_1}$$

Using the last result and Eq. (3.5.5), we can write the solution $p^{(1)}$ for the case of the liquid layer over the liquid halfspace in the form

$$p^{(1)}_\infty = -\frac{1}{8\pi}\int_{c_z^-}\left\{\frac{\gamma_1\cos\gamma_1(h_1 - z_s) - i\delta_1\gamma_2\sin\gamma_1(h_1 - z_s)}{\gamma_1\cos\gamma_1 h_1 - i\delta_1\gamma_2\sin\gamma_1 h_1}\right\}$$

$$\cdot \frac{\sin(\gamma_1 z)}{\gamma_1} H_0^{(1)}(\sqrt{\lambda}\,r)\,d\lambda \quad (3.5.7)$$

146 PROPAGATION IN THE OCEAN: THE HOMOGENEOUS LAYERED MODEL

Inspection of the integral of Eq. (3.5.7) shows that, while it is still an even function of γ_1, it is no longer an even function of γ_2. Hence, it now makes a difference as to which of the two branches of

$$\gamma_2 = \pm(k_2^2 - \lambda)^{1/2}$$

as given by Eq. (3.3.11) we choose. We will shortly see that the appearance of a branch point in the integral introduces a continuous spectrum.

But first, let us obtain an expression for $p^{(2)}$, the solution in layer 2. The depth-dependent Green's function $G_z^{(2)}$ for this layer is given by Eq. (3.3.34), namely,

$$G_z^{(2)} = -\frac{1}{M}\rho_2 \sin\gamma_1 z_s \cos\gamma_2(d_2 - z)$$

Taking the limit of $G_z^{(2)}$ as $d_2 \to \infty$ gives

$$\lim_{d_2 \to \infty} G_z^{(2)} = -\frac{\sin(\gamma_1 z_s)e^{i\gamma_2(z-h_1)}}{i\delta_1\gamma_2 \sin\gamma_1 h_1 - \gamma_1 \cos\gamma_1 h_1}$$

Consequently, the solution for layer 2, $p^{(2)}$ given by Eq. (3.3.42), becomes for the case of a homogeneous layer over a homogeneous halfspace

$$p_\infty^{(2)} = \frac{1}{4\pi}\int_{C_z^-}\left\{\frac{\sin\gamma_1 z_s}{i\delta_1\gamma_2 \sin\gamma_1 h_1 - \gamma_1 \cos\gamma_1 h_1}\right\}$$

$$\cdot H_0^{(1)}(\sqrt{\lambda}\,r)e^{i\gamma_2(z-h_1)}\,d\lambda \qquad (3.5.8)$$

We now want to apply the residue theorem to evaluate the contour integral Eq. (3.5.7). For convenience, let

$$F(\gamma_1, \gamma_2) = -\frac{1}{8\pi}\left\{\frac{\gamma_1 \cos\gamma_1(h_1 - z_s) - i\delta_1\gamma_2 \sin\gamma_1(h_1 - z_s)}{\gamma_1 \cos\gamma_1 h_1 - i\delta_1\gamma_2 \sin\gamma_1 h_1}\right\}$$

$$\cdot \frac{\sin\gamma_1 z}{\gamma_1} \qquad (3.5.9)$$

so that we can write Eq. (3.5.7) compactly as

$$p_\infty^{(1)} = \int_{C_z^-} F(\gamma_1, \gamma_2) H_0^{(1)}(\sqrt{\lambda}\,r)\,d\lambda \qquad (3.5.10)$$

The residue theorem requires the integrand to be analytic on and within the closed contour C_z^- except for a finite number of singular points interior to the

COMPLEX GREEN'S FUNCTION WITH A DISCRETE PLUS CONTINUOUS SPECTRUM 147

contour. Since the integrand contains two branch points, it is not analytic on the entire λ plane. The integrand will need to be made analytic by choosing two branch cuts.

The first branch point that we need to be concerned with is due to the radical $\sqrt{\lambda}$ in the argument of the radial Green's function G_r or the Hankel function $H_0^{(1)}(\sqrt{\lambda}\, r)$. The radical $\sqrt{\lambda}$ has a branch point at

$$\lambda = 0 \qquad (3.5.11)$$

Mathematically, any branch cut will do; however, physical considerations dictate a particular branch cut. Let

$$\kappa = \sqrt{\lambda} \equiv \kappa' + i\kappa'' \qquad (3.5.12)$$

Substituting this expression for κ into the asymptotic expression for the Hankel function as given by Eq. (3.1.32), we get

$$e^{i\kappa r} = e^{i\kappa' r} e^{-\kappa'' r}$$

In order for this expression to vanish as $r \to \infty$, we must require

$$\kappa'' = \text{Im}\sqrt{\lambda} > 0 \qquad (3.5.13)$$

Thus, we must choose the two branches of $\sqrt{\lambda}$ so that on one branch $\text{Im}\sqrt{\lambda} > 0$ and on the other branch $\text{Im}\sqrt{\lambda} < 0$. Thus, the branch cut must be taken as

$$\text{Im}\,\kappa = \text{Im}\sqrt{\lambda} = 0 \qquad (3.5.14)$$

We note that if we let

$$\kappa^2 = \delta^2$$

so that

$$\kappa = \delta$$

Then the condition $\text{Im}\,\kappa = 0$ is automatically satisfied on the branch cut. Thus κ^2 must be real and positive. Now let

$$\kappa^2 = \lambda \equiv \alpha + i\beta \qquad (3.5.15)$$

So the equations describing the branch cut are

$$\beta = 0 \qquad (3.5.16a)$$

$$\alpha > 0 \qquad (3.5.16b)$$

Thus the appropriate branch cut runs from the origin along the positive real axis.

The second branch point is due to the radical

$$\gamma_2 = (k_2^2 - \lambda)^{1/2} \qquad (3.5.17)$$

As in Section 3.1, we give k_1 and k_2 a small, positive imaginary part, say

$$k_1 = k_1' + i\varepsilon \frac{k_2'}{k_1'} \qquad (3.5.18)$$

$$k_2 = k_2' + i\varepsilon \qquad (3.5.19)$$

The branch point of the radical γ_2 is at

$$\lambda = k_2^2 = k_2'^2 - \varepsilon^2 + 2ik_2'\varepsilon \qquad (3.5.20)$$

Using Eqs. (3.5.15), (3.5.17), and (3.5.19), we get

$$\gamma_2^2 = k_2^2 - \lambda = (k_2'^2 - \varepsilon^2 - \alpha) + i(2k_2'\varepsilon - \beta) \qquad (3.5.21)$$

Further, let

$$\gamma_2 = \gamma_2' + i\gamma_2''$$

and consider the exponential factor $e^{i\gamma_2 z}$ in Eq. (3.5.8), the solution $p_\infty^{(2)}$. Then

$$e^{i\gamma_2 z} = e^{i\gamma_2' z} e^{-\gamma_2'' z}$$

In order that the radiation field in layer 2 vanish as $z \to \infty$, we must require

$$\mathrm{Im}\, \gamma_2 > 0 \qquad (3.5.22)$$

Thus, we will choose the two branches of γ_2 so that on one branch $\mathrm{Im}\, \gamma_2 > 0$ and on the other branch $\mathrm{Im}\, \gamma_2 < 0$. The branch must then be given by

$$\mathrm{Im}\, \gamma_2 = 0 \qquad (3.5.23)$$

We note that if we let

$$\gamma_2^2 = \Delta^2$$

so that

$$\gamma_2 = \Delta$$

then the condition $\mathrm{Im}\, \gamma_2 = 0$ is automatically satisfied on the branch cut. Thus γ_2^2 must be real and positive. Therefore, from Eq. (3.5.21) the equations defining the branch cut are

$$\beta = 2k_2'\varepsilon \qquad (3.5.24a)$$

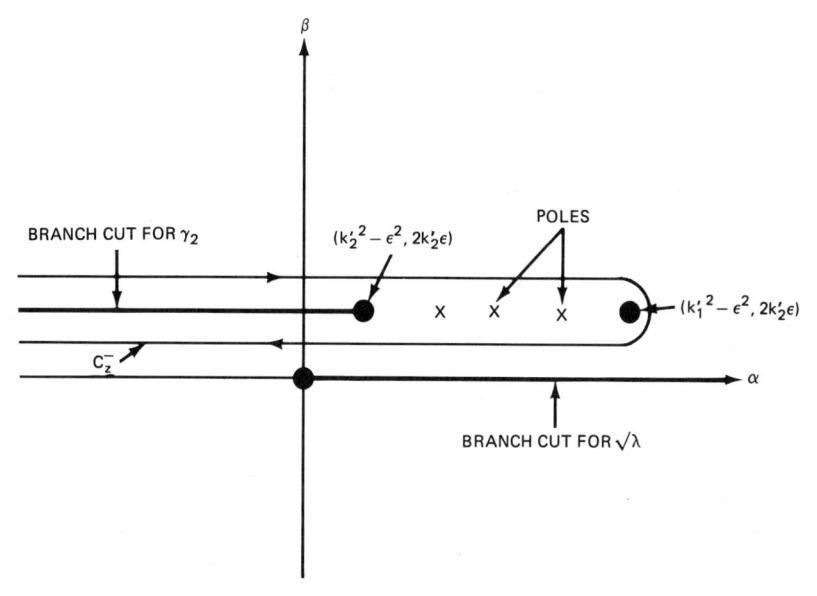

COMPLEX GREEN'S FUNCTION WITH A DISCRETE PLUS CONTINUOUS SPECTRUM   149

FIGURE 3.5.2. Branch cut for radicals γ_2 and $\sqrt{\lambda}$.

which is a horizontal line, and

$$\alpha < k_2'^2 - \varepsilon^2 \tag{3.5.24b}$$

which describes which part of the line is to be used.

The cuts for $\sqrt{\lambda}$ and γ_2 are shown in Figure 3.5.2.

Next we need to determine on which portions of the branch $\operatorname{Im}\gamma_2 > 0$ we have $\operatorname{Re}\gamma_2 > 0$ and $\operatorname{Re}\gamma_2 < 0$. To illustrate the technique, consider two points P_1 and P_2 shown in Figure 3.5.3. Point P_1 is assumed to lie on the upper edge of the cut, and point P_2 is assumed to be on the lower edge of the cut after we have taken the limit $\varepsilon \to 0$.

Now,

$$\begin{aligned}
\gamma_2 &= \left(k_2^2 - \lambda\right)^{1/2} \\
&= \left[e^{i\pi}(\lambda - k_2^2)\right]^{1/2} \\
&= \left[e^{i\pi}Re^{i\phi}\right]^{1/2} \\
&= R^{1/2}e^{i(\phi+\pi)/2} \\
&= R^{1/2}\left\{\cos\left(\frac{\phi+\pi}{2}\right) + i\sin\left(\frac{\phi+\pi}{2}\right)\right\}
\end{aligned}$$

150 PROPAGATION IN THE OCEAN: THE HOMOGENEOUS LAYERED MODEL

Here we have put

$$(\lambda - k_2^2) = Re^{i\phi}$$

For the point P_1, $\phi = \pi$, so $\cos(\phi + \pi)/2 = -1$. Hence, for points on the upper edge of the cut we have

$$\text{Re } \gamma_2 < 0$$

Similarly, for the P_2, $\phi = -\pi$. So $\cos(\phi + \pi)/2 = 1$. Hence, for points on the lower edge of the cut, we have

$$\text{Re } \gamma_2 > 0$$

This argument can be generalized to show that Re $\gamma_2 < 0$ in quadrants I and II and Re $\gamma_2 > 0$ in quadrants III and IV as shown in Figure 3.5.3. It is also easy to show that on the portion of the positive real axis running from $\lambda = k_2^2$ to positive infinity we have Re $\gamma_2 = 0$.

We must now examine the poles given by the characteristic Eq. (3.5.6), which can be written in the form

$$\tan \gamma_1 h_1 = \frac{\gamma_1}{i\delta_1 \gamma_2}$$

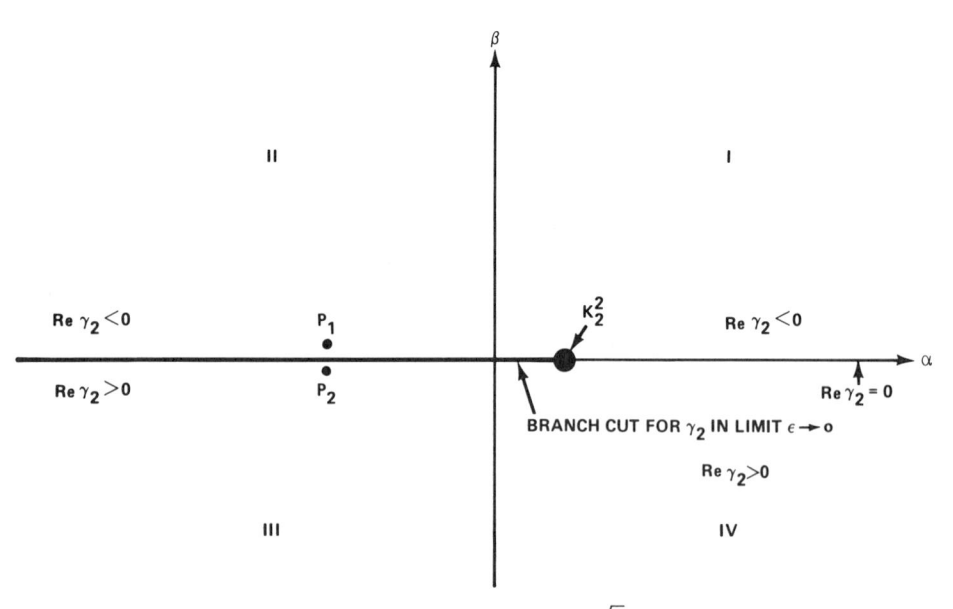

FIGURE 3.5.3. Branch cut for radicals γ_2 and $\sqrt{\lambda}$ in limit $\varepsilon \to 0$.

COMPLEX GREEN'S FUNCTION WITH A DISCRETE PLUS CONTINUOUS SPECTRUM

or, using Eqs. (3.3.10) and (3.5.3),

$$\tan(k_1^2 - \lambda)^{1/2} h_1 = \frac{-(k_1^2 - \lambda)^{1/2}}{\delta_1(\lambda - k_2^2)^{1/2}} \qquad (3.5.25)$$

First, let us examine the real roots of this equation, taking k_1 and k_2 to be real positive quantities. We see that this equation only has a finite number of solutions for real values of λ that satisfy

$$k_2^2 \leq \lambda \leq k_1^2 \qquad (3.5.26)$$

If $k_1 < k_2$ or $c_2 < c_1$, there are no real solutions of the characteristic equation and hence there is no discrete spectrum. So let us always assume that $c_2 > c_1$ so a discrete spectrum exists.

Now let us examine the characteristic equation for complex roots. Let

$$\gamma_1 h_1 = p_1 + i q_1 \qquad (3.5.27a)$$

$$\gamma_2 h_1 = p_2 + i q_2 \qquad (3.5.27b)$$

Because of the choice for the branch cut of γ_2, we have

$$q_2 > 0$$

$$p_2 < 0, \text{ Quadrants I and II}$$

$$p_2 > 0, \text{ Quadrants III and IV} \qquad (3.5.28)$$

Since the integrand of Eq. (3.5.10) was an even function of γ_1, we were not forced into any particular branch cuts. So let us choose a branch that is similar to the branch chosen for γ_2. That is, we choose a branch of γ_1, so that on this branch

$$q_1 > 0$$

$$p_1 < 0, \text{ Quadrants I and II}$$

$$p_1 > 0, \text{ Quadrants III and IV} \qquad (3.5.29)$$

Now, let us substitute Eq. (3.5.27) into Eq. (3.5.25), use the well-known relations for complex z_1, z_2, and real y given by

$$\tan(z_1 + z_2) = \frac{\tan z_1 + \tan z_2}{1 - \tan z_1 \tan z_2}$$

and
$$\tan(iy) = i\tanh y$$

and then separate the resulting equation into real and imaginary parts.

$$\tan(p_1 + iq_1) = \frac{p_1 + iq_1}{i\delta_1(p_2 + iq_2)}$$

$$\frac{\tan p_1 + \tan iq_1}{1 - \tan p_1 \tan iq_1} = \frac{p_1 + ip_2}{i\delta_1(p_2 + iq_2)}$$

$$\frac{\tan p_1 + i\tanh q_1}{1 - \tan p_1 \tanh q_1} = \frac{p_1 + ip_2}{i\delta_1(p_2 + iq_2)}$$

The real part is

$$\frac{\tan p_1(1 - \tanh^2 q_1)}{1 + \tan^2 p_1 \tanh^2 q_1} = \frac{p_2 q_1 - p_1 q_2}{\delta_1(p_2^2 + q_2^2)}$$

The imaginary part is

$$\frac{\tanh q_1(1 + \tan^2 p_1)}{1 + \tan^2 p_1 \tanh^2 q_1} = -\frac{(p_1 p_2 + q_1 q_2)}{\delta_1(p_2^2 + q_2^2)} \tag{3.5.30}$$

Consider Eq. (3.5.30). Because of the conditions (3.5.28) and (3.5.29), Eq. (3.5.30) cannot be satisfied. The left side is always positive and the right side is always negative. Thus, there are no complex roots on the permissible Riemann sheet.

Because the branch cut for γ_2 and the real roots of the characteristic equation overlap the branch cut for the radial Green's function, we must give k_1 and k_2 a small imaginary part before performing the integration. We will give k_1 and k_2 an imaginary part as indicated in Eqs. (3.5.18) and (3.5.19). We then choose the contour C_z^- as shown in Figure 3.5.2. C_z^- lies on the edges of the cut. After integrating around the contour, we allow $\varepsilon \to 0$. Thus, using the residue theorem, we find

$$p_\infty^{(1)} = \int_{C_z^-} F(\gamma_1, \gamma_2) H_0^{(1)}(\sqrt{\lambda}\, r)\, d\lambda$$

$$= \int_{-\infty}^0 F(\gamma_1^-, \gamma_2^-) H_0^{(1)}(\sqrt{\lambda}\, r)\, d\lambda$$

$$+ \int_0^{k_2^2} F(\gamma_1^-, \gamma_2^-) H_0^{(1)}(\sqrt{\lambda}\, r)\, d\lambda + \int_{k_2^2}^0 F(\gamma_1^+, \gamma_2^+) H_0^{(1)}(\sqrt{\lambda}\, r)\, d\lambda$$

$$+ \int_0^{-\infty} F(\gamma_1^+, \gamma_2^+) H_0^{(1)}(\sqrt{\lambda}\, r)\, d\lambda + 2\pi i \sum_{n=1}^N \text{residues}$$

Rewritten,

$$p_\infty^{(1)} = \int_{-\infty}^{0} [F(\gamma_1^-, \gamma_2^-) - F(\gamma_1^+, \gamma_2^+)] H_0^{(1)}(\sqrt{\lambda}\, r)\, d\lambda$$

$$+ \int_0^{k_2^2} [F(\gamma_1^-, \gamma_2^-) - F(\gamma_1^+, \gamma_2^+)] H_0^{(1)}(\sqrt{\lambda}\, r)\, d\lambda$$

$$+ 2\pi i \sum_{n=1}^{N} \text{residues} \qquad (3.5.31)$$

Before discussing the physical interpretation of Eq. (3.5.31), let us put it into a different form. Let us change the variable of integration from λ to γ_2. Using the relation

$$\gamma_2 = (k_2^2 - \lambda)^{1/2}$$

Eq. (3.5.31) can be written

$$p_\infty^{(1)} = 2\int_0^\infty [F(\gamma_1^-, \gamma_2^-) - F(\gamma_1^+, \gamma_2^+)] H_0^{(1)} \gamma_2\, d\gamma_2$$

$$+ 2\pi i \sum_{n=1}^{N} \text{residues} \qquad (3.5.32)$$

Now let us give a physical interpretation to the discrete and continuous spectrum in terms of rays. We will need the result given in Eq. (5.2.17). Since this result is independent of the WKB approximation, the first part of Section 5.2 could be read at this time.

Recall from Eq. (3.5.12)

$$\kappa_n = \sqrt{\lambda}$$

From an examination of the characteristic equation, Eq. (3.5.25), we know that for a discrete spectrum to exist we must have $c_1 < c_2$. Let us assume that this is true.

Let c_p be the phase velocity associated with a mode, that is, let

$$c_{pn} = \frac{\omega}{\kappa_n} \qquad (3.5.33)$$

Since we are considering a homogeneous layer, the ray will propagate through the entire layer at a single grazing angle θ. This angle θ is related to the phase velocity of the mode, c_p, and the speed of sound in the layer, c_1, by Eq. (5.2.17):

$$\cos \theta_n = \frac{c_1}{c_{pn}} \qquad (3.5.34)$$

154 PROPAGATION IN THE OCEAN: THE HOMOGENEOUS LAYERED MODEL

From Eq. (3.5.26), we see that the range of values of c_p corresponding to the discrete spectrum is

$$c_1 \leq c_p \leq c_2 \qquad (3.5.35)$$

From Eq. (3.5.34), the corresponding range of values for θ is

$$0 \leq \theta \leq \theta_c \qquad (3.5.36)$$

where

$$\theta_c = \cos^{-1}\frac{c_1}{c_2}$$

θ_c is the critical angle. Thus the discrete spectrum corresponds to rays which undergo total internal reflection upon striking the interface between layers 1 and 2. Recall total reflection means no energy is transmitted into layer 2.

Now consider one of the integrals that constitute the continuous spectrum, namely, the second integral in Eq. (3.5.31) which has 0 to k_2^2 as its limits of integration. In terms of the phase velocity c_p, this range of values for λ corresponds to

$$c_2 \leq c_p < \infty \qquad (3.5.37)$$

From Eq. (3.5.34), this range of phase velocities corresponds to rays that strike the bottom from the critical angle θ_c to $\theta = 90°$. According to ray theory, when a ray strikes an interface at angles greater than critical, part of the energy is reflected and part of the energy is transmitted into the second layer and is radiated to infinity.

Lastly, there is the first integral in Eq. (3.5.31) whose limits of integration are from $-\infty$ to 0. This corresponds to imaginary values of κ, that is, κ goes from $i\infty$ to 0.

The asymptotic expansion of the Hankel function appearing in this integral is

$$H_0^{(1)}(\kappa r) = \left(\frac{2}{\pi \kappa r}\right)^{1/2} e^{i(\kappa r - \pi/4)}$$

For imaginary values of κ, say $\kappa = i\alpha$, and for real positive α, the Hankel function becomes

$$H_0^{(1)}(i\alpha r) = \left(\frac{2}{\pi i \alpha r}\right)^{1/2} e^{-\alpha r - i\pi/4}$$

Thus, this range of values for κ corresponds to continuous modes that are attenuated in range, that is, to nonpropagating modes of the continuum. While these modes do not correspond to real rays, they are necessary to make the set of depth eigenfunctions complete.

APPENDIX 3.A. DIFFRACTION THEORY AND THE SOMMERFELD RADIATION CONDITION

We wish to consider solutions of the inhomogeneous Helmholtz equations

$$\nabla^2 \psi(\mathbf{r}) + k^2 \psi(\mathbf{r}) = -4\pi\sigma(\mathbf{r}) \tag{3.A.1}$$

subject to arbitrary boundary conditions on the closed surface S.

The required Green's function is the solution of the inhomogeneous Helmholtz equation

$$\nabla^2 G(\mathbf{r},\mathbf{r}_0) + k^2 G(\mathbf{r},\mathbf{r}_0) = -4\pi\delta(\mathbf{r} - \mathbf{r}_0) \tag{3.A.2}$$

for a point source at $\mathbf{r}_0$ and subject to prescribed boundary conditions. First, let us interchange $\mathbf{r}$ and $\mathbf{r}_0$ to give

$$\nabla_0^2 \psi(\mathbf{r}_0) + k^2 \psi(\mathbf{r}_0) = -4\pi\sigma(\mathbf{r}_0) \tag{3.A.3}$$

$$\nabla_0^2 G(\mathbf{r}_0,\mathbf{r}) + k^2 G(\mathbf{r}_0,\mathbf{r}) = -4\pi\delta(\mathbf{r} - \mathbf{r}_0) \tag{3.A.4}$$

where ∇_0 operates on $\mathbf{r}_0$ and ∇ on $\mathbf{r}$.

Now multiply Eq. (3.A.3) by $G(\mathbf{r}_0,\mathbf{r})$ and Eq. (3.A.4) by $\psi(\mathbf{r}_0)$, and subtract Eq. (3.A.4) from Eq. (3.A.3) to give

$$G(\mathbf{r}_0,\mathbf{r})\nabla_0^2\psi(\mathbf{r}_0) + k^2 G(\mathbf{r}_0,\mathbf{r}) - \psi(\mathbf{r}_0)\nabla_0^2 G(\mathbf{r}_0,\mathbf{r})$$

$$- k^2 \psi(\mathbf{r}_0)G(\mathbf{r}_0,\mathbf{r}) = -4\pi\sigma(\mathbf{r}_0)G(\mathbf{r}_0,\mathbf{r}) + 4\pi\delta(\mathbf{r}_0 - \mathbf{r})\psi(\mathbf{r}_0)$$

or

$$4\pi\delta(\mathbf{r}_0 - \mathbf{r})\psi(\mathbf{r}_0) = 4\pi\sigma(\mathbf{r}_0)G(\mathbf{r},\mathbf{r}_0)$$

$$+ G(\mathbf{r},\mathbf{r}_0)\nabla_0^2\psi(\mathbf{r}_0) - \psi(\mathbf{r}_0)\nabla_0^2 G(\mathbf{r},\mathbf{r}_0) \tag{3.A.5}$$

when we have used the symmetry of the Green's function

$$G(\mathbf{r}_0,\mathbf{r}) = G(\mathbf{r},\mathbf{r}_0) \tag{3.A.6}$$

Let V_0 be the volume inside the closed surface S_0. Here S_0 is the boundary surface expressed in the coordinates $\mathbf{r}_0$. The same surface expressed in the coordinates $\mathbf{r}$ will be labeled S.

Integrating Eq. (3.A.5) over V_0 and assuming $\mathbf{r} \in V_0$ gives

$$\psi(\mathbf{r}) = \iiint_{V_0} \sigma(\mathbf{r}_0)G(\mathbf{r},\mathbf{r}_0)\,dV_0$$

$$+ \frac{1}{4\pi} \iiint_{V_0} \{G(\mathbf{r},\mathbf{r}_0)\nabla_0^2\psi(\mathbf{r}_0) - \psi(\mathbf{r}_0)\nabla_0^2 G(\mathbf{r},\mathbf{r}_0)\}\,dV_0$$

156 PROPAGATION IN THE OCEAN: THE HOMOGENEOUS LAYERED MODEL

or, upon using Gauss' theorem,

$$\psi(\mathbf{r}) = \iiint_{V_0} \sigma(\mathbf{r}_0) G(\mathbf{r}, \mathbf{r}_0) \, dV_0$$

$$+ \frac{1}{4\pi} \iint_{S_0} \{ G(\mathbf{r}, \mathbf{r}_0) \nabla_0 \psi(\mathbf{r}_0) - \psi(\mathbf{r}_0) \nabla_0 G(\mathbf{r}, \mathbf{r}_0) \} \cdot \hat{n}_0 \, dS_0$$

(3.A.7)

where $\hat{n}_0$ is the unit, outward normal to S_0.

The general solution of Eq. (3.A.1) has the form

$$\psi = \psi_i + \psi_s \qquad (3.A.8)$$

where ψ_i is the particular solution of Eq. (3.A.1) due to the inhomogeneous term $\sigma(\mathbf{r})$, and ψ_s is the complementary solution of the corresponding homogeneous equations that must satisfy the imposed boundary conditions. Comparing Eqs. (3.A.7) and (3.A.8) gives

$$\psi_i(\mathbf{r}) = \iiint_{V_0} \sigma(\mathbf{r}_0) G(\mathbf{r}, \mathbf{r}_0) \, dV_0 \qquad (3.A.9)$$

$$\psi_s(\mathbf{r}) = \frac{1}{4\pi} \iint_{S_0} \{ G(\mathbf{r}, \mathbf{r}_0) \nabla_0 \psi(\mathbf{r}_0) - \psi(\mathbf{r}_0) \nabla_0 G(\mathbf{r}, \mathbf{r}_0) \} \cdot \hat{n}_0 \, dS$$

(3.A.10)

Note that ψ under the integral in Eq. (3.A.10) is still the total field.

The general solution of Eq. (3.A.2) can also be written as the sum of a particular solution $g(\mathbf{r}, \mathbf{r}_0)$ and a complementary solution $F(\mathbf{r}, \mathbf{r}_0)$:

$$G(\mathbf{r}, \mathbf{r}_0) = g(\mathbf{r}, \mathbf{r}_0) + F(\mathbf{r}, \mathbf{r}_0) \qquad (3.A.11)$$

The particular solution g satisfies Eq. (3.A.2) and, in the absence of boundaries, g is the appropriate Green's function. For three dimensions, it is of the form

$$g(\mathbf{r}, \mathbf{r}_0) = \frac{e^{ik|\mathbf{r} - \mathbf{r}_0|}}{|\mathbf{r} - \mathbf{r}_0|} \qquad (3.4.12a)$$

and, for two dimensions, it is

$$g(\mathbf{r}, \mathbf{r}_0) = H_0^{(1)}(k|r - r_0|) \qquad (3.A.12b)$$

for a time factor $e^{-i\omega t}$.

DIFFRACTION THEORY AND THE SOMMERFELD RADIATION CONDITION

The introduction of a boundary causes reflections that must be added to the wave developed by the source to give the total pressure. In Eq. (3.A.11) F represents the boundary effects and satisfies

$$\nabla^2 F(\mathbf{r}, \mathbf{r}_0) + k^2 F(\mathbf{r}, \mathbf{r}_0) = 0 \tag{3.A.13}$$

throughout V; that is, F has no singularities in the volume V. At this point the only condition placed on F is that it be a solution of Eq. (3.A.13). Thus we are free to choose F to satisfy any boundary conditions we might want to impose on G in order to simplify the surface integral appearing in Eq. (3.A.7).

Now let us apply these results to the diffraction problem. We will do the problem in two dimensions. We want to find a solution for the wave equation exterior to a surface Σ_0 (a curve in two dimensions). It is assumed that the source is inside Σ_0. The geometry is shown in Figure 3.A.1.

The surface Σ_0 is usually taken to be a wave front at a fixed time that is generated by the source.

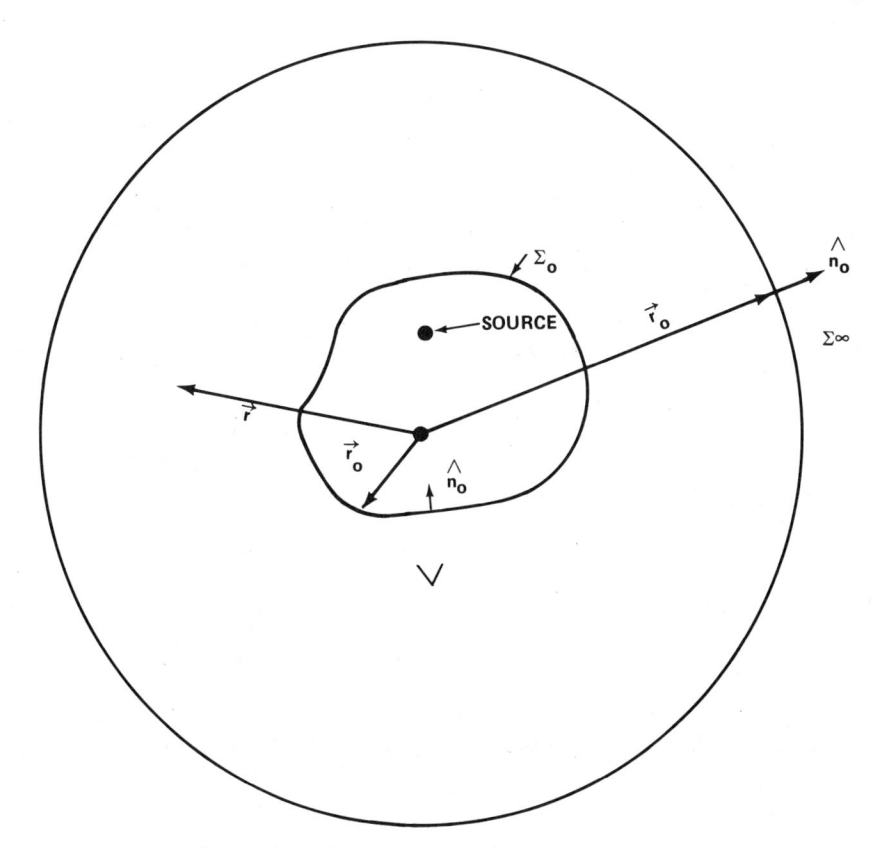

FIGURE 3.A.1. Example of a possible diffraction geometry.

158 PROPAGATION IN THE OCEAN: THE HOMOGENEOUS LAYERED MODEL

Recall that Gauss' divergence theorem only applies to a volume completely enclosed by a surface. Hence, we enclose Σ_0 by a large sphere Σ_∞ (a circle in two dimensions) and apply Gauss' theorem to the volume V_0 (area in two dimensions) enclosed by $S_0 = \Sigma_0 + \Sigma_\infty$, where S_0 is the surface of integration in Eq. (3.A.7). Let the source be located as shown in Figure 3.A.1.

Since $\sigma(\mathbf{r}_0) = 0$ for $\mathbf{r}_0 \in V_0$, Eq. (3.A.7) becomes

$$\psi(\mathbf{r}) = \frac{1}{4\pi} \iint_{\Sigma_0} \{G(\mathbf{r},\mathbf{r}_0)\nabla_0\psi(\mathbf{r}_0) - \psi(\mathbf{r}_0)\nabla_0 G(\mathbf{r},\mathbf{r}_0)\} \cdot \hat{n}_0 \, d\Sigma_0$$

$$+ \frac{1}{4\pi} \iint_{\Sigma_\infty} \{G(\mathbf{r},\mathbf{r}_0)\nabla_0\psi(\mathbf{r}_0) - \psi(\mathbf{r}_0)\nabla_0 G(\mathbf{r},\mathbf{r}_0)\} \cdot \hat{n}_0 \, d\Sigma_\infty$$

(3.A.14)

Now let us choose $F = 0$ in V and V_0, and first evaluate the integral over Σ_∞ in Eq. (3.A.14). Calling this term I_∞ we have

$$I_\infty = \frac{1}{4\pi} \iint_{\Sigma_\infty} \{G(\mathbf{r},\mathbf{r}_0)\nabla_0\psi(\mathbf{r}_0) - \psi(\mathbf{r}_0)\nabla_0 G(\mathbf{r},\mathbf{r}_0)\} \cdot \hat{n}_0 \, d\Sigma_\infty$$

(3.A.15)

Since we are assuming r_0 is very large and actually will let it go to infinity in the end, we can use the asymptotic expansion of the Hankel function given in Eq. (3.A.12b). Thus we choose for the Green's function here

$$G(\mathbf{r},\mathbf{r}_0) = \sqrt{\frac{2}{\pi}} \frac{e^{ik|\mathbf{r}-\mathbf{r}_0|-i(\pi/4)}}{\sqrt{k|\mathbf{r}-\mathbf{r}_0|}}$$

(3.A.16)

Let

$$R = |\mathbf{r} - \mathbf{r}_0|$$

(3.A.17)

Then we get

$$I_\infty = \sqrt{\frac{1}{8\pi^3}} e^{-i(\pi/4)} \int_{\Sigma_\infty} \left\{ \frac{e^{ikR}}{\sqrt{kR}} \nabla_0\psi(\mathbf{r}_0) - \psi(\mathbf{r}_0)\nabla_0 \frac{e^{ikR}}{\sqrt{kR}} \right\} \cdot \hat{n}_0 \, d\Sigma_\infty$$

(3.A.18)

Let $\hat{r}_0 = \mathbf{r}_0/r_0$ and $\hat{r} = \mathbf{r}/r$ where $r_0 = |\mathbf{r}_0|$ and $r = |\mathbf{r}|$.
On Σ_∞ we have

$$\hat{n}_0 \cdot \nabla_0 = \hat{r}_0 \cdot \nabla_0 = \frac{\partial}{\partial r_0}$$

DIFFRACTION THEORY AND THE SOMMERFELD RADIATION CONDITION 159

The expression for I_∞ becomes

$$I_\infty = \left(\frac{1}{8\pi^3}\right)^{1/2} e^{-i(\pi/4)} \int_{\Sigma_\infty} \left\{\frac{\partial \psi}{\partial r_0} - \left(ik - \frac{1}{2R}\right)\psi \frac{\partial R}{\partial r_0}\right\} \frac{e^{ikR}}{\sqrt{kR}} d\Sigma_\infty \quad (3.A.19)$$

Now

$$R = [r^2 + r_0^2 - 2rr_0\cos(\hat{r}, \hat{r}_0)]^{1/2}$$

$$= r_0\left\{1 - \frac{2r}{r_0}\cos(\hat{r}, \hat{r}_0) + \frac{r^2}{r_0^2}\right\}^{1/2}$$

$$= r_0\left\{1 - \frac{r}{r_0}\cos(\hat{r}, \hat{r}_0) + o\left(\frac{1}{r_0^2}\right)\right\} \quad (3.A.20)$$

So

$$\frac{\partial R}{\partial r_0} = 1 + o\left(\frac{1}{r_0^2}\right) \quad (3.A.21)$$

For $r_0 \gg 1$, we take

$$R \approx r_0$$

$$\frac{\partial R}{\partial r_0} \approx 1$$

and neglect the $1/(2r_0)$ term in Eq. (3.A.19). Hence, we get

$$I_\infty = \left(\frac{1}{8\pi^3}\right)^{1/2} e^{-i(\pi/4)} \int_{\Sigma_\infty} \left\{\frac{\partial \psi}{\partial r_0} - ik\psi\right\} \frac{e^{ikr_0}}{\sqrt{kr_0}} d\Sigma_\infty$$

In polar coordinates $d\Sigma_\infty = r_0 \, d\theta_0$. So we get

$$I_\infty = \left(\frac{1}{8\pi^3 k}\right)^{1/2} e^{-(\pi/4)i} \int_{\Sigma_\infty} \sqrt{r_0}\left[\frac{\partial \psi}{\partial r_0} - ik\psi\right] e^{ikr_0} d\theta_0 \quad (3.A.22)$$

In order for this integral to vanish as $r_0 \to \infty$, we must require

$$\lim_{r_0 \to \infty} \sqrt{r_0}\left(\frac{\partial \psi}{\partial r_0} - ik\psi\right) = 0 \quad (3.A.23)$$

This is the Sommerfeld radiation condition.

The solution for ψ given by Eq. (3.A.14) then becomes

$$\psi(\mathbf{r}) = \left(\frac{1}{8\pi^3 k}\right)^{1/2} e^{-i\pi/4} \int \left\{ \frac{e^{ikR}}{\sqrt{R}} \nabla_0 \psi - \psi \nabla_0 \left(\frac{e^{ikR}}{\sqrt{R}}\right) \right\} \cdot \hat{n}_0 \, d\Sigma_0$$

(3.A.24)

APPENDIX 3.B. EVALUATION OF THE COMPLEX INTEGRAL SOLUTION OF THE WAVE EQUATION USING THE CONTOUR C_r^+

We saw in Section 3.1 that the general solution of the wave equation could be represented in the form given by Eq. (3.1.28):

$$p(r, r_s, z, z_s) = -\frac{1}{2\pi i} \int_{C_\lambda} G_r(r, r_s, \lambda) G_z(z, z_s, \lambda) \, d\lambda \qquad (3.B.1)$$

Throughout Chapter 3 we used the contour $C_\lambda = C_z^-$ to evaluate this integral. We will now evaluate it using the contour C_r^+ and show that it leads to the well-known Hankel transform solution of the wave equation. We will also see that the eigenfunction expansion representation is much more difficult to obtain using the contour C_r^+.

We saw that for all the cases we considered, the radial Green's function was given by Eq. 3.1.34:

$$G_r = \frac{i}{4} H_0^{(1)}(\sqrt{\lambda}\, r) \qquad (3.B.2)$$

Thus Eq. (3.B.1) becomes

$$p = -\frac{1}{8\pi} \int_{C_r^+} G_z H_0^{(1)}(\sqrt{\lambda}\, r) \, d\lambda \qquad (3.B.3)$$

The singularities of G_r consist of only one branch point at the origin as we discussed in Section 3.1. Using the same technique as was used in Section 3.5, we can show that $\text{Re}\sqrt{\lambda} > 0$ on the upper edge of the cut and $\text{Re}\sqrt{\lambda} < 0$ on the lower edge of the cut. This is shown in Figure 3.B.1.

Recall the contour C_r^+ encloses all the λ-plane singularities of G_r but none of the singularities of G_z and is taken in the counterclockwise sense. We saw that by adding a small imaginary part to the wave number k, we could always displace the singularities of G_z off the real axis, so we can assume we have already done that. The contour C_r^+ in Figure 3.B.1 is exaggerated for clarity. It should run right along the upper and lower edges of the cut.

INTEGRAL SOLUTION OF THE WAVE EQUATION USING THE CONTOUR C_r^+

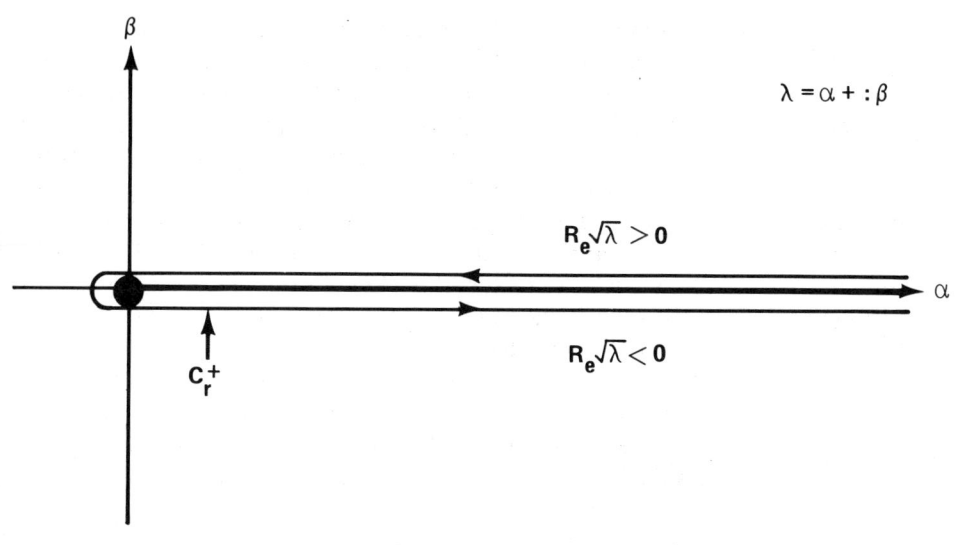

FIGURE 3.B.1. Contour and branch cuts for C_r^+.

Evaluating the integral in Eq. (3.B.3) along this contour gives

$$p = -\frac{1}{8\pi}\int_\infty^0 G_z H_0^{(1)}(\sqrt{\lambda}\,r)\,d\lambda$$

$$-\frac{1}{8\pi}\int_0^\infty G_z H_0^{(1)}(-\sqrt{\lambda}\,r)\,d\lambda \qquad (3.B.4)$$

Now let us introduce a new variable ζ defined by

$$\zeta \equiv \sqrt{\lambda} \qquad (3.B.5)$$

Then Eq. (3.B.4) becomes with the variable ζ

$$p = -\frac{1}{4\pi}\int_\infty^0 G_z H_0^{(1)}(\zeta r)\zeta\,d\zeta - \frac{1}{4\pi}\int_0^\infty G_z H_0^{(1)}(-\zeta r)\zeta\,d\zeta$$

Using the substitution $\zeta' = -\zeta$ in this last integral, we get

$$p = -\frac{1}{4\pi}\int_\infty^0 G_z H_0^{(1)}(\zeta r)\zeta\,d\zeta - \frac{1}{4\pi}\int_0^{-\infty} G_z H_0^{(1)}(\zeta' r)\zeta'\,d\zeta'$$

or, since ζ' is a dummy variable of integration, we get the final result as

$$p = \frac{1}{4\pi}\int_{-\infty}^\infty G_z H_0^{(1)}(\zeta r)\zeta\,d\zeta \qquad (3.B.6)$$

This is also the result that would be obtained by applying the Hankel transform directly to the wave equation. For those who are unfamiliar with using transform techniques to solve partial differential equations, an excellent introduction can be found in Sneddon [3.4].

The solution to the wave equation represented by Eq. (3.B.6) is now a real integral. In order to evaluate this by the theory of residues, we must take a closed contour that extends along the real axis and is closed by a semicircle at infinity. That is, we must apply the standard technique used to evaluate real infinite integrals. We then have to show that the integral vanishes along the semicircle. If there are branch cuts needed for G_z, we must go around these. Thus we see that by using the counter C_r^+, we still have a very difficult contour integral to evaluate after we have evaluated it for C_r^+. On the other hand, using the contour C_z^-, we arrive at the series expansion directly and very easily.

The form of the solution given by Eq. (3.B.6) is that used by Ewing, Jardetzky, and Press [3.5].

REFERENCES

3.1. L. E. Kinsler and A. R. Frey, *Fundamentals of Acoustics*, (Wiley, New York, 1962), pp. 128–130.

3.2. See Kinsler and Frey, pp. 163–165.

3.3. A. Sommerfeld, *Partial Differential Equations in Physics*, (Academic Press, New York, 1967).

3.4. I. N. Sneddon, *Elements of Partial Differential Equations*, (McGraw-Hill, New York, 1957), pp. 126–129.

3.5. W. M. Ewing, W. S. Jardetzky, and F. Press, *Elastic Waves in Layered Media*, (McGraw-Hill, New York, 1962).

4 PROPAGATION IN THE OCEAN AS A BOUNDARY VALUE PROBLEM: THE INHOMOGENEOUS LAYERED MODEL

4.0. INTRODUCTION

Now that we have acquired some dexterity in solving the wave equation for some simple models, we are in a position to treat the oceanic waveguide in a more realistic manner. In Section 4.1 we solve the wave equation for a model of the ocean consisting of N inhomogeneous layers where the sound velocity and density can be discontinuous from layer to layer. This property is useful for representing the multilayered structure of the bottom sediments.

This normal mode model is a slightly modified version of the model published by Stickler [4.1]. This model, which is used extensively at The Johns Hopkins University Applied Physics Laboratory, differs from Stickler's model in two respects. First, the final boundary or floor at the bottom of the last sediment layer is assumed to be rigid. This is so that we do not have to deal with the continuous spectrum. Second, there is attenuation in both the water and sediments in this model that was missing from the original model published by Stickler. The major difference between these two models, which we will not discuss here, is the numerical implementation of the equations.

4.1. A NORMAL MODE MODEL FOR N-INHOMOGENEOUS LAYERS WITH DISCONTINUOUS PROPERTIES

4.1.1. Formulation of the Problem

We will consider a horizontally stratified oceanic wave guide as shown in Figure 4.1.1. The model will consist of N horizontally stratified layers with $c_\alpha(z)$ and ρ_α representing the sound speed and density in the αth layer where

164 PROPAGATION IN THE OCEAN: THE INHOMOGENEOUS LAYERED MODEL

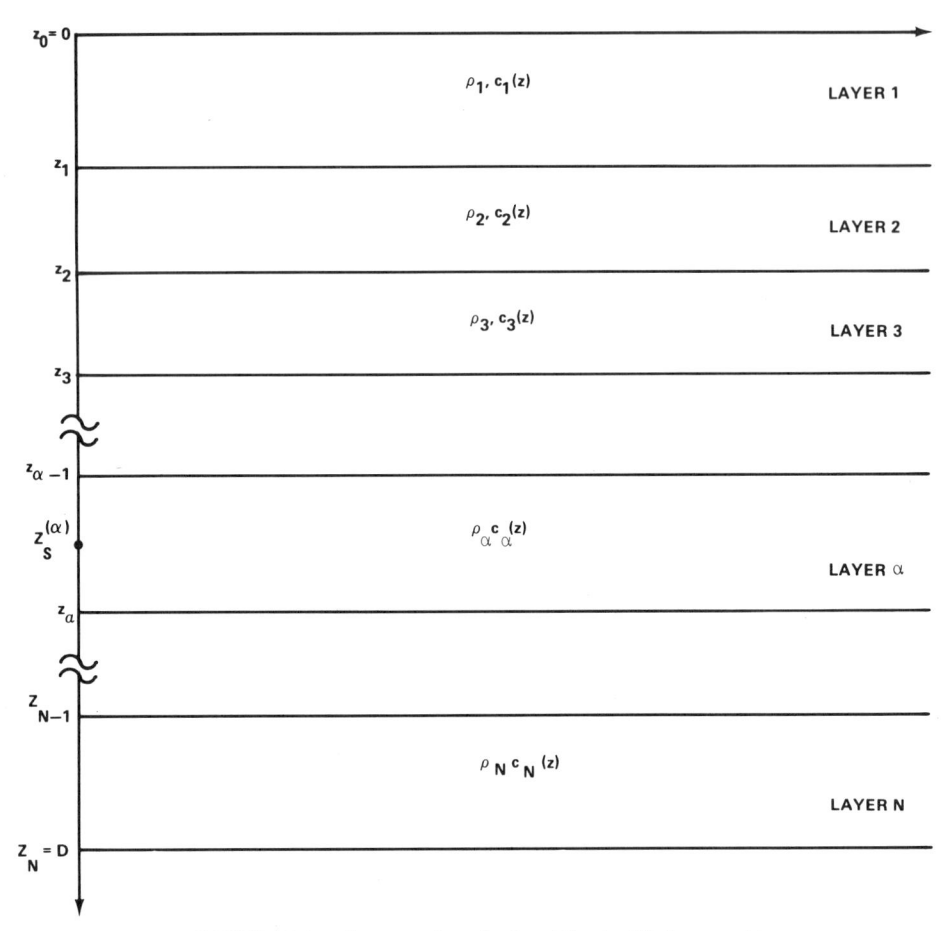

FIGURE 4.1.1. Geometry for a horizontally stratified waveguide.

$\alpha = 1, \ldots, N$. The density is constant in each layer, but its value can differ from layer to layer. The bottom of the Nth layer will be assumed a rigid boundary. A rigid boundary is used for the sole reason of avoiding the continuous spectrum. By confining the waveguide between a pressure-release surface and a rigid lower boundary, we are guaranteed an infinite discrete spectrum that is mathematically rigorous. We can use the upper layers to represent the water column and as many of the lower layers as we wish to represent the bottom sediments. If desired, the Nth layer could be taken to be a nearly homogeneous layer and made as thick as we please in order to approximate the continuous spectrum as closely as we please. Terminating an ocean waveguide model by a homogeneous, infinite halfspace may not be more realistic than this model, which terminates the bottom of the sediment layers by a rigid boundary.

A NORMAL MODE MODEL FOR N-INHOMOGENEOUS LAYERS

The ocean will be assumed cylindrically symmetric and a time factor $e^{-i\omega t}$ will be suppressed.

The wave equation for the acoustic pressure p in cylindrical coordinates r and z is

$$\frac{\partial^2 p}{\partial r^2} + \frac{1}{r}\frac{\partial p}{\partial r} + \frac{\partial^2 p}{\partial z^2} - \frac{1}{\rho(z)}\frac{d\rho}{dz}\frac{\partial p}{\partial z} + k^2(z)p = -\frac{1}{2\pi r}\delta(r)\delta(z - z_s) \qquad (4.1.1)$$

Here

$$\rho(z) = \rho_\alpha \quad \text{for} \quad z_{\alpha-1} < z < z_\alpha \quad \text{and} \quad \alpha = 1, \ldots, N \qquad (4.1.2)$$

and $k(z)$ is the local wave number defined by

$$k(z) = \frac{\omega}{c(z)} \qquad (4.1.3)$$

where the sound speed $c(z)$ is defined by

$$c(z) = c_\alpha(z), \quad z_{\alpha-1} < z < z_\alpha \quad \alpha = 1, \ldots, N \qquad (4.1.4)$$

We are also assuming there is a source at $r = 0$, $z = z_s^{(\alpha)}$. We define the index of refraction $n(z)$ as

$$n(z) = \frac{1}{c(z)} \qquad (4.1.5)$$

If, for a single mode $p_n(r, z)$ of the pressure field, we assume a separated solution of the form

$$p_n(r, z) = \phi_n(r)\psi_n(z) \qquad (4.1.6)$$

we find that $\psi_n(z)$ satisfies the separated ordinary differential equation

$$\frac{d}{dz}\left[\frac{1}{\rho(z)}\frac{d\psi_n}{dz}\right] + \left[\frac{k^2(z)}{\rho(z)} - \frac{\kappa_n^2}{\rho(z)}\right]\psi_n = 0 \qquad (4.1.7)$$

where κ_n^2 is the separation constant.

We know from the boundary conditions already stated, that is, that the oceanic waveguide is confirmed between a pressure-release surface and a rigid lower boundary, that the set of eigenvalues that are solutions of Eq. (4.1.7) form an infinite discrete spectrum and the corresponding eigenfunctions form a complete set. Thus we can expand the general solution of Eq. (4.1.1) in terms of the $\psi_n(z)$ as

$$p(r, z) = \sum_{n=1}^{\infty} \phi_n(r)\psi_n(z) \qquad (4.1.8)$$

4.1.2. Description of the Vertical Sound Speed Profile

In each layer, $n^2(z)$ is assumed to be a linear function of z, that is,

$$n^2 = a'_\alpha z + b'_\alpha, \qquad \alpha = 1,\ldots,N \qquad (4.1.9)$$

Consider the αth layer shown in Figure 4.1.2. Here $c_{T,\alpha}$ is the sound speed at the top of the αth layer, $C_{B,\alpha}$ is the sound speed at the bottom of the αth layer, and l_α is the thickness of the αth layer.

From Eqs. (4.1.5) and (4.1.9),

$$\frac{1}{c_{T,\alpha}^2} = a'_\alpha z_{\alpha-1} + b'_\alpha$$

$$\frac{1}{c_{B,\alpha}^2} = a'_\alpha z_\alpha + b'_\alpha$$

Solving these equations for a'_α and b'_α results in

$$a'_\alpha = \frac{1}{l_\alpha}\left(\frac{1}{c_{B,\alpha}^2} - \frac{1}{c_{T,\alpha}^2}\right)$$

$$b'_\alpha = \frac{1}{c_{T,\alpha}^2} - \frac{z_{\alpha-1}}{l_\alpha}\left(\frac{1}{c_{B,\alpha}^2} - \frac{1}{c_{T,\alpha}^2}\right)$$

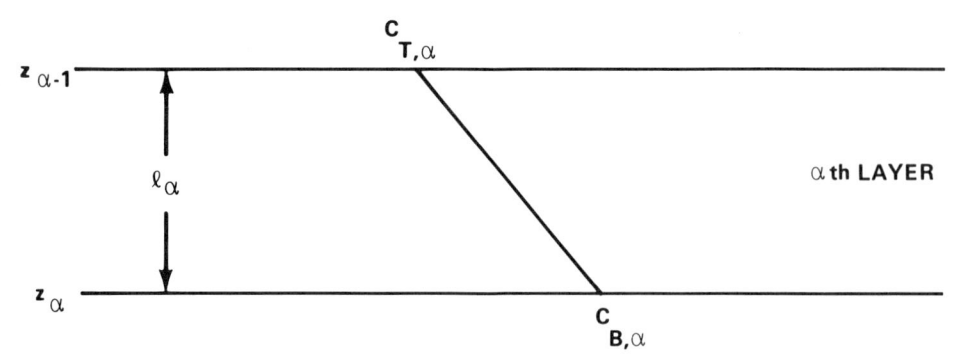

FIGURE 4.1.2. Geometry for a single layer.

Combining this result with Eq. (4.1.9) gives

$$n^2 = \frac{1}{c_{T,\alpha}^2} - \left(\frac{1}{c_{T,\alpha}^2} - \frac{1}{c_{B,\alpha}^2}\right)\frac{(z - z_{\alpha-1})}{l_\alpha}$$

Let

$$a_\alpha = \frac{1}{c_{T,\alpha}^2} \tag{4.1.10}$$

$$b_\alpha = \frac{1}{l_\alpha}\left(\frac{1}{c_{T,\alpha}^2} - \frac{1}{c_{B,\alpha}^2}\right) \tag{4.1.11}$$

so that we can write n^2 in the form

$$n^2 = a_\alpha + b_\alpha z_{\alpha-1} - b_\alpha z \tag{4.1.12}$$

4.1.3. Solution of the Depth Equation

Using Eqs. (4.1.3), (4.1.5), and (4.1.12), the depth equation (4.1.7) becomes for the αth layer

$$\frac{d^2\psi_n^{(\alpha)}}{dz^2} + \left(\omega^2 a_\alpha + \omega^2 b_\alpha z_{\alpha-1} - \kappa_n^2 - \omega^2 b_\alpha z\right)\psi_n^{(\alpha)} = 0 \tag{4.1.13}$$

Note ρ_α has been dropped from Eq. (4.1.13) because it is constant in each layer. Let

$$Z_\alpha = -\left(\omega^2 b_\alpha\right)^{-2/3}\left[\omega^2 a_\alpha + \omega^2 b_\alpha z_{\alpha-1} - \kappa_n^2 - \omega^2 b_\alpha z\right] \tag{4.1.14}$$

Using Eq. (4.1.14), we obtain

$$\frac{d\psi_n^{(\alpha)}}{dz} = \frac{d\psi_n^{(\alpha)}}{dZ_\alpha}\frac{dZ_\alpha}{dz} = \left(\omega^2 b_\alpha\right)^{1/3}\frac{d\psi_n^{(\alpha)}}{dZ_\alpha}$$

and

$$\frac{d^2\psi_n^{(\alpha)}}{dz^2} = \frac{d}{dz}\left[\left(\omega^2 b_\alpha\right)^{1/3}\frac{d\psi_n^{(\alpha)}}{dZ_\alpha}\right]$$

$$= \frac{d}{dZ_\alpha}\left[\left(\omega^2 b_\alpha\right)^{1/3}\frac{d\psi_n^{(\alpha)}}{dZ_\alpha}\right]\frac{dZ_\alpha}{dz}$$

$$= \left(\omega^2 b_\alpha\right)^{2/3}\frac{d^2\psi_n^{(\alpha)}}{dZ_\alpha^2}$$

168 PROPAGATION IN THE OCEAN: THE INHOMOGENEOUS LAYERED MODEL

Using these results, Eq. (4.1.13) becomes

$$\frac{d^2\psi_n^{(\alpha)}}{dZ_\alpha^2} - Z_\alpha \psi_n^{(\alpha)} = 0 \qquad (4.1.15)$$

This is Airy's differential equation which was discussed in Section 2.10. The general solution of Eq. (4.1.15) is

$$\psi_n^{(\alpha)} = A_\alpha Ai[Z_\alpha(z)] + B_\alpha Bi[Z_\alpha(z)] \qquad (4.1.16)$$

where Ai and Bi are the Airy functions defined by Eqs. (2.10.5) and (2.10.6).

The boundary conditions are:

(Z.1) $\quad\quad\quad \psi_n^{(1)}(0) = 0$

(Z.2) $\quad\quad\quad \psi_n^{(\alpha)}(z_\alpha) = \psi_n^{(\alpha+1)}(z_\alpha), \quad \alpha = 1, \ldots, N-1$

(Z.3) $\quad\quad\quad \dfrac{1}{\rho_\alpha}\dfrac{\partial \psi_n^{(\alpha)}}{\partial z}(z_\alpha) = \dfrac{1}{\rho_{\alpha+1}}\dfrac{\partial \psi_n^{(\alpha+1)}}{\partial z}(z_\alpha), \quad \alpha = 1, \ldots, N-1$

(Z.4) $\quad\quad\quad \dfrac{\partial \psi_n^{(N)}}{\partial z}(z_N) = 0$

Applying the boundary conditions,

$$A_1 Ai[Z_1(0)] + B_1 Bi[Z_1(0)] = 0 \qquad (4.1.17)$$

$$\vdots$$

$$Ai(\alpha+1, \alpha)A_{\alpha+1} + Bi(\alpha+1, \alpha)B_{\alpha+1} = Ai(\alpha, \alpha)A_\alpha + Bi(\alpha, \alpha)B_\alpha \qquad (4.1.18)$$

$$\frac{b_{\alpha+1}^{1/3}}{\rho_{\alpha+1}} Ai'(\alpha+1, \alpha)A_{\alpha+1} + \frac{b_{\alpha+1}^{1/3}}{\rho_{\alpha+1}} Bi'(\alpha+1, \alpha)B_{\alpha+1}$$

$$= \frac{b_\alpha^{1/3}}{\rho_\alpha} Ai'(\alpha, \alpha)A_\alpha + \frac{b_\alpha^{1/3}}{\rho_\alpha} Bi'(\alpha, \alpha)B_\alpha \qquad (4.1.19)$$

$$\vdots$$

$$Ai'(N, N)A_N + Bi'(N, N)B_N = 0 \qquad (4.1.20)$$

where we have used the notation

$$Ai(\alpha, \beta) = Ai[Z_\alpha(z_\beta)] \qquad (4.1.21)$$

A NORMAL MODE MODEL FOR N-INHOMOGENEOUS LAYERS

and primes on Ai and Bi indicate differentiation with respect to the argument Z_α.

Let us write Eqs. (4.1.18) and (4.1.19) in the form

$$\alpha_3 A_{\alpha+1} + \beta_3 B_{\alpha+1} = \alpha_1 A_\alpha + \beta_1 B_\alpha$$

$$\alpha_4 A_{\alpha+1} + \beta_4 B_{\alpha+1} = \alpha_2 A_\alpha + \beta_2 B_\alpha \qquad (4.1.22)$$

In matrix notation, Eq. (4.1.22) can be written

$$\begin{bmatrix} \alpha_3 & \beta_3 \\ \alpha_4 & \beta_4 \end{bmatrix} \begin{bmatrix} A_{\alpha+1} \\ B_{\alpha+1} \end{bmatrix} = \begin{bmatrix} \alpha_1 & \beta_1 \\ \alpha_2 & \beta_2 \end{bmatrix} \begin{bmatrix} A_\alpha \\ B_\alpha \end{bmatrix}$$

or

$$\begin{bmatrix} A_{\alpha+1} \\ B_{\alpha+1} \end{bmatrix} = M_\alpha \begin{bmatrix} A_\alpha \\ B_\alpha \end{bmatrix}$$

where

$$M_\alpha = \begin{bmatrix} \alpha_3 & \beta_3 \\ \alpha_4 & \beta_4 \end{bmatrix}^{-1} \begin{bmatrix} \alpha_1 & \beta_1 \\ \alpha_2 & \beta_2 \end{bmatrix}$$

$$= \frac{1}{\Delta} \begin{bmatrix} \beta_4 & -\beta_3 \\ -\alpha_4 & \alpha_3 \end{bmatrix} \begin{bmatrix} \alpha_1 & \beta_1 \\ \alpha_2 & \beta_2 \end{bmatrix}$$

where

$$\Delta = \alpha_3 \beta_4 - \alpha_4 \beta_3 \qquad (4.1.23)$$

Thus

$$M_\alpha = \frac{1}{\Delta} \begin{bmatrix} \alpha_1 \beta_4 - \alpha_2 \beta_3 & \beta_4 \beta_1 - \beta_2 \beta_3 \\ \alpha_2 \alpha_3 - \alpha_1 \alpha_4 & \alpha_3 \beta_2 - \alpha_4 \beta_1 \end{bmatrix} \qquad (4.1.24)$$

Let

$$M_\alpha = \begin{bmatrix} M_{11}^{(\alpha)} & M_{12}^{(\alpha)} \\ M_{21}^{(\alpha)} & M_{22}^{(\alpha)} \end{bmatrix} \qquad (4.1.25)$$

Then, using Eqs. (4.1.18), (4.1.19), (4.1.22), and (4.1.23), we get

$$\Delta = Ai(\alpha+1, \alpha) \frac{b_{\alpha+1}^{1/3}}{\rho_{\alpha+1}} Bi'(\alpha+1, \alpha) - \frac{b_{\alpha+1}^{1/3}}{\rho_{\alpha+1}} Ai'(\alpha+1, \alpha) Bi(\alpha+1, \alpha)$$

$$\Delta = \frac{b_{\alpha+1}^{1/3}}{\pi \rho_{\alpha+1}} \qquad (4.1.26)$$

where we have used the Wronskian

$$W = Ai(z)Bi'(z) - Bi(z)Ai'(z) = \frac{1}{\pi} \quad (4.1.27)$$

Further, we obtain for the elements of the matrix M_α

$$M_{11}^{(\alpha)} = \pi\{Ai(\alpha,\alpha)Bi'(\alpha+1,\alpha) - T_\alpha Ai'(\alpha,\alpha)Bi(\alpha+1,\alpha)\}$$
$$(4.1.28)$$

$$M_{12}^{(\alpha)} = \pi\{Bi'(\alpha+1,\alpha)Bi(\alpha,\alpha) - T_\alpha Bi'(\alpha,\alpha)Bi(\alpha+1,\alpha)\}$$
$$(4.1.29)$$

$$M_{21}^{(\alpha)} = \pi\{T_\alpha Ai'(\alpha,\alpha)Ai(\alpha+1,\alpha) - Ai(\alpha,\alpha)Ai'(\alpha+1,\alpha)\}$$
$$(4.1.30)$$

$$M_{22}^{(\alpha)} = \pi\{T_\alpha Ai(\alpha+1,\alpha)Bi'(\alpha,\alpha) - Ai'(\alpha+1,\alpha)Bi(\alpha,\alpha)\}$$
$$(4.1.31)$$

where

$$T_\alpha = \frac{\rho_{\alpha+1}}{\rho_\alpha}\left(\frac{b_\alpha}{b_{\alpha+1}}\right)\left(\frac{b_{\alpha+1}^2}{b_\alpha^2}\right)^{1/3} \quad (4.1.32)$$

and so our set of Eqs. (4.1.18) and (4.1.19) can be written

$$\begin{bmatrix} A_{\alpha+1} \\ B_{\alpha+1} \end{bmatrix} = M_\alpha \begin{bmatrix} A_\alpha \\ B_\alpha \end{bmatrix} \quad (4.1.33)$$

The amplitudes A_α and B_α and the eigenvalues κ_n are now obtained by the following procedure. Eq. (4.1.17) gives the solution for A_1 and B_1. This solution is simply

$$A_1 = N_n Bi[Z_1(0)] \quad (4.1.34)$$

$$B_1 = -N_n Ai[Z_1(0)] \quad (4.1.35)$$

where N_n is a normalizing factor.

Eq. (4.1.33) is then used repeatedly to solve the systems of equations resulting from application of the boundary condition, two at a time. We thus arrive at A_N, B_N by the cascaded matrix product

$$\begin{bmatrix} A_N \\ B_N \end{bmatrix} = M_{N-1} M_{N-2} \ldots M_2 M_1 \begin{bmatrix} A_1 \\ A_2 \end{bmatrix} \quad (4.1.36)$$

This result for A_N and B_N from Eq. (4.1.36) is substituted into Eq. (4.1.20), which then becomes the characteristic equation for the eigenvalues

$$Ai'(N, N)A_N + Bi'(N, N)B_N = 0 \qquad (4.1.37)$$

According to the general theory of eigenfunctions discussed in Section 2.2, we can then choose the arbitrary normalizing factor N_n so that the depth functions form an orthonormal set relative to the weight function ρ^{-1}. This orthonormality is expressed by the condition

$$\int_0^d \frac{1}{\rho(z)} \psi_n(z) \psi_m(z)\, dz = \delta_{nm} \qquad (4.1.38)$$

4.1.4. Attenuation

Attenuation effects are included by using first-order perturbation theory. It is assumed that the attenuation effects of the medium may be treated by introducing a small imaginary component of the wave number; that is, $k(z)$ becomes

$$k(z) + i\alpha(z) \qquad (4.1.39)$$

where we assume $\alpha \ll k$. The parameter α is the attenuation as a function of depth in units of nepers per unit length.

Using Eq. (4.1.39), the depth equation (4.1.7) becomes

$$\frac{d}{dz}\left(\frac{1}{\rho}\frac{d\psi_n}{dz}\right) + \left(\frac{k^2}{\rho} - \frac{\lambda_n^2}{\rho} + \frac{2i\alpha k}{\rho}\right)\psi_n = 0 \qquad (4.1.40)$$

where we have denoted the eigenvalue here by λ_n instead of κ_n and have neglected the term of order α^2.

Let ψ_n^0 and κ_n^2 be the unperturbed eigenfunction and corresponding eigenvalue. Then they satisfy

$$\frac{d}{dz}\left(\frac{1}{\rho}\frac{d\psi_n^0}{dz}\right) + \left(\frac{k^2}{\rho} - \frac{\kappa_n^2}{\rho}\right)\psi_n^0 = 0 \qquad (4.1.41a)$$

or

$$\frac{d^2\psi_n^0}{dz^2} - \frac{\rho'}{\rho}\frac{d\psi_n^0}{dz} + (k^2 - \kappa_n^2)\psi_n^0 = 0 \qquad (4.1.41b)$$

To a first approximation we write

$$\psi_n = \psi_n^0 + \varepsilon \psi_n^1 \qquad (4.1.42)$$

$$\lambda_n^2 = \kappa_n^2 + i\varepsilon\delta_n \qquad (4.1.43)$$

172 PROPAGATION IN THE OCEAN: THE INHOMOGENEOUS LAYERED MODEL

where the small parameter ε is introduced only to facilitate comparison of quantities as to order of magnitude and does not enter into any calculations. Note we must use $k + i\varepsilon\alpha$ in Eq. (4.1.39).

Let us write Eq. (4.1.40) in the form

$$\frac{d^2\psi_n}{dz^2} - \frac{\rho'}{\rho}\frac{d\psi_n}{dz} + k^2\psi_n + 2i\varepsilon\alpha k\psi_n = \lambda_n^2\psi_n \qquad (4.1.44)$$

Now substitute Eqs. (4.1.42) and (4.1.43) into Eq. (4.1.44):

$$\frac{d^2\psi_n^0}{dz^2} + \varepsilon\frac{d^2\psi_n^1}{dz^2} - \frac{\rho'}{\rho}\frac{d\psi_n^0}{dz} - \varepsilon\frac{\rho'}{\rho}\frac{d\psi_n^1}{dz} + k^2\psi_n^0 + \varepsilon k^2\psi_n^1 + 2i\varepsilon\alpha k\psi_n^0$$

$$= \kappa_n^2\psi_n^0 + \varepsilon\kappa_n^2\psi_n^1 + \varepsilon i\delta_n\psi_n^0$$

$$\qquad (4.1.45)$$

where terms of order ε^2 have been neglected.

Using Eq. (4.1.41b), Eq. (4.1.45) becomes

$$\frac{d^2\psi_n^1}{dz^2} - \frac{\rho'}{\rho}\frac{d\psi_n^1}{dz} + k^2\psi_n^1 + 2i\alpha k\psi_n^0 = \kappa_n^2\psi_n^1 + i\delta_n\psi_n^0 \qquad (4.1.46)$$

Multiply Eq. (4.1.46) by $\rho^{-1}\psi_n^0$, integrate over z from 0 to d, and use Eq. (4.1.38) to obtain

$$\int_0^d \rho^{-1}\psi_n^0 \frac{d^2\psi_n^1}{dz^2}\,dz - \int_0^d \rho^{-1}\psi_n^0 \frac{\rho'}{\rho}\frac{d\psi_n^1}{dz}\,dz$$

$$+ \int_0^d k^2\rho^{-1}\psi_n^0\psi_n^1\,dz + 2i\int_0^d \rho^{-1}\psi_n^0\alpha k\psi_n^0\,dz \qquad (4.1.47)$$

$$= \int_0^d \kappa_n^2\rho^{-1}\psi_n^0\psi_n^1\,dz + i\delta_n$$

Combining the first two terms on the left side results in

$$\int_0^d \psi_n^0 \frac{d}{dz}\left(\frac{1}{\rho}\frac{d\psi_n^1}{dz}\right)dz + \int_0^d k^2\rho^{-1}\psi_n^0\psi_n^1\,dz + 2i\int_0^d \rho^{-1}\psi_n^0\alpha k\psi_n^0\,dz$$

$$= \int_0^d \kappa_n^2\rho^{-1}\psi_n^0\psi_n^1\,dz + i\delta_n$$

$$\qquad (4.1.48)$$

A NORMAL MODE MODEL FOR N-INHOMOGENEOUS LAYERS

Let us integrate the first term twice by parts.

$$\int_0^d \psi_n^0 \frac{d}{dz}\left(\frac{1}{\rho}\frac{d\psi_n^1}{dz}\right) dz = \left[\rho^{-1}\psi_n^0 \frac{d\psi_n^1}{dz}\right]_0^d - \left[\rho^{-1}\psi_n^1 \frac{d\psi_n^0}{dz}\right]_0^d$$

$$= \int_0^d \psi_n^1 \frac{d}{dz}\left(\frac{1}{\rho}\frac{d\psi_n^0}{dz}\right) dz$$

The integrated terms vanish because of boundary conditions. Thus Eq. (4.1.48) becomes

$$\int_0^d \psi_n^1 \frac{d}{dz}\left(\frac{1}{\rho}\frac{d\psi_n^0}{dz}\right) dz + \int_0^d \psi_n^1 \rho^{-1} k^2 \psi_n^0 \, dz$$

$$- \int_0^d \psi_n^1 \rho^{-1} \kappa_n^2 \psi_n^0 \, dz + 2i \int_0^d \rho^{-1} \alpha k \psi_n^0 \psi_n^0 \, dz = i\delta_n$$

Using Eq. (4.1.41a), this last equation becomes

$$\delta_n = 2\int_0^d \frac{\alpha(z)}{\rho(z)} k(z) \psi_n^0(z) \psi_n^0(z) \, dz \qquad (4.1.49)$$

Hence, the new eigenvalues of the problem become

$$\kappa_n^2 + i\delta_n = \kappa_n^2 \left(1 + i\frac{\delta_n}{\kappa_n^2}\right)$$

The correction to κ_n can be found by expansion, and it is seen to be

$$\left[\kappa_n^2\left(1 + i\frac{\delta_n}{\kappa_n^2}\right)\right]^{1/2} \approx \kappa_n\left(1 + i\frac{\delta_n}{2\kappa_n^2}\right)$$

$$= \kappa_n + \frac{i\delta_n}{2\kappa_n} \qquad (4.1.50)$$

4.1.5. Solution of the Radial Equation

In order to obtain the radial equation for $\phi_n(r)$, we substitute the series expansion for the acoustic pressure p given by Eq. (4.1.8) into the wave equation (4.1.1). After carrying out the indicated differentiations, making use of the depth equation (4.1.7), multiplying by $\rho^{-1}\psi_m$, integrating over the interval $z = 0$ to $z = d$, and making use of the orthonormality relation (4.1.38), we obtain the desired radial equation

$$\frac{d^2\phi_m}{dr^2} + \frac{1}{r}\frac{d\phi_m}{dr} + \kappa_m^2 \phi_m = \frac{-\psi_m^{(\alpha)}(z_s^{(\alpha)})}{2\pi r \rho_\alpha} \delta(r) \qquad (4.1.51)$$

174 PROPAGATION IN THE OCEAN: THE INHOMOGENEOUS LAYERED MODEL

The boundary conditions are given by R.1 and R.2 in Section 3.1. Consequently, the solution of Eq. (4.1.51) subject to these boundary conditions is

$$\phi_m = \frac{\psi_m^{(\alpha)}(z_s^{(\alpha)})}{\rho_\alpha} \left\{ \frac{i}{4} H_0^{(1)}(\kappa_m r) \right\} \tag{4.1.52}$$

The procedure for obtaining this solution is identical to the procedure in Section 3.2.

Assuming $\kappa_m r \gg 1$, we can use the asymptotic expansion of the Hankel function given by Eq. (2.9.51). This is

$$H_0^{(1)}(\kappa_m r) \approx \sqrt{\frac{2}{\pi \kappa_m r}} e^{i(\kappa_m r - (\pi/4))}$$

Thus, asymptotically, Eq. (4.1.52) becomes

$$\phi_m = \frac{i}{4} \left(\frac{2}{\pi \kappa_m r} \right)^{1/2} \frac{\psi_m^{(\alpha)}(z_s^{(\alpha)})}{\rho_\alpha} e^{i(\kappa_m r - (\pi/4))} \tag{4.1.53}$$

4.1.6. Solution of the Wave Equation

The general solution of Eq. (4.1.8) becomes

$$p(r, z) = \frac{i}{4\rho_\alpha} \left(\frac{2}{\pi r} \right)^{1/2} e^{-i(\pi/4)} \sum_{m=1}^{\infty} \frac{1}{\sqrt{\kappa_m}} \psi_m^{(\alpha)}(z_s) \psi_m(z) e^{i\kappa_m r} \tag{4.1.54}$$

When attenuation is added via Eq. (4.1.50), the solution becomes

$$p(r, z) = \frac{1}{\rho_\alpha} \left(\frac{1}{8\pi r} \right)^{1/2} e^{i(\pi/4)} \sum_{m=1}^{\infty} \frac{1}{\sqrt{\kappa_m}} \psi_m^{(\alpha)}(z_s) \psi_m(z) e^{(i\kappa_m - (\delta_m/2\kappa_m))r}$$

$$\tag{4.1.55}$$

Notice that we used the unperturbed depth eigenfunctions.

4.1.7. Transmission Loss

The quantity of greatest interest in acoustic propagation problems is the transmission loss (TL). This is defined by the relation

$$\text{TL} = -10 \log \frac{|p|^2}{|p_0|^2} \tag{4.1.56}$$

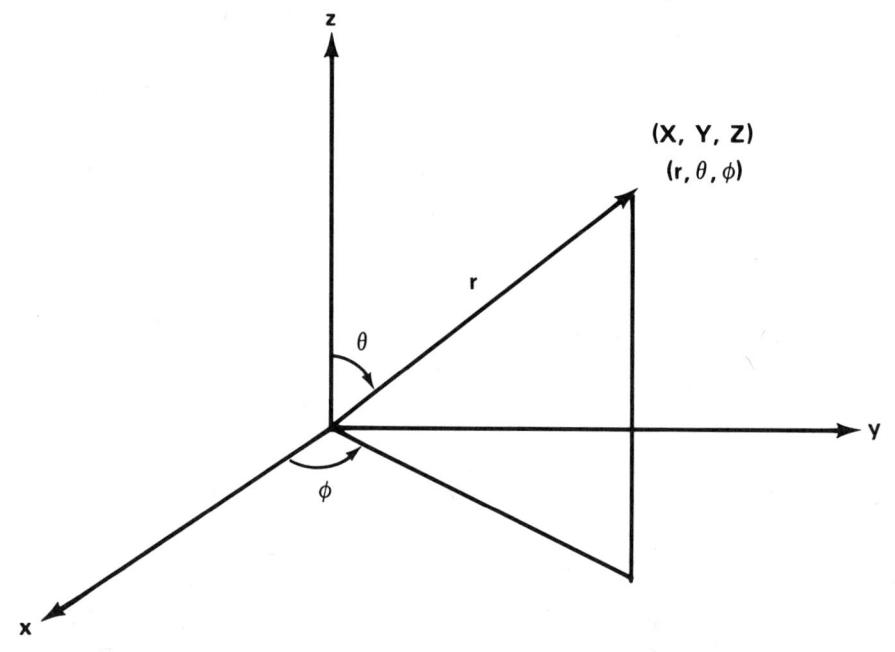

A NORMAL MODE MODEL FOR N-INHOMOGENEOUS LAYERS 175

FIGURE 4.1.3. Relation between Cartesian coordinates and spherical coordinates.

where $p(r, z)$ is the pressure at a range r from the source and p_0 is the pressure at a distance of 1 meter from the source. Thus transmission loss is a measure of the power lost in signal transmission at a range r relative to a reference distance of 1 meter from the source.

The pressure $p(r, z)$ is calculated from Eq. (4.1.55). We now need to determine p_0.

We will assume we are close enough to the point source at 1 meter so that the radiation has not interacted with any boundaries. Anticipating that a point source radiates spherical waves, we will solve the wave equation in spherical coordinates. Figure 4.1.3 shows the geometry of our spherical coordinate system which we have centered on the source.

The transformation equations between Cartesian coordinates x, y, and z, and spherical coordinates r, θ, and ϕ are:

$$x = r \sin \theta \cos \phi$$
$$y = r \sin \theta \sin \phi$$
$$z = r \sin \theta \tag{4.1.57}$$

The Jacobian of this transformation is

$$J = r^2 \sin \theta \tag{4.1.58}$$

176 PROPAGATION IN THE OCEAN: THE INHOMOGENEOUS LAYERED MODEL

Using Eq. (3.1.11), we can express the delta function in spherical coordinates as

$$\delta(\mathbf{r} - \mathbf{r}_s) = \frac{1}{r^2 \sin\theta} \delta(r - r_s)\delta(\phi - \phi_s)\delta(\theta - \theta_s) \qquad (4.1.59)$$

The wave equation (3.1.1), namely,

$$\nabla^2 p_0 + k^2 p_0 = -\delta(\mathbf{r} - \mathbf{r}_s)$$

becomes in spherical coordinates

$$\frac{1}{r}\frac{\partial^2}{\partial r^2}(rp_0) + \frac{1}{r^2 \sin\theta}\frac{\partial}{\partial \theta}\left(\sin\theta \frac{\partial p_0}{\partial \theta}\right) + \frac{1}{r^2 \sin^2\theta}\frac{\partial^2 p_0}{\partial \phi^2} + k^2 p_0$$

$$= -\frac{1}{r^2 \sin\theta}\delta(r - r_s)\delta(\phi - \phi_s)\delta(\theta - \theta_s) \qquad (4.1.60)$$

Since we are dealing with a point source and we are close to it, we can assume the radiation field has a spherical symmetry; that is, we can assume $p_0(r, \theta, \phi)$ is independent of θ and ϕ. This gives for Eq. (4.1.60)

$$\frac{d^2 p_0}{dr^2} + \frac{2}{r}\frac{dp_0}{dr} + k^2 p_0 = -\frac{1}{r^2 \sin\theta}\delta(r - r_s)\delta(\phi - \phi_s)\delta(\theta - \theta_s)$$

$$(4.1.61)$$

We must eliminate the variables θ and ϕ from the right side of this equation. We do this by integrating Eq. (4.1.61) over the full solid angle; that is, we multiply Eq. (4.1.61) by the solid angle $d\Omega = \sin\theta\, d\theta\, d\phi$ and integrate over θ from 0 to π and ϕ from 0 to 2π:

$$\int_0^{2\pi} d\phi \int_0^{\pi} \sin\theta\, d\theta \left\{ \frac{d^2 p_0}{dr^2} + \frac{2}{r}\frac{dp_0}{dr} + k^2 p_0 \right\}$$

$$= -\frac{\delta(r - r_s)}{r^2} \int_0^{2\pi} d\phi \int_0^{\pi} \sin\theta\, d\theta \left\{ \frac{1}{\sin\theta}\delta(\phi - \phi_s)\delta(\theta - \theta_s) \right\}$$

or

$$4\pi \left\{ \frac{d^2 p_0}{dr^2} + \frac{2}{r}\frac{dp_0}{dr} + k^2 p_0 \right\} = -\frac{\delta(r - r_s)}{r^2}$$

or finally,

$$\frac{d^2 p_0}{dr^2} + \frac{2}{r}\frac{dp_0}{dr} + k^2 p_0 = -\frac{1}{4\pi r^2}\delta(r) \qquad (4.1.62)$$

since $r_s = 0$.

The Green's function solution for Eq. (4.2.62) that satisfies the condition of outgoing waves only is

$$p_0 = \frac{1}{4\pi r} e^{ikr} \qquad (4.1.63)$$

Consequently,

$$|p_0| = \frac{1}{4\pi r} \qquad (4.1.64)$$

Substituting Eq. (4.1.64) back into Eq. (4.1.56), we get for $r = 1$ meter:

$$\begin{aligned} TL &= -10 \log|p|^2 - 20 \log(4\pi r) \\ &= -10 \log|p|^2 - 20 \log 4\pi \\ &= -10 \log|p|^2 - 21.98 \, dB \end{aligned} \qquad (4.1.65)$$

REFERENCE

4.1. D. C. Stickler, "Normal Mode Program with Both the Discrete and Branch Line Contributions," *J. Acoust. Soc. Am.* **57** 856–861 (1975).

5 APPROXIMATE SOLUTIONS OF THE WAVE EQUATION

5.0. INTRODUCTION

There exists a hierarchy of approximate solutions to the wave equation. We shall study three levels of approximations: ray theory, ray theory with asymptotic corrections for smooth caustics, and the WKB approximation.

In Section 5.1 we shall derive the ray theory approximation under the assumption that the wave front is locally a plane wave. We have chosen this approach first as a means of examining the physical nature of the ray theory approximation, because this approach allows for the simplest physical interpretation of the approximation.

In Sections 5.2 and 5.3 we discuss the WKB approximation, which is not as severe an approximation as the ray theory approximation. The application of the WKB method to a wave guide is discussed in Section 5.4. In Section 5.5 we derive the WKB Green's function for an unbounded medium. We will need this result in Section 5.6, where we show that the ray theory approximation can be arrived at through a combination of the WKB method and the method of steepest descent.

In Section 5.7 we discuss the physical basis for the formation of smooth caustics and briefly discuss cusp caustics. Finally, in Section 5.8 we derive the mathematical correction to ray theory which accounts for smooth caustics.

5.1. RAY EQUATIONS AS A QUASI-PLANE WAVE APPROXIMATION

A plane wave is characterized by the property that its direction of propagation and amplitude are the same everywhere. Arbitrary acoustic waves do not have this property. Nevertheless, a great many acoustic waves, which are not plane,

RAY EQUATIONS AS A QUASI-PLANE WAVE APPROXIMATION 179

have the property that within each small region of space they can be considered to be plane. For this, it is clearly necessary that the amplitude and direction of the wave remain practically constant over distances of the order of the wavelength. If this condition is satisfied, we can introduce the so-called wave surface, that is, a surface such that the phase of the wave is the same at all points at a given time.

Let us consider the case in which we are in a region of space that has no source, so that $Q = 0$ and the density ρ_0 is constant.

In the notation of Eq. (1.8.19), we will assume the field has the form

$$\mathscr{P} = A(x_1, x_2, x_3) e^{i[k_0 L(x_1, x_2, x_3) - \omega t]} \tag{5.1.1}$$

where, for convenience, we have put $x_1 = x$, $x_2 = y$, and $x_3 = z$. That is, we assume the field to consist of a single-frequency (ω) time harmonic wave that is characterized by a constant wave number k_0. The wave fronts or surfaces of constant phase are given by the equation

$$L(x_1, x_2, x_3) = \text{constant}$$

Under all these assumptions, the wave equation, Eq. (1.7.30), then becomes in component form

$$\frac{\partial^2 \mathscr{P}}{\partial x_i \, \partial x_i} = \frac{1}{c^2} \frac{\partial^2 \mathscr{P}}{\partial t^2} \tag{5.1.2}$$

Now carrying out the differentiation indicated in Eq. (5.1.2) on $\mathscr{P}$ as given by Eq. (5.1.1), we get

$$\frac{\partial \mathscr{P}}{\partial x_i} = \frac{\partial A}{\partial x_i} e^{i(k_0 L - \omega t)} + ik_0 A \frac{\partial L}{\partial x_i} e^{i(k_0 L - \omega t)}$$

$$\frac{\partial}{\partial x_i} \frac{\partial \mathscr{P}}{\partial x_i} = \frac{\partial^2 A}{\partial x_i^2} e^{i(k_0 L - \omega t)} + ik_0 \frac{\partial A}{\partial x_i} \frac{\partial L}{\partial x_i} e^{i(k_0 L - \omega t)}$$

$$+ ik_0 \frac{\partial A}{\partial x_i} \frac{\partial L}{\partial x_i} e^{i(k_0 L - \omega t)} + ik_0 A \frac{\partial^2 L}{\partial x_i^2} e^{i(k_0 L - \omega t)}$$

$$- k_0^2 A \frac{\partial L}{\partial x_i} \frac{\partial L}{\partial x_i} e^{i(k_0 L - \omega t)}$$

$$\frac{\partial \mathscr{P}}{\partial t} = -i\omega A e^{i(k_0 L - \omega t)}$$

$$\frac{\partial^2 \mathscr{P}}{\partial t^2} = -\omega^2 A e^{i(k_0 L - \omega t)}$$

180 APPROXIMATE SOLUTIONS OF THE WAVE EQUATION

Using these results, Eq. (5.1.2) becomes

$$\frac{\partial^2 A}{\partial x_i^2} + 2ik_0 \frac{\partial A}{\partial x_i}\frac{\partial L}{\partial x_i} + ik_0 A \frac{\partial^2 L}{\partial x_i^2} - k_0^2 A \left(\frac{\partial L}{\partial x_i}\right)^2 + \frac{\omega^2}{c^2}A = 0 \quad (5.1.3)$$

Separating Eq. (5.1.3) into real and imaginary parts gives

$$\frac{1}{k_0^2 A}\frac{\partial^2 A}{\partial x_i^2} - \frac{\partial L}{\partial x_i}\frac{\partial L}{\partial x_i} + \frac{\omega^2}{k_0^2 c^2} = 0 \quad (5.1.4)$$

and

$$2\frac{\partial A}{\partial x_i}\frac{\partial L}{\partial x_i} + A\frac{\partial^2 L}{\partial x_i^2} = 0 \quad (5.1.5)$$

Let

$$c_0 = \omega/k_0 \quad (5.1.6)$$

$$n = c_0/c \quad (5.1.7)$$

Then Eq. (5.1.4) can be written

$$\left(\frac{\partial L}{\partial x}\right)^2 + \left(\frac{\partial L}{\partial y}\right)^2 + \left(\frac{\partial L}{\partial z}\right)^2 - n^2 = \frac{\lambda_0^2}{4\pi^2 A}\left(\frac{\partial^2 A}{\partial x^2} + \frac{\partial^2 A}{\partial y^2} + \frac{\partial^2 A}{\partial z^2}\right) \quad (5.1.8)$$

where we have put

$$k_0 = \frac{2\pi}{\lambda_0} \quad (5.1.9)$$

Now in the limit of zero wavelength, L satisfies the following equation; that is, for $\lambda_0 = 0$, Eq. (5.1.8) becomes

$$\left(\frac{\partial L}{\partial x}\right)^2 + \left(\frac{\partial L}{\partial y}\right)^2 + \left(\frac{\partial L}{\partial z}\right)^2 = n^2 \quad (5.1.10)$$

This is called the iconal equation, and L is called the iconal. If L is a solution of the iconal equation, it will be a good approximation to the wave equation if the last expression of Eq. (5.1.8) is small compared with the sum of the first three terms. However, it is not sufficient to say that this will be true for small λ_0; we should like to know how small λ_0 must be with regard to the physical conditions of a particular problem. Let L be an exact solution of the iconal equation. Then $\mathscr{P}$ in the form of Eq. (5.1.1) will be a good approximation to the solution of the wave equation if the following condition is met:

$$\frac{\lambda_0^2}{4\pi^2 A} \nabla^2 A \ll \left(\frac{\partial L}{\partial x}\right)^2 + \left(\frac{\partial L}{\partial y}\right)^2 + \left(\frac{\partial L}{\partial z}\right)^2 \qquad (5.1.11)$$

A physical interpretation of Eq. (5.1.11) can be obtained from order of magnitude considerations. Let a prime denote any space derivative. Then Eq. (5.1.11) can be written:

$$\lambda_0^2 \frac{A''}{A} \ll (L')^2 \qquad (5.1.12)$$

Using the identity

$$\frac{A''}{A} = (\ln A)'^2 + (\ln A)'' \qquad (5.1.13)$$

Eq. (5.1.12) becomes

$$\lambda_0^2 (\ln A)'^2 + \lambda_0^2 (\ln A)'' \ll (L')^2 \qquad (5.1.14)$$

This will be satisfied if

$$\lambda_0^2 (\ln A)'^2 \ll (L')^2 \qquad (5.1.15)$$

and

$$\lambda_0^2 (\ln A)'' \ll (L')^2 \qquad (5.1.16)$$

Or rewritting these two equations, we have

$$\lambda_0 (\ln A)' \ll L' \qquad (5.1.17)$$

$$\lambda_0^2 (\ln A)'' \ll (L')^2 \qquad (5.1.18)$$

From Eq. (5.1.10),

$$(L')^2 \sim n^2 \qquad (5.1.19)$$

or

$$L' \sim n \qquad (5.1.20)$$

From Eq. (5.1.5),

$$L'' \sim L'(\ln A)' \qquad (5.1.21)$$

182 APPROXIMATE SOLUTIONS OF THE WAVE EQUATION

Using Eq. (5.1.17), Eq. (5.1.21) implies

$$\lambda_0 L'' \sim \lambda_0 L'(\ln A)' \ll L'L' \quad (5.1.22)$$

Combining Eq. (5.1.22) and (5.1.19),

$$\lambda_0 L'' \ll n^2 \quad (5.1.23)$$

In the ocean $n > 1$. So Eq. (5.1.23) becomes

$$\lambda_0 L'' \ll 1 \quad (5.1.24)$$

This states that the first spatial derivative of L must not change much over a distance of one wavelength.

The first spatial derivatives of L gives the direction of the rays, while the second derivatives, yielding the rate of change of ray direction, give the curvature of the rays. Therefore, Eq. (5.1.24) becomes the following. The direction of the ray must not change much over the distance of one wavelength. In regions where the ray curves very strongly, ray acoustics cannot be applied safely.

Differentiating Eq. (5.1.9) gives

$$L'L'' \sim nn' \quad (5.1.25)$$

because of Eq. (5.1.20),

$$L'' \sim n' \quad (5.1.26)$$

In view of Eq. (5.1.23), this means

$$\lambda_0 n' \ll n^2 \sim 1 \quad (5.1.27)$$

Thus, the index of refraction must not change much over the distance of one wavelength.

Using Eq. (5.1.17) and (5.1.20) implies

$$\lambda_0 (\ln A)' \ll 1 \quad (5.1.28)$$

This means that $\ln A$ must not change much over one wavelength. Since

$$\lambda_0 (\ln A)' = \lambda_0 \frac{A'}{A} \quad (5.1.29)$$

Eq. (5.1.28) means that the fractional change in A over one wavelength must be small.

RAY EQUATIONS AS A QUASI-PLANE WAVE APPROXIMATION

Thus, the following three statements are equivalent conditions [all derived from Eq. (5.1.17)] for the iconal equation to be a good approximation:

1. The direction or curvature of the rays must not change much over the distance of one wavelength.
2. The index of refraction must not change much over one wavelength.
3. The fractional change in A must be small over one wavelength.

A second condition comes from Eq. (5.1.18). It states that the change in the derivative of $\ln A$ also must be small over a wavelength.

We must now ask the question as to what is the direction of the energy flow relative to the wave front. The answer to this question leads to the concept of rays. Again using the notation in Eq. (1.8.19) and (1.8.20), Eq. (1.7.11) becomes for our assumptions of ρ_0 constant, $Q_1 = 0$, and neglecting the force of gravity,

$$\rho_0 \frac{\partial \mathscr{V}_i}{\partial t} = -\frac{\partial \mathscr{P}}{\partial x_i}$$

or since we are considering a time harmonic wave

$$\mathscr{V}_i = \frac{1}{i\omega\rho_0} \frac{\partial \mathscr{P}}{\partial x_i}$$

Using Eq. (5.1.6), this last expression can be written

$$\mathscr{V}_i = \frac{1}{ik_0 c_0 \rho_0} \frac{\partial \mathscr{P}}{\partial x_i} \tag{5.1.30}$$

Using Eq. (5.1.1) and carrying out the indicated differentiation, Eq. (5.1.30) becomes

$$\mathscr{V}_i = \frac{1}{ik_0 c_0 \rho_0} \frac{\partial A}{\partial x_i} e^{i(k_0 L - \omega t)} + \frac{A}{\rho_0 c_0} \frac{\partial L}{\partial x_i} e^{i(k_0 L - \omega t)} \tag{5.1.31}$$

Using the ray theory approximation $k_0 \to \infty$, Eq. (5.1.31) becomes

$$\mathscr{V}_i = \frac{A}{\rho_0 c_0} \frac{\partial L}{\partial x_i} e^{i(k_0 L - \omega t)}$$

or

$$\mathscr{V}_i = \frac{\mathscr{P}}{\rho_0 c_0} \frac{\partial L}{\partial x_i} \tag{5.1.32}$$

where use was made of Eq. (5.1.1). Substituting Eq. (5.1.32) into Eq. (1.8.21)

184 APPROXIMATE SOLUTIONS OF THE WAVE EQUATION

gives

$$\langle \mathbf{J} \rangle = \tfrac{1}{2} \text{Re}\{ \mathcal{P} \mathcal{V}^* \}$$

$$= \text{Re}\left\{ \frac{\mathcal{P}\mathcal{P}^*}{2\rho_0 c_0} \right\} \nabla L$$

$$= \frac{|p|^2}{2\rho_0 c_0} n \frac{\nabla L}{n}$$

$$= \frac{|p|^2}{2\rho_0 c} \hat{s}$$

$$= I\hat{s} \tag{5.1.33}$$

where n is given by Eq. (5.1.7) and

$$I = \frac{|p|^2}{2\rho_0 c} \tag{5.1.34}$$

and

$$\hat{s} = \frac{\nabla L}{n} \tag{5.1.35}$$

From Eq. (5.1.34) we see that the magnitude I of the average intensity vector $\langle \mathbf{J} \rangle$ is just the plane wave intensity, and the direction of the average intensity vector is in the direction of ∇L, which is orthogonal to the wave front $L = $ constant. Also, we note that $\hat{s}$ is a unit vector because of the iconal equation (5.1.10).

The fact that energy flows in a direction perpendicular to a wave front leads to the idea of rays. A ray is a curve that is everywhere orthogonal to the wave fronts. Thus, we can say that the average intensity propagates in the directions of the rays.

Now take a narrow bundle of rays or a ray tube formed by all the rays proceeding from an element dS_1 of a wave front $L = a_1$ (a_1 being a constant) and denote by dS_2 the corresponding element in which these rays intersect a second wave front $L = a_2$ (Figure 5.1.1).

Recall from Chapter 1 that the conservation law for the average intensity vector, Eq. (1.8.18), is

$$\nabla \cdot \langle \mathbf{J} \rangle = 0$$

On using Eq. (5.1.33), this becomes

$$\nabla \cdot (I\hat{s}) = 0 \tag{5.1.36}$$

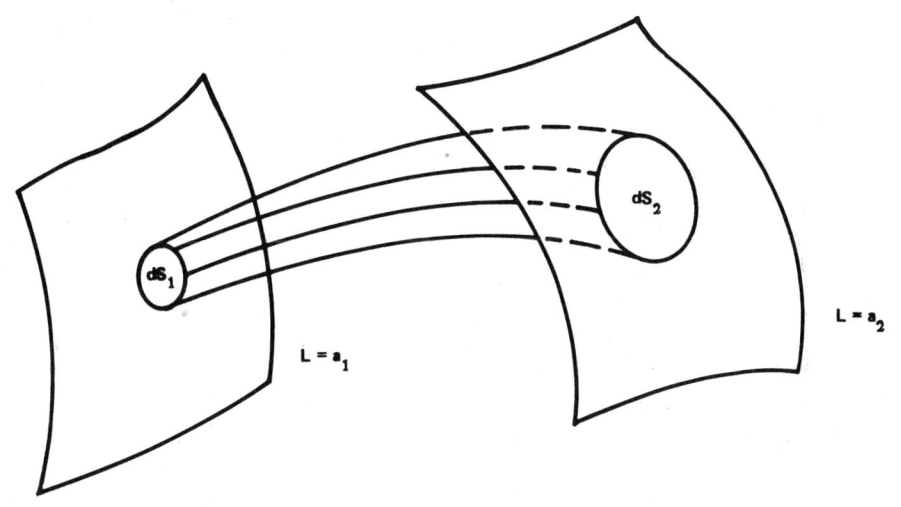

RAY EQUATIONS AS A QUASI-PLANE WAVE APPROXIMATION

FIGURE 5.1.1. Energy confined to a ray bundle.

Integrating Eq. (5.1.36) throughout the volume of the tube and applying the divergence theorem, we obtain

$$\iiint \nabla \cdot (I\hat{s}) \, dV = \iint I\hat{s} \cdot \hat{n} \, dS = 0 \tag{5.1.37}$$

where $\hat{n}$ is the outward normal to the surface. Now

$$\hat{s} \cdot \hat{n} = 1 \quad \text{on } dS_2$$
$$= -1 \quad \text{on } dS_1$$
$$= 0 \quad \text{elsewhere}$$

so that Eq. (5.1.37) reduces to

$$I_1 \, dS_1 = I_2 \, dS_2 \tag{5.1.38}$$

where I_1 and I_2 denote the intensity on dS_1 and dS_2, respectively. Hence, $I \, dS$ remains constant along a tube of rays. This result expresses the intensity law of geometrical acoustics. Note that if dS_2 approaches zero, then I_2 must approach infinity so that $I \, dS$ remains constant. A point where I becomes infinite is called a focal point. There exists surfaces such that each point on the surface is a focal point. Such focal surfaces are called caustics. We will have a lot more to say about caustics later in Sections 5.7 and 5.8. In the ray theory approximation, energy is confined to the ray tube. Leakage of energy out of the ray tube is called diffraction.

Using Eq. (5.1.38), let us derive an expression for the ray theory intensity for a medium in which the speed of sound c varies only in the z direction. We

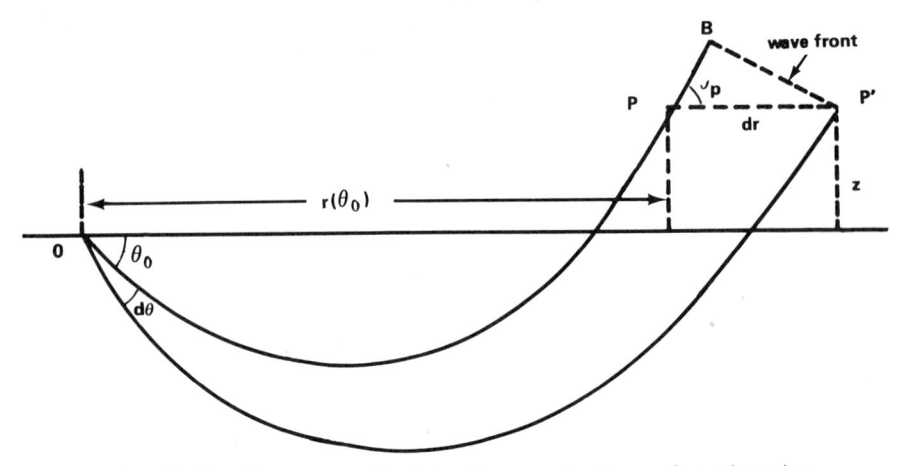

186 APPROXIMATE SOLUTIONS OF THE WAVE EQUATION

FIGURE 5.1.2. Geometry used to derive the expression for ray theory intensity.

want to find an expression for I at an arbitrary point P. The geometry is shown in Figure 5.1.2. We wish to determine the energy which leaves the source at the point 0 and is confined between the rays which leave the source at the angles θ_0 and $\theta_0 + d\theta$. Let θ be the angle of an arbitrary ray which lies between θ_0 and $\theta_0 + d\theta$. Denote by $r(\theta)$ the horizontal distance covered by the ray, leaving the source at angle θ, in traveling to any point lying at the same level z as P. Let θ_p be the grazing angle of the ray at the point P.

The cross section of the ray tube in the plane of the figure, that is bounded by the rays that leave the source at angles θ_0 and $\theta_0 + d\theta$, is

$$BP' = PP'\sin\theta_p = (dr)\sin\theta_p$$

$$= \left(\frac{\partial r}{\partial \theta}\right)_{\theta=\theta_0} d\theta \sin\theta_p \qquad (5.1.39)$$

Taking the cylindrical symmetry of the problem into account, the area of the wave front contained between these two rays is

$$dS_2 = 2\pi r \sin\theta_p \left(\frac{\partial r}{\partial \theta}\right)_{\theta_0} d\theta \qquad (5.1.40)$$

Let W be the total power emitted by the source. We need to calculate the power radiated into the interval $d\theta$. Consider Figure 5.1.3, which is in the vicinity of the source. Let a be a small distance from the source. The area of the wave front cut off by the ray bundle $d\theta$ near the source is

$$dS_1 = (a\,d\theta)(2\pi b)$$

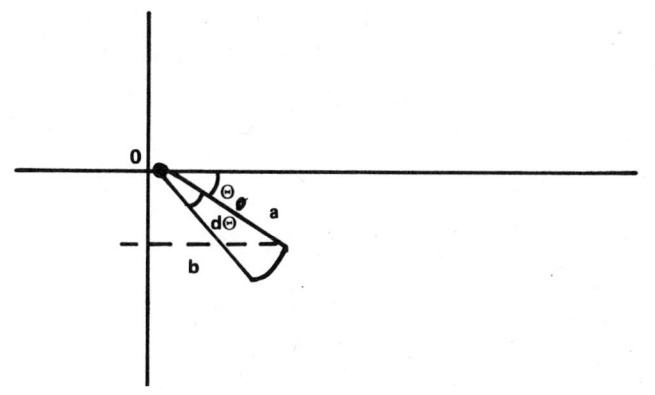

RAY EQUATIONS AS A QUASI-PLANE WAVE APPROXIMATION

FIGURE 5.1.3. Geometry to calculate power injected at the source.

where cylindrical symmetry has been taken into account. But

$$a \cos \theta_0 = b$$

so

$$dS_1 = 2\pi a^2 \cos \theta_0 \, d\theta \tag{5.1.41}$$

The total power radiated per unit area at a distance a from the source is $W/4\pi a^2$. Therefore, the total power radiated into the interval $d\theta$ is dW, where

$$dW = \frac{W}{4\pi a^2} dS_1$$

$$= \frac{W}{4\pi a^2} 2\pi a^2 \cos \theta_0 \, d\theta$$

$$= \tfrac{1}{2} W \cos \theta_0 \, d\theta \tag{5.1.42}$$

Consequently, the acoustic intensity at the point P is

$$I = \frac{dW}{dS_2}$$

$$I = \frac{\left(\tfrac{1}{2} W \cos \theta_0 \, d\theta\right)}{2\pi r \sin \theta_p \left(\frac{\partial r}{\partial \theta}\right)_{\theta_0} d\theta}$$

$$I = \frac{W}{4\pi} \left\{ \frac{\cos \theta_0}{r \sin \theta_p \left(\frac{\partial r}{\partial \theta}\right)_{\theta_0}} \right\} \tag{5.1.43}$$

where $W/4\pi$ is the total power per unit solid angle.

188   APPROXIMATE SOLUTIONS OF THE WAVE EQUATION

Next, we want to derive the equations for the ray paths. Recall that a ray is a curve orthogonal to a wave front and a wave front is a surface of constant phase; that is, a wave surface satisfies the equation

$$L = \text{constant}$$

From vector analysis we know that ∇L is a vector orthogonal to the surface $L = \text{constant}$, so that it is in the direction of a ray. Let $\mathbf{r}$ be the position vector of a point on a ray, and let s be the arc length along the ray. Then $d\mathbf{r}/ds$ is the unit tangent vector to the curve.

Since ∇L and $d\mathbf{r}/ds$ are in the same direction they must be proportional; that is, we must have

$$\nabla L = a \frac{d\mathbf{r}}{ds} \tag{5.1.44}$$

where a is a constant of proportionality. In component form, Eq. (5.1.44) becomes

$$\frac{\partial L}{\partial x} = a \frac{dx}{ds} \tag{5.1.45}$$

$$\frac{\partial L}{\partial y} = a \frac{dy}{ds} \tag{5.1.46}$$

$$\frac{\partial L}{\partial z} = a \frac{dz}{ds} \tag{5.1.47}$$

Squaring and adding Eq. (5.1.45) to Eq. (5.1.47) gives

$$\left(\frac{\partial L}{\partial x}\right)^2 + \left(\frac{\partial L}{\partial y}\right)^2 + \left(\frac{\partial L}{\partial z}\right)^2 = a^2 \left\{ \left(\frac{dx}{ds}\right)^2 + \left(\frac{dy}{ds}\right)^2 + \left(\frac{dz}{ds}\right)^2 \right\}$$

By definition

$$ds^2 = dx^2 + dy^2 + dz^2$$

and, using Eq. (5.1.10), we get

$$n^2 = a^2$$

Thus Eqs. (5.1.45) through (5.1.47) become

$$n \frac{dx}{ds} = \frac{\partial L}{\partial x} \tag{5.1.48}$$

$$n \frac{dy}{ds} = \frac{\partial L}{\partial y} \tag{5.1.49}$$

$$n \frac{dz}{ds} = \frac{\partial L}{\partial z} \tag{5.1.50}$$

Next we need to eliminate the iconal L from these equations.

RAY EQUATIONS AS A QUASI-PLANE WAVE APPROXIMATION 189

Consider Eq. (5.1.48). Differentiating this equation with respect to arc length gives

$$\frac{d}{ds}\left(n\frac{dx}{ds}\right) = \frac{d}{ds}\left(\frac{\partial L}{\partial x}\right)$$

$$= \frac{\partial}{\partial x}\left(\frac{\partial L}{\partial x}\right)\frac{dx}{ds} + \frac{\partial}{\partial y}\left(\frac{\partial L}{\partial x}\right)\frac{dy}{ds} + \frac{\partial}{\partial z}\left(\frac{\partial L}{\partial x}\right)\frac{dz}{ds}$$

$$= \frac{\partial}{\partial x}\left\{\frac{\partial L}{\partial x}\frac{dx}{ds} + \frac{\partial L}{\partial y}\frac{dy}{ds} + \frac{\partial L}{\partial z}\frac{dz}{ds}\right\}$$

$$= \frac{\partial}{\partial x}\left\{n\left(\frac{dx}{ds}\right)^2 + n\left(\frac{dy}{ds}\right)^2 + n\left(\frac{dz}{ds}\right)^2\right\}$$

$$= \frac{\partial n}{\partial x}$$

where use was made of Eq. (5.1.48). Eqs. (5.1.49) and (5.1.50) can be treated similarly. Thus Eqs. (5.1.48) through (5.1.50) become

$$\frac{d}{ds}\left(n\frac{dx}{ds}\right) = \frac{\partial n}{\partial x} \qquad (5.1.51)$$

$$\frac{d}{ds}\left(n\frac{dy}{ds}\right) = \frac{\partial n}{\partial y} \qquad (5.1.52)$$

$$\frac{d}{ds}\left(n\frac{dz}{ds}\right) = \frac{\partial n}{\partial z} \qquad (5.1.53)$$

These are the differential equations for the ray paths.

Now let us assume that the velocity is a function of z only, that is, let

$$n = n(z)$$

Then Eqs. (5.1.51) through (5.1.53) reduce to

$$n\frac{dx}{ds} = \text{constant}$$

$$n\frac{dy}{ds} = \text{constant}$$

$$\frac{d}{ds}\left(n\frac{dz}{ds}\right) = \frac{dn}{dz}$$

The first two equations imply that the ray path is confined to a plane normal to

190 APPROXIMATE SOLUTIONS OF THE WAVE EQUATION

the xy plane. Let us take this plane to be the xz plane. Then the first and third equations preceding become

$$n\frac{dx}{ds} = \text{constant}$$

$$\frac{d}{ds}\left(n\frac{dz}{ds}\right) = \frac{dn}{dz} \tag{5.1.54}$$

The direction cosines are given by Figure 5.1.4:

$$\frac{dx}{ds} = \cos\theta$$

$$\frac{dz}{ds} = \sin\theta \tag{5.1.55}$$

Combining the first equation of Eq. (5.1.54) with the first equation in (5.1.55) gives

$$n\frac{dx}{ds} = \text{constant}$$

$$\frac{\cos\theta}{c} = K = \frac{1}{c_n} = \text{constant} \tag{5.1.56}$$

Note: c_n = velocity when $\theta = 0$ or ray becomes horizontal.

This is Snell's law. The ray parameter K is constant along a given ray but varies from ray to ray. It is sometimes useful to state this relation in terms of

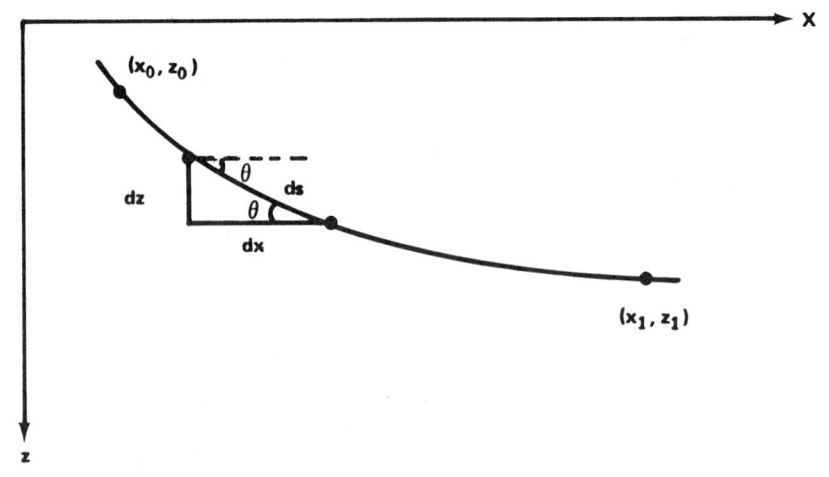

FIGURE 5.1.4. Geometry for a single ray.

RAY EQUATIONS AS A QUASI-PLANE WAVE APPROXIMATION

the grazing angle θ_s and velocity c_s of the ray at its source,

$$\frac{\cos\theta}{c} = \frac{\cos\theta_s}{c_s} \tag{5.1.57}$$

From the second equation of Eq. (5.1.54) and the second equation of Eq. (5.1.55), we get

$$\frac{dn}{dz} = \frac{d}{ds}(n\sin\theta)$$

$$= n\cos\theta\frac{d\theta}{ds} + \sin\theta\frac{dn}{dz}\frac{dz}{ds}$$

$$= n\cos\theta\frac{d\theta}{ds} + \sin^2\theta\frac{dn}{dz}$$

and solving for $d\theta/ds$,

$$\frac{d\theta}{ds} = \frac{\cos\theta}{n}\frac{dn}{dz}$$

$$\frac{d\theta}{ds} = -\frac{\cos\theta}{c}\frac{dc}{dz}$$

$$\frac{d\theta}{ds} = -K\frac{dc}{dz} \tag{5.1.58}$$

This states that the curvature of a ray, $d\theta/ds$, is directly proportional to the velocity gradient dc/dz.

From Figure 5.1.4,

$$\frac{dx}{dz} = \operatorname{ctn}\theta \tag{5.1.59}$$

so,

$$x_1 - x_0 = \int_{z_0}^{z_1} \operatorname{ctn}\theta\, dz \tag{5.1.60a}$$

The travel time is given by

$$dt = \frac{ds}{c}$$

$$t_1 - t_0 = \int_{s_0}^{s_1} \frac{ds}{c}$$

$$t_1 - t_0 = \int_{z_0}^{z_1} c^{-1}\frac{dz}{\sin\theta} \tag{5.1.60b}$$

192 APPROXIMATE SOLUTIONS OF THE WAVE EQUATION

and the path length along a ray is given by

$$ds = \frac{dz}{\sin \theta}$$

$$s_1 - s_0 = \int_{z_0}^{z_1} \frac{dz}{\sin \theta} \tag{5.1.60c}$$

From Eq. (5.1.56),

$$\cos \theta = Kc$$

$$\sin \theta = \sqrt{1 - K^2 c^2}$$

$$\operatorname{ctn} \theta = \frac{Kc}{\sqrt{1 - K^2 c^2}}$$

Using these results, we get from Eqs. (5.1.60a), (5.1.60b), and (5.1.60c)

$$x_1 - x_0 = \int_{z_0}^{z_1} \frac{Kc \, dz}{\sqrt{1 - K^2 c^2}} \tag{5.1.61a}$$

$$t_1 - t_0 = \int_{z_0}^{z_1} \frac{dz}{c\sqrt{1 - K^2 c^2}} \tag{5.1.61b}$$

$$s_1 - s_0 = \int_{z_0}^{z_1} \frac{dz}{\sqrt{1 - K^2 c^2}} \tag{5.1.61c}$$

Now take the case of a constant velocity gradient. Assume

$$c = c_m + gz \tag{5.1.62}$$

From Eq. (5.1.58),

$$\frac{d\theta}{ds} = -Kg$$

This states that the curvature along any given ray is constant so the ray is the arc of a circle. The radius of curvature R_c is

$$R_c = \frac{1}{|d\theta/ds|} = \frac{1}{|Kg|}$$

The equations for the horizontal range, travel time, and path length can be obtained conveniently in terms of θ. From Eq. (5.1.62),

$$\frac{dc}{dz} = c' = g$$

RAY EQUATIONS AS A QUASI-PLANE WAVE APPROXIMATION

From Eq. (5.1.56),

$$K = \frac{\cos \theta_0}{c_0} = \frac{\cos \theta_1}{c_1}$$

where c_0, θ_0 are the sound velocity and angle at (x_0, z_0) and c_1, θ_1 are the sound velocity and angle at (x_1, z_1). From Eq. (5.1.60a),

$$x_1 - x_0 = \int_{z_0}^{z_1} \cos \theta \frac{dz}{\sin \theta}$$

$$= -\int_{\theta_0}^{\theta_1} \frac{\cos \theta \, d\theta}{Kc'}$$

where we have used

$$dz/\sin \theta = -d\theta/Kc'$$

obtained by differentiating Snell's law, Eq. (5.1.56).

The integral now becomes

$$x_1 - x_0 = -\int_{\theta_0}^{\theta_1} \frac{\cos \theta \, d\theta}{Kg}$$

$$= -\frac{1}{Kg}(\sin \theta)\Big|_{\theta_0}^{\theta_1}$$

$$= \frac{1}{Kg}(\sin \theta_0 - \sin \theta_1)$$

$$= \frac{c_0}{g}\left(\frac{\sin \theta_0 - \sin \theta_1}{\cos \theta_0}\right)$$

Similarly, from Eq. (5.1.60b),

$$t_1 - t_0 = \int_{z_0}^{z_1} \frac{dz}{c \sin \theta}$$

$$= -\int_{\theta_0}^{\theta_1} \frac{d\theta}{g \cos \theta}$$

$$= -\frac{1}{g} \ln \left\{ \frac{\tan\left(\frac{\theta_1 + \pi/2}{2}\right)}{\tan\left(\frac{\theta_0 + \pi/2}{2}\right)} \right\}$$

$$= -\frac{1}{g} \ln \left\{ \frac{\cos \theta_0 (1 + \sin \theta_1)}{\cos \theta_1 (1 + \sin \theta_0)} \right\}$$

194 APPROXIMATE SOLUTIONS OF THE WAVE EQUATION

and, from Eq. (5.1.60c),

$$s_1 - s_0 = \int_{z_0}^{z_1} \frac{dz}{\sin\theta}$$

$$= -\int_{\theta_0}^{\theta_1} \frac{d\theta}{Kg}$$

$$= \frac{1}{Kg}(\theta_0 - \theta_1)$$

$$= \frac{c_0}{g\cos\theta_0}(\theta_0 - \theta_1)$$

Summarizing, for the case of a constant gradient,

$$x_1 - x_0 = \frac{c_0}{g}\left(\frac{\sin\theta_0 - \sin\theta_1}{\cos\theta_0}\right) \quad (5.1.63a)$$

$$t_1 - t_0 = -\frac{1}{g}\ln\left\{\frac{\cos\theta_0(1 + \sin\theta_1)}{\cos\theta_1(1 + \sin\theta_0)}\right\} \quad (5.1.63b)$$

$$s_1 - s_0 = \frac{c_0}{g\cos\theta_0}(\theta_0 - \theta_1) \quad (5.1.63c)$$

An equation for the change in depth of the ray can easily be derived from Eq. (5.1.55). Using Eq. (5.1.62), namely,

$$z = \frac{c - c_m}{g}$$

we have directly

$$z_1 - z_0 = \left(\frac{c_1 - c_m}{g}\right) - \left(\frac{c_0 - c_m}{g}\right)$$

$$= \frac{c_1 - c_0}{g}$$

$$= \frac{1}{g}\left(c_0\frac{\cos\theta_1}{\cos\theta_0} - c_0\right)$$

$$= \frac{c_0}{g\cos\theta_0}(\cos\theta_1 - \cos\theta_0) \quad (5.1.64)$$

The manner in which the equation for the ray paths are used is to first divide the ocean into horizontal layers as shown in Figure 4.1.1. The thickness of each layer is determined by the requirement that the actual variation of sound speed in that layer can be approximated by a known curve. Usually we choose that either the sound speed c or the refractive index n be a polynomial in z. We have found that the linear fit is the most convenient to work with. Thus, in each layer we let the sound speed be represented by Eq. (5.1.62). Thus we approximate the speed of sound in the ocean by a series of straight lines. While Pedersen [5.1] and Pedersen and Gordon [5.2] have discussed the problem that can arise with the linear fit to the sound speed, we have found the linear fit to be very accurate when we have compared the ray model with experimental data. If a nonlinear fit is used, many more ocean layers are needed, and the increase in computer time is not offset by the small increase in accuracy. If Eq. (5.1.62) is used to approximate the speed of sound in each layer, then Snell's law as well as Eqs. (5.1.63a) to (5.1.63c) and Eq. (5.1.64) are used to trace the ray through each layer.

For a given variation of sound speed there is a simple way to remember in what direction a ray is bent or refracted. Consider, for example, an oceanic waveguide of depth d in which the sound speed decreases monotonically from the surface to the bottom as shown in Figure 5.1.5a. Let a point source be located at a depth z_s. As shown in Figure 5.1.5b, let OA be a ray emitted from the source in an upward direction. At point A on the ray, consider a small section of the wave front which is perpendicular to the ray. The top of this wave front is at a depth where the sound speed is greater than the sound speed at the bottom of the wave front. Hence, the top of the wave front is moving at a greater speed than the bottom, causing the ray which is always perpendicular to the wave front to bend or refract downward as is shown by the portion of the ray denoted AB. The same reasoning can be applied to the ray OC which is

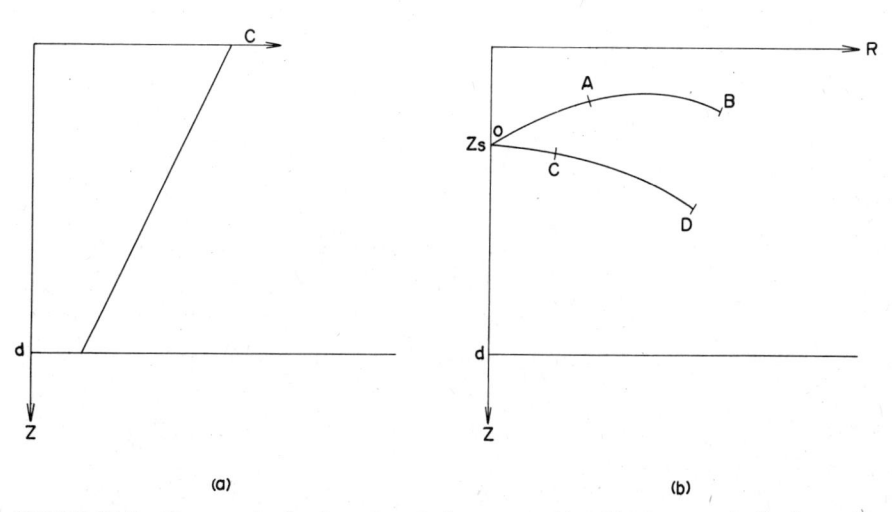

FIGURE 5.1.5. Downward refraction of a ray in a waveguide with a monotonically decreasing sound speed.

196 APPROXIMATE SOLUTIONS OF THE WAVE EQUATION

emitted from the source in a downward direction. Again the ray is refracted downward as shown by the portion of the ray CD. Hence, when a ray is traveling through a waveguide in which the sound velocity gradient is negative, the ray is refracted downward.

Now consider an oceanic waveguide of depth d in which the sound speed increases monotonically from the surface to the bottom as shown in Figure 5.1.6a. Now suppose the source is at the depth z_s as shown in Figure 5.1.6b. Let OA be a ray emitted in the upward direction, and again consider a small section of a wave front at the point A on the ray. This time the top of the wave front is at a depth where the sound speed is less than the sound speed at the bottom of the wave front. Hence, the bottom of the wave front is moving at a faster rate than the top, causing the ray which must remain perpendicular to the wave front to refract upward, as is shown by the portion of the ray AB. Again the same reasoning can be applied to the ray emitted from the source in a downward direction. This ray is also refracted upward as shown by the portion of the ray CD. Hence, when a ray is traveling through a waveguide in which the sound velocity gradient is positive, the ray is refracted upward.

Let us consider an oceanic waveguide with a bilinear profile as shown in Figure 5.1.7a. The sound speed has a constant negative gradient down to a depth z_m and then a constant positive gradient from z_m to the bottom. Consider a source located at a depth z_s as shown in Figure 5.1.7b. To illustrate how a ray cycles down the waveguide, we will consider only the ray that leaves the source in a horizontal direction. Using the reasoning that we applied to Figure 5.1.5, we see that initially the ray will be refracted downward. The ray will continue to be refracted downward until the depth z_m is reached. Once the ray enters into the lower portion of the waveguide below z_m, the situation is the

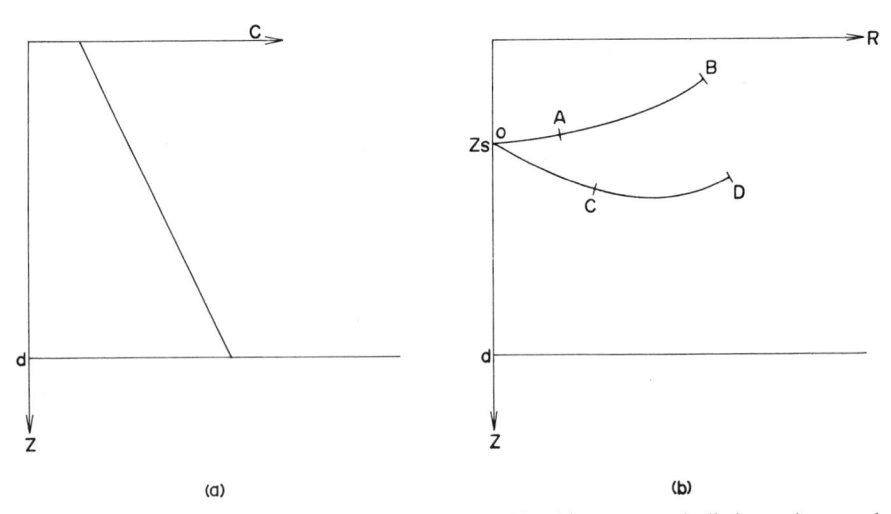

FIGURE 5.1.6. Upward refraction of a ray in a waveguide with a monotonically increasing sound speed.

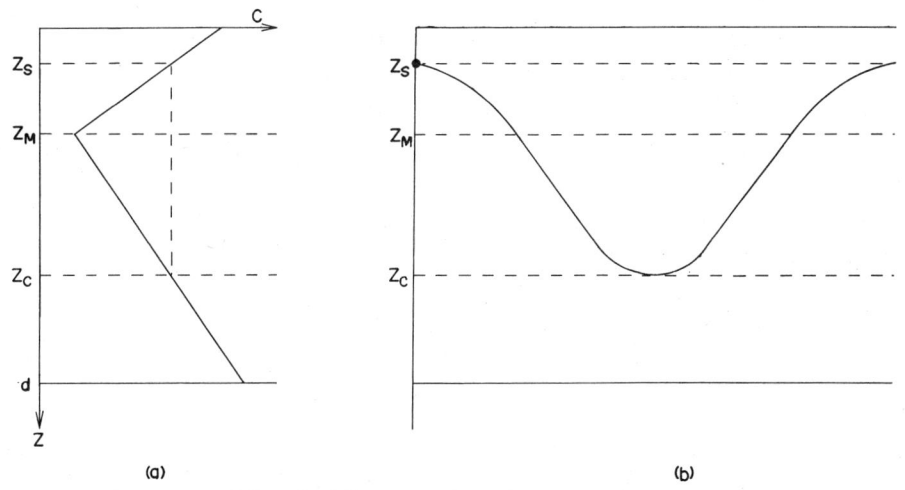

FIGURE 5.1.7. Effect of a bilinear sound speed profile on the refraction of a ray.

same as that shown in Figure 5.1.6. Now the ray will be refracted upwards. Snell's law states that this particular ray will turn over at a depth z_c where the sound speed equals the sound speed at the source depth. After the ray turns over, it will continue to be refracted upward until this depth z_m is reached. Above z_m the ray is again refracted downward until it turns over at source depth. The ray will continue to cycle down the waveguide in this manner, refracting back and forth between the two turning points z_s and z_c. The path of any other ray can be discerned in a similar manner, using Snell's law to find the turning points if any.

5.2. WKB APPROXIMATION

Recall from Section 2.1 that the Helmholtz equation with no source term [Eq. (2.1.13)] reduced to two separated, ordinary differential equations. The ordinary differential equation describing the variation of the acoustic pressure field with range was Eq. (2.1.17):

$$\frac{d^2R}{dr^2} + \frac{1}{r}\frac{dR}{dr} + k_0^2 \xi^2 R = 0 \qquad (5.2.1)$$

The ordinary differential equation describing the variations of the acoustic pressure field with depth was Eq. (2.1.18):

$$\frac{d^2\psi}{dz^2} + k_0^2(n^2 - \xi^2)\psi = 0 \qquad (5.2.2)$$

198 APPROXIMATE SOLUTIONS OF THE WAVE EQUATION

Eq. (5.2.1) is independent of any explicit properties of the medium and, as we have seen in Section 3.1, it is Bessel's equation. The solution is known and in fact is given by Eq. (3.1.34). However, Eq. (5.2.2) does depend explicitly on the properties of the medium through the index of refraction $n(z)$. Depending on how the refractive index varies with z, Eq. (5.2.2) may or may not have a known solution. In Chapter 4, we saw that when we chose n^2 linear in z, Eq. (5.2.2) was Airy's differential equation. Since Eq. (5.2.2) is in general very difficult to solve, it is advantageous to seek an approximate solution of it. One such asymptotic solution is the WKB solution.

Before proceeding to derive the WKB solution, let us give a physical interpretation to the separation parameter ξ. We start with the asymptotic solution of Bessel's equation, Eq. (5.2.1):

$$R = A\left(\frac{2}{\pi k'r}\right)^{1/2} e^{i(k'r - \omega t - (\pi/4))} \tag{5.2.3}$$

where A is an arbitrary constant and, for convenience, we have put

$$k' = k_0 \xi \tag{5.2.4}$$

As in the case of interpreting a plane-wave solution, the term

$$\left(\frac{2}{\pi k'r}\right)^{1/2} e^{i(k'r - \omega t) - i(\pi/4)}$$

represents a wave propagating in the positive r direction.

A wave front or a surface of constant phase at a fixed time t_1 is given by

$$k'r - \omega t_1 - \pi/4 = \text{constant}$$

This equation is equivalent to

$$r = \text{constant}$$

However, this equation describes a cylindrical surface in the (r, z)-coordinate system. Hence, the wave fronts are cylinders, and the above solution represents cylindrical waves. Note that the amplitude of the waves is decreasing as $(1/r)^{1/2}$. Since the wave fronts are cylindrical, this is called cylindrical spreading.

Now consider the case in which we have a single source situated at the origin. Then we would expect acoustic waves to propagate only in the positive r direction because there is no source at infinity by assumption, and there is no mechanism to reflect the outgoing waves because of our assumptions. Thus, for the entire field, we have

$$\mathcal{P} = A\psi(z)\left(\frac{2}{\pi k'r}\right)^{1/2} e^{i(k'r - \omega t - (\pi/4))} \tag{5.2.5}$$

From examining Eq. (5.2.5), we see that the parameter k' is to be interpreted as a horizontal wave number, that is, a wave number associated with the r direction. Thus, we can associate a horizontal phase velocity c_r of the wave front by the usual relation

$$c_r = \frac{\omega}{k'} \tag{5.2.6}$$

The phase velocity c_r is the speed that the wave front (surface of constant phase, hence, the name phase velocity) propagates in the r direction.

We can now associate the parameter ξ with a particular ray. At this point, this ray is a mathematical construct that is to be used to give one a "physical feeling" for the solution Eq. (5.2.5). The ray that we will associate with the wave solution is not physically equivalent to it because the wave solution given by Eq. (5.2.5) is at this point an exact solution of the wave equation, and it includes full diffraction effects that are not properly accounted for by ray theory. Consider Figure 5.2.1. In the figure a ray is shown leaving the source at the origin and arriving at a point R at an arbitrary depth z. At R, let us construct a small section of a wave front PQ that is, of course, perpendicular to the ray at the point R. At a depth z, the wave front in the immediate vicinity of R is moving with a speed $c(z)$ in the direction of the normal vector to the wave front. We can decompose this motion into a wave motion in the horizontal

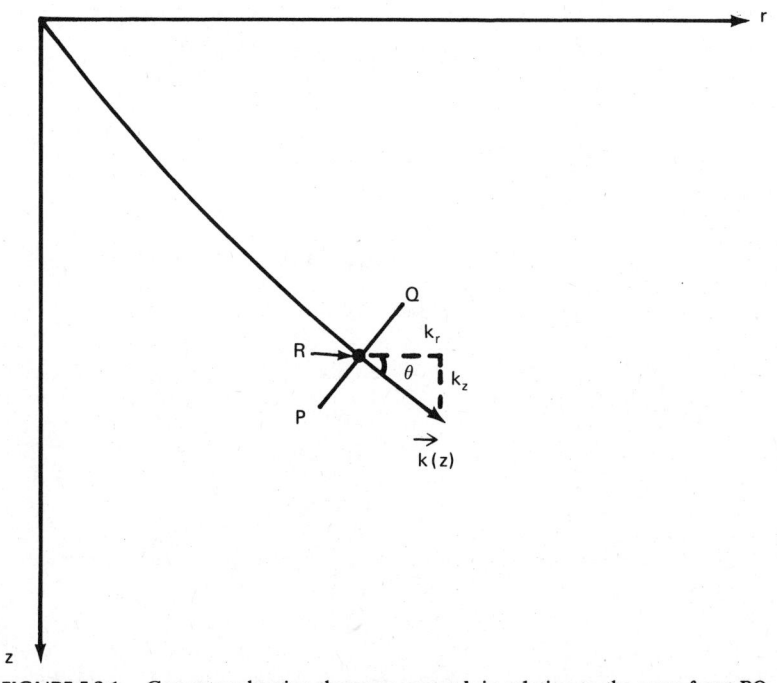

FIGURE 5.2.1. Geometry showing the wave vector **k** in relation to the wave front PQ.

200 APPROXIMATE SOLUTIONS OF THE WAVE EQUATION

direction and a wave motion in the vertical direction by the following procedure. Recall from the iconal equation, Eq. (5.1.10), that we have

$$\nabla L = n\hat{s} \tag{5.2.7}$$

where $\hat{s}$ is the unit vector given by Eq. (5.1.35) that is normal to the wave front. Using Eqs. (2.1.8) and (2.1.9), we can write Eq. (5.2.7) in the form

$$\nabla L = \frac{k(z)}{k_0}\hat{s} \tag{5.2.8}$$

where we have put

$$k(z) = \frac{\omega}{c(z)} \tag{5.2.9}$$

Rearranging Eq. (5.2.8), we get

$$\nabla(k_0 L) = \mathbf{k} \tag{5.2.10}$$

where we have put

$$\mathbf{k} = k(z)\hat{s} \tag{5.2.11}$$

The vector $\mathbf{k}$ is a generalized wave vector in the direction $\hat{s}$, which as we know is perpendicular to the wave front. This is indicated in Figure 5.2.1. We can now resolve $\mathbf{k}$ into vertical and horizontal components. If θ is the angle between the vector $\mathbf{k}$ and the horizontal, we get

$$k_r = k \cos \theta \tag{5.2.12}$$

$$k_z = k \sin \theta \tag{5.2.13}$$

We can now define a horizontal ray velocity V_r by using the usual method,

$$V_r = \frac{\omega}{k_r} \tag{5.2.14}$$

or, by using Eq. (5.2.12),

$$V_r = \frac{\omega}{k \cos \theta} \tag{5.2.15}$$

and finally using Eq. (5.2.9),

$$V_r = \frac{c(z)}{\cos \theta} \tag{5.2.16}$$

To associate a particular ray with the wave solution, we equate the horizontal component of the ray speed to the horizontal component of the wave speed to get

$$c_r = V_r$$

or

$$c_r = \frac{c(z)}{\cos \theta} \qquad (5.2.17)$$

From Snell's law, Eq. (5.1.56), we see that c_r is the value of the sound speed at the depth z_r, where the ray becomes horizontal or turns over. From Eq. (5.2.4), (5.2.6), (5.2.9), and (5.2.17), we easily obtain the desired result

$$\xi = n \cos \theta \qquad (5.2.18)$$

Recall that even though $n(z)$ and $\theta(z)$ are functions of z, by Snell's law ξ is a constant.

If we take as a reference velocity c_0, the speed of sound at the source depth, we can then write Eq. (5.2.18) in the form

$$\xi = \cos \theta_s \qquad (5.2.19)$$

where θ_s is the exit angle of the ray of the source. Thus, we associate with the full wave solution, Eq. (5.2.5), the ray that exits the source at an angle θ_s, which gives the ray a turning point at the depth where the speed of sound is equal to the horizontal phase speed of the outgoing cylindrical wave. It must be kept in mind that this "equivalent ray" is only a heuristic concept.

The limitations of ray theory were discussed in the last section. In many important applications these conditions fail to be met. One example is the failure of ray theory when diffraction effects are important. Diffraction can be important even up to 1000 Hz for some types of deep ocean propagation, such as the energy diffracted by the formation of caustics to be discussed in Section 5.7. At lower frequencies, diffraction effects can be important in surface duct propagation which we will discuss in Sections 5.3 and 6.4, and in energy leaking into shadow zones formed in the region of the main thermocline to be discussed in Section 6.2. A second example of the failure of ray theory is the formation of caustics themselves.

Hence, it is desirable to try to find an approximate solution which, at least to some order, includes diffraction effects and caustics, but which simplifies the solution of Eq. (5.2.2). The WKB solution does just this. So now let us derive the WKB solution.

Let us assume a solution of the form

$$\psi = e^{ik_0 S} \qquad (5.2.20)$$

202 APPROXIMATE SOLUTIONS OF THE WAVE EQUATION

Then,
$$\psi' = ik_0 S' e^{ik_0 S}$$
$$\psi'' = ik_0 S'' e^{ik_0 S} - k_0^2 S'^2 e^{ik_0 S}$$

Substituting these expressions in Eq. (5.2.21) results in

$$ik_0 S'' e^{ik_0 S} - k_0^2 S'^2 e^{ik_0 S} + k_0^2 (n^2 - \xi^2) e^{ik_0 S} = 0$$

$$ik_0 S'' - k_0^2 S'^2 + k_0^2 (n^2 - \xi^2) = 0$$

$$iS'' - k_0 S'^2 + k_0 (n^2 - \xi^2) = 0 \qquad (5.2.21)$$

Now, let us expand the function S in reciprocal powers of the wave number k_0.

$$S = S_0 + \frac{1}{k_0} S_1 + \frac{1}{k_0^2} S_2 + \cdots \qquad (5.2.22)$$

Keeping only terms of order $1/k_0$ and substituting Eq. (5.2.22) into Eq. (5.2.21), we get

$$i\left(S_0'' + \frac{1}{k_0} S_1''\right) - k_0 \left(S_0' + \frac{1}{k_0} S_1'\right)^2 + k_0 (n^2 - \xi^2) = 0$$

$$iS''_0 + \frac{i}{k_0} S_1'' - k_0 S_0'^2 - 2 S_0' S_1' - \frac{1}{k_0} S_1'^2 + k_0 (n^2 - \xi^2) = 0$$

$$i \frac{S_0''}{k_0} + i \frac{S_1''}{k_0^2} - S_0'^2 - \frac{2}{k_0} S_0' S_1' - \frac{S_1'^2}{k_0^2} + (n^2 - \xi^2) = 0$$

Thus, equating like powers of $1/k_0$, we get

$$-S_0'^2 + (n^2 - \xi^2) = 0 \qquad (5.2.23)$$

$$i \frac{S_0''}{k_0} - \frac{2}{k_0} S_0' S_1' = 0 \qquad (5.2.24)$$

for the zeroth and first-order approximation. From Eq. (5.2.23),

$$S_0'^2 = n^2 - \xi^2$$

$$S_0' = \pm \sqrt{n^2 - \xi^2}$$

$$S_0 = \pm \int \sqrt{n^2 - \xi^2} \, dz \qquad (5.2.25)$$

From Eq. (5.2.24),

$$2S_1'S_0' = iS_0''$$

$$S_1' = i\frac{S_0''}{2S_0'}$$

$$S_1 = \frac{i}{2}\int \frac{S_0''}{S_0'} dz$$

$$= \frac{i}{2}\ln S_0'$$

$$= i\ln(n^2 - \xi^2)^{1/4} \qquad (5.2.26)$$

where we have used the expression for S_0 given by Eq. (5.2.25). Since

$$S = S_0 + \frac{1}{k_0}S_1$$

we have, from Eq. (5.2.20),

$$\psi = e^{ik_0 S_0 + iS_1}$$

$$\psi = e^{ik_0 S_0 - \ln(n^2 - \xi^2)^{1/4}}$$

$$\psi = \frac{1}{(n^2 - \xi^2)^{1/4}} e^{\pm ik_0 \int \sqrt{n^2 - \xi^2}\, dz} \qquad (5.2.27)$$

This is the WKB approximation. If we use Eq. (5.2.18), this solution can be put into the form

$$\psi(z) = \frac{1}{(n\sin\theta)^{1/2}} e^{\pm ik_0 \int n\sin\theta\, dz} \qquad (5.2.28)$$

We note from this form of the solution that when $\theta = 0$, the amplitude becomes infinite and the solution does not exist. When $\theta = 0$, the "equivalent ray" becomes horizontal and consequently turns over. Hence, this point is called a "turning point." Thus, we say that the WKB solution breaks down at a turning point even though the WKB solution is not the ray solution. The WKB solution as presented here does account for first-order diffraction effects and caustics whereas the ray solution does not. In Section 5.3, 5.4, and 5.5, we will continue to interpret the WKB solution in terms of the "equivalent ray," but bear in mind the WKB solution is not the ray solution.

204 APPROXIMATE SOLUTIONS OF THE WAVE EQUATION

Using Eq. (5.2.13), we can put Eq. (5.2.28) in yet another form:

$$\psi(z) = \left(\frac{k_0}{k_z}\right)^{1/2} e^{\pm i\int k_z \, dz} \qquad (5.2.29)$$

This puts the solution to the depth equation in terms of quantities associated entirely with the z direction. Thus, $\int k_z \, dz$ is simply the vertical component of the phase along the "ray."

5.3. CORRECTION FOR TURNING POINTS IN THE WKB METHOD

We have seen in the last section that the WKB solution fails at a turning point. It is clear from the nature of the singularity in Eq. (5.2.28) that the WKB solution is inapplicable not only at the turning point, but in some small layer, say $\pm \Delta$, about the turning point. In order to illustrate how the WKB solution is corrected for a turning point, we will consider the case when the speed of sound increases monotonically with depth as shown in Figure 5.3.1a. However, the technique applies to an arbitrary variation of sound speed with depth.

Figure 5.3.1b shows a particular "equivalent ray" which turns over or is totally reflected at a depth z_m. We shall see that this means that no acoustic energy is radiated in the positive z direction for value of $z > z_m$. At points which the WKB solution is applicable, the wave propagates without reflection. This is shown in a paper by Bremmer [5.3]. Consequently, the entire reflection process must take place in the region $z_m - \Delta < z < z_m + \Delta$. We will assume that the density gradient can be neglected so that this problem is reduced to the solution of the Helmholtz equation:

$$\nabla^2 p + k^2(z) p = 0$$

The procedure that we will follow in the remainder of this section is that due to Brekhovskikh [5.4].

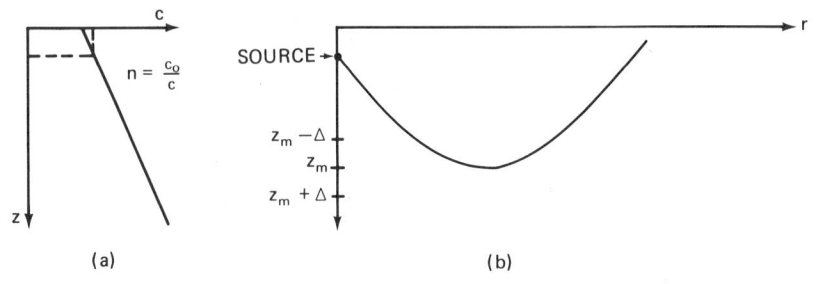

FIGURE 5.3.1. Total reflection of a ray due to a monotonically increasing sound speed.

CORRECTION FOR TURNING POINTS IN THE WKB METHOD

We have seen in Section 5.2, from Eqs. (5.2.2) and (5.2.18), that the depth equation can be represented in the form

$$\frac{d^2\psi}{dz} + \left[k_0^2 n^2(z) - k_0^2 \cos^2\theta_s\right]\psi = 0 \tag{5.3.1}$$

where $n = c_0/c(z)$, c_0 being the speed of sound at source depth, and all symbols are defined there. We will assume that $(k_0^2 n^2 - k_0^2 \cos^2\theta_s)$ can be expanded in a series of power of $z - z_m$ and that over the entire region $z_m - \Delta < z < z_m + \Delta$ we need only consider the linear term in the expansion. Then,

$$n^2 \cos^2\theta_s = -a(z - z_m) \tag{5.3.2}$$

where a is a coefficient with the dimensions of inverse length. Since we are assuming that the properties of the medium vary only slightly over a wavelength, we have in particular

$$a\lambda \ll 1 \tag{5.3.3}$$

because a is the coefficient of the highest order term in the series expansion that we kept.

Furthermore, we assume that the retention of only the linear terms in expansion of Eq. (5.3.2) is justified over the entire region of interest to us if we impose the requirement

$$a\Delta \ll 1 \tag{5.3.4}$$

Eq. (5.3.1) can now be written

$$\frac{d^2\psi}{dz^2} + k_0^2 a(z_m - z)\psi = 0 \tag{5.3.5}$$

This is just Airy's differential equation. We saw in Section 2.10 that solutions of this equation are Bessel function of order 1/3. Taking Hankel functions of the first and second kind as linearly independent solutions, we obtain the following general solution:

$z < z_m$:

$$\psi = Av^{1/2} H_{1/3}^{(1)}\left(\frac{2}{3}v^{2/3}\right) + Bv^{1/2} H_{1/3}^{(2)}\left(\frac{2}{3}v^{3/2}\right)$$

where $\quad v = (ak_0^2)^{1/3}(z_m - z) \tag{5.3.6}$

$z > z_m$:

$$\psi = Cu^{1/2} H_{1/3}^{(1)}\left(\frac{2i}{3}u^{3/2}\right) + Du^{1/2} H_{1/3}^{(2)}\left(\frac{2i}{3}u^{3/2}\right)$$

where $\quad u = (ak_0^2)^{1/3}(z - z_m) \tag{5.3.7}$

Because of the asymptotic nature of $H^{(2)}_{1/3}$ in Eq. (5.3.7), it represents a wave propagating from $z = \infty$ toward z_m. For exactly the same reasons discussed for propagation in the radial directions in Section 3.1, we must require $D = 0$. We shall now show that for sufficiently large $z_m - z$, the term in Eq. (5.3.6) containing the constant B degenerates into a wave incident on the turning point z_m, and the term containing the constant A degenerates into a wave reflected from the region containing the turning point z_m.

Using the arbitrariness available to us in the choice of Δ, we choose it in such a way that the condition

$$\tfrac{2}{3}v^{3/2} = \tfrac{2}{3}k_0 a^{1/2}\Delta^{3/2} \gg 1 \tag{5.3.8}$$

is satisfied. Note $(2/3) v^{3/2}$ is the argument of the Hankel functions. The conditions given by Eqs. (5.3.3) and (5.3.8) do not contradict one another, and it is always possible to choose a Δ such that they are satisfied simultaneously. Let $\delta = a\Delta$. According to Eq. (5.3.4), $\delta \ll 1$. Since $\lambda \ll \Delta$, we can choose Δ such that a $\lambda \sim \delta^2$. Then according to Eq. (5.3.8), $k_0 a^{1/2}\Delta^{3/2} \sim \delta^{-1/2} \gg 1$. As a result, all the requirements are compatible.

We will now make use of the asymptotic expressions for the Hankel functions for large values of the argument given by Eqs. (2.9.51) and (2.9.52):

$$H^{(1)}_{1/3}(x) = \sqrt{\frac{2}{\pi x}}\, e^{i(x - (5/12)\pi)} \tag{5.3.9}$$

$$H^{(2)}_{1/3}(x) = \sqrt{\frac{2}{\pi x}}\, e^{-i(x - (5\pi/12))} \tag{5.3.10}$$

Using these expressions in Eq. (5.3.6), we obtain for $(z_m - z) \sim \Delta$:

$$\psi(z) = \left(\frac{3}{\pi}\right)^{1/2} v^{-1/4} \left\{ A e^{i((2/3)v^{3/2} - (5\pi/12))} + B e^{-i((2/3)v^{3/2} - (5\pi/12))} \right\} \tag{5.3.11}$$

This expression can also be written in a form which is typical of the WKB solution given by Eq. (5.2.28). In fact, according to Eq. (5.3.2),

$$v^{-1/4} = \left(ak_0^2\right)^{-1/12}(z_m - z)^{-1/4}$$

$$= \left(\frac{k_0}{a}\right)^{-1/6}\left(n^2 - \cos^2\theta_s\right)^{-1/4}$$

$$= \left(\frac{k_0}{a}\right)^{-1/6}(n\sin\theta)^{-1/2} \tag{5.3.12}$$

Moreover, we also have the identity

$$\frac{2}{3}v^{3/2} = \int_0^v v^{1/2}\, dv$$

$$= -k_0 \int_{z_m}^z n(z)\sin\theta\, dz \tag{5.3.13}$$

As a result, Eq. (5.3.11) becomes

$$\psi(z) = \left(\frac{a}{k_0}\right)^{1/6} \left(\frac{3}{\pi n \sin\theta}\right)^{1/2} \left\{ Be^{ik_0 \int_{z_m}^z n \sin\theta\, dz + (i5\pi/12)} \right.$$

$$\left. + Ae^{-ik_0 \int_{z_m}^z n \sin\theta\, dz - (i5\pi/12)} \right\} \tag{5.3.14}$$

which has the same structure as the WKB solution (5.2.28). The first term corresponds to a wave propagating in the positive z direction, and the second term to a wave in the negative z direction.

Eq. (5.3.14) which we have obtained from Eq. (5.3.6) is valid in the region $z_m - \Delta < z < z_m + \Delta$. However, since the WKB solution is undoubtedly applicable beyond the limits of this region, Eq. (5.3.14) will describe the field for any value of $z < z_m - \Delta$. Thus, we see the constant B represents the amplitude of the incident wave and consequently can be prescribed arbitrarily.

The constants A and C can be found from the continuity conditions on ψ and $d\psi/dz$ at $z = z_m$. To analyze the field at the point, we need the following formulas:

$$H_{1/3}^{(1)} = \frac{i}{\sin(\pi/3)} \left\{ e^{-i\pi/3} J_{1/3} - J_{-1/3} \right\}$$

$$H_{1/3}^{(2)} = \frac{-i}{\sin(\pi/3)} \left\{ e^{+i\pi/3} J_{1/3} + J_{-1/3} \right\} \tag{5.3.15}$$

These formulas can be obtained simply from Eqs. (2.9.36), (2.9.53), and (2.9.54). It is expedient to expand the Bessel functions in powers of their arguments. Using Eqs. (2.9.25) and (2.9.26), we have

$$u^{1/2} J_{1/3}\left(\frac{2i}{3} u^{3/2}\right) = e^{i\pi/6} \frac{1}{\sqrt[3]{3}\, \Gamma(4/3)} P(u)$$

$$u^{1/2} J_{-1/3}\left(\frac{2i}{3} u^{3/2}\right) = e^{-i\pi/6} \frac{\sqrt[3]{3}}{\Gamma(2/3)} Q(u) \tag{5.3.16}$$

where

$$P(u) = u\left[1 + \frac{u^3}{12} + \frac{u^6}{504} + \cdots\right]$$

$$Q(u) = 1 + \frac{u^3}{6} + \frac{u^6}{180} + \cdots \tag{5.3.17}$$

208 APPROXIMATE SOLUTIONS OF THE WAVE EQUATION

From Eqs. (5.3.15) and (5.3.16), we obtain

$$u^{1/2}H^{(1)}_{1/3}\left(\frac{2i}{3}u^{3/2}\right) = \frac{2}{\sqrt{3}}e^{\pi i/3}[C_1 P(u) - C_2 Q(u)] \qquad (5.3.18)$$

where we have used the notation

$$C_1 = \frac{1}{\sqrt[3]{3}\,\Gamma(4/3)}, \qquad C_2 = \frac{\sqrt[3]{3}}{\Gamma(2/3)} \qquad (5.3.19)$$

Furthermore, from Eqs. (2.9.25) and (2.9.26), we find

$$v^{1/2}J_{1/3}\left(\frac{2}{3}v^{3/2}\right) = C_1 v\left[1 - \frac{v^3}{12} + \frac{v^6}{504} + \cdots\right] = -C_1 P(u)$$

$$v^{1/2}J_{-1/3}\left(\frac{2}{3}v^{3/2}\right) = C_2\left[1 - \frac{v^3}{6} + \frac{v^6}{180} + \cdots\right] = C_2 Q(u) \qquad (5.3.20)$$

As a result, according to Eq. (5.3.15), we obtain

$$v^{1/2}H^{(1)}_{1/3}\left(\frac{2}{3}v^{2/3}\right) = \frac{-2i}{\sqrt{3}}\left\{e^{-(i\pi/3)}C_1 P(u) + C_2 Q(u)\right\}$$

$$v^{1/2}H^{(2)}_{1/3}\left(\frac{2}{3}v^{3/2}\right) = \frac{2i}{\sqrt{3}}\left\{e^{i\pi/3}C_1 P(u) + C_2 Q(u)\right\} \qquad (5.3.21)$$

To determine ψ and $d\psi/dz$ at the point $u = v = 0$, we limit ourselves to the first terms in Eq. (5.3.17), which gives $P = u$, $Q = 1$.

Then, from Eqs. (5.3.6), (5.3.7), (5.3.18), and (5.3.21), we obtain $z < z_m$:

$$\psi = -\frac{2i}{\sqrt{3}}A\left\{e^{-(i\pi/3)}C_1 u + C_2\right\} + \frac{2i}{\sqrt{3}}B\left\{e^{i\pi/3}C_1 + u + C_2\right\} \qquad (5.3.22)$$

and $z > z_m$:

$$\psi = \frac{2}{\sqrt{3}}Ce^{i\pi/3}(C_1 u - C_2) \qquad (5.3.23)$$

where

$$u = (ak_0^2)^{1/2}(z - z_m) \qquad (5.3.24)$$

The continuity on ψ and $d\psi/dz$ at $z = z_m$ yields two equations:

$$B - A = iCe^{i\pi/3}$$

$$Be^{i\pi/3} - Ae^{-i\pi/3} = -iCe^{i\pi/3} \qquad (5.3.25)$$

where we find

$$A = Be^{i\pi/3}, \qquad C = ie^{i\pi/3}B \tag{5.3.26}$$

Using these relations in Eqs. (5.3.6) and (5.3.7), we obtain $z_m - \Delta < z < z_m$:

$$\psi = B\sqrt{v}\left\{ H^{(2)}_{1/3}\left(\tfrac{2}{3}v^{3/2}\right) + e^{i\pi/3}H^{(1)}_{1/3}\left(\tfrac{2}{3}v^{3/2}\right)\right\} \tag{5.3.27}$$

and $z_m < z < z_m + \Delta$:

$$\psi = iBe^{i\pi/3}\sqrt{u}\, H^{(1)}_{1/3}\left(\frac{2i}{3}u^{3/2}\right) \tag{5.3.28}$$

Since $|A| = |B|$ according to Eq. (5.3.26), the amplitude of the reflected wave is equal to the amplitude of the incident wave at every point. As regards the phase of the waves, we take Eq. (5.3.26) into account and obtain the following expression for the phase difference between the incident and the reflected waves at an arbitrary point z according to Eq. (5.3.14):

$$\begin{aligned}\psi &= -k_0\int_{z_m}^{z} n\sin\theta\, dz - \frac{5\pi}{12} + \frac{\pi}{3} - k_0\int_{z_m}^{z} n\sin\theta\, dz - \frac{5\pi}{12}\\ &= -\frac{\pi}{2} - 2k_0\int_{z_m}^{z} n\sin\theta\, dz\end{aligned} \tag{5.3.29}$$

If, in spite of the inapplicability of the WKB solution near the point $z = z_m$, we had used it to calculate the phase change over the entire path from z to z_m, we would have obtained from Eq. (5.2.28):

$$\phi_{\text{WKB}} = -2\int_{z_m}^{z} n\sin\theta\, dz \tag{5.3.30}$$

This differs from Eq. (5.3.29) only by the term $-\pi/2$. It is often convenient to regard the reflection as taking place at the plane $z = z_m$ in which case the quantity $-\pi/2$ must be interpreted as the phase of the reflection coefficient.

We can now use our solution to investigate wave leakage through a layer. We will examine the field in the region $z > z_m$ for values of $z - z_m$, which are sufficiently large to permit the use of the asymptotic representation Eq. (5.3.9) for the function $H^{(1)}_{1/3}$ in Eq. (5.3.28). As a result we obtain for $z > z_m$:

$$\psi = \sqrt{\frac{3}{\pi}}\, Bu^{-1/4}e^{i\pi/6}e^{-(2/3)u^{3/2}}$$

where

$$u = \left(ak_0^2\right)^{1/3}(z - z_m) \gg 1 \tag{5.3.31}$$

210 APPROXIMATE SOLUTIONS OF THE WAVE EQUATION

Thus as $z - z_m$ increases, the field decreases. Eq. (5.3.31) can also be represented in a form similar to the WKB solution. In fact, remembering the definition of u and using Eq. (5.3.2), we obtain for $z_m < z < z_m + \Delta$:

$$u^{-1/4} = \left(\frac{a}{k_0}\right)^{1/6} \frac{1}{\sqrt[4]{(\cos^2\theta_0 - n^2)}}$$

$$\frac{2}{3}u^{3/2} = \int_0^u u^{1/2}\,du = k_0 \int_{z_m}^z \left[\cos^2\theta_0 - n^2\right]^{1/2} dz$$

where
$$\cos^2\theta_0 - n^2 > 0$$

As a result, Eq. (5.3.31) is written

$$\psi = \left(\frac{3}{\pi}\right)^{1/2} B e^{i\pi/6} \left(\frac{a}{k_0}\right)^{1/6} \frac{1}{\left[\cos^2\theta_s - n^2\right]^{1/4}}$$

$$\cdot e^{-k_0 \int_{z_m}^z [\cos^2\theta_s - n^2]^{1/2}\,dz} \qquad (5.3.32)$$

This last expression represents a nonpropagating wave with an amplitude decreasing as z increases.

Now consider the following situation shown in Figure 5.3.2. Since $\cos\theta_s = c_0/c(z_m) = n(z_m)$, we see from Figure 5.3.2 that when $z_m < z < z'_m$,

$$n(z) < n(z_m) = \cos\theta_s$$

Hence, for $z_m < z < z'_m$, $[\cos\theta_s - n(z)] > 0$ and the solution Eq. (5.3.32) is exponentially decaying. That is, the acoustic field in the layer $z_m < z < z'_m$ is

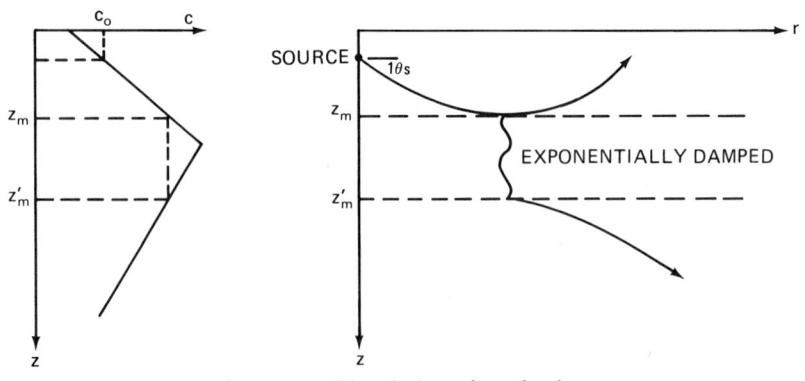

FIGURE 5.3.2. Wave leakage through a layer.

not a radiation field, but represents diffraction leakage or "tunneling" out of the duct.

As z increases, we again find some point $z = z'_m$ at which $n(z'_m) = \cos \theta_s$ and for values of $z > z'_m$ we have

$$n(z) > \cos \theta_s$$

So

$$\cos \theta_s - n(z) < 0$$

and the exponential factor in Eq. (5.3.32) becomes

$$e^{-k_0 i \int_{z_m}^{z'} [n^2 - \cos^2 \theta_s]^{1/2} \, dz}$$

which represents a traveling wave. This wave will carry a definite part of the energy beyond the boundaries of the layer (z_m, z'_m) and so the reflection cannot be complete.

Near $z = z'_m$, the solution Eq. (5.3.32) in the WKB approximation will again be inapplicable since the term in the denominator of the amplitude, namely, $(\cos^2 \theta_s - n^2)^{1/2}$ again becomes zero. A rigorous examination of the behavior of a wave in this region can be carried out by analogy with the method used above; the quantity $\cos^2 \theta_s - n^2(z)$ is expanded in a series of power of $(z - z'_m)$.

Without carrying out such a detailed investigation it can be shown that there will be no reflection at the level $z = z_m$, and the wave leaves the layer (z_m, z'_m) with the same amplitude with which it reached the level z'_m. For sufficiently large values of $z - z_m$, the geometrical acoustics solution will be

$$\psi = \left(\frac{3}{\pi}\right)^{1/2} e^{i\pi/6} \left(\frac{a}{k_0}\right)^{1/6} \frac{B}{(n \sin \theta)^{1/2}}$$

$$e^{-k_0 \int_{z_m}^{z'} [\cos^2 \theta_s - n^2(z)]^{1/2} \, dz} \left\{ e^{i k_0 \int_{z_m}^{z} n \sin \theta \, dz} \right\} \quad (5.3.33)$$

The amplitude of the transmitted wave will, as a rule, be very small since the factor

$$e^{-k_0 \int_{z_m}^{z'_m} (\cos^2 \theta_s - n^2)^{1/2} \, dz}$$

is small.

5.4. WKB SOLUTION IN A WAVEGUIDE

We will show in this section how the WKB approximation can be used to simplify the solution of the wave equation for an inhomogeneous waveguide such as that described in Chapter 4. There are two ways in which the exact solution can be simplified. The first is to replace the depth modes in Eq.

(4.1.54) by the WKB solution for the depth modes. The second is to derive a characteristic equation which is more numerically tractable than Eq. (4.1.37).

In Section 5.2 we derived the WKB approximation to the depth modes so we need not repeat that here. We will only derive the characteristic equation in the WKB approximation and, for simplicity, we will consider a restricted case. However, it will be clear how the derivation of the characteristic equation can be generalized to any situation.

The general solution of Eq. (5.2.2) in the WKB approximation is

$$\psi = \frac{A}{\beta^{1/2}} \exp\left\{ ik_0 \int_{z_0}^{z} \beta \, dz \right\} + \frac{B}{\beta^{1/2}} \exp\left\{ -ik_0 \int_{z_0}^{z} \beta \, dz \right\} \quad (5.4.1)$$

where $\quad \beta(z) = (n^2 - \xi^2)^{1/2} \quad$ (5.4.2)

and A, B, and z_0 are constants. Only two of the constants are independent and the third is determined by the normalization condition Eq. (4.1.38).

Recall the physical meaning of the exponent in Eq. (5.4.1) when β is real. The exponent $k_1 \int_{z_0}^{z} \beta \, dz$ is the change of phase of a wave which has traveled from an arbitrary point z_0 to the point of observation z. Since the suppressed time factor is $\exp(-i\omega t)$, the first term in Eq. (5.4.1) represents a wave propagating in the positive z direction, and the second term represents a wave propagating in the negative z direction.

Now consider an inhomogeneous waveguide of depth d as shown in Figure 5.4.1. Let z_0 be a reference level. We will only derive the characteristic equation for those modes which are trapped between two turning points z_u and z_L. When you have read Section 6.2, you will see that for this example, these modes correspond to the WKB approximation for modes 1 through 8 for the modes listed in Table 6.2.2 that correspond to the sound velocity profile shown in Figure 6.2.1. We begin by breaking the waveguide up into two regions, namely, the region $0 \leq z \leq z_0$ and the region $z_0 \leq z \leq d$. Let ψ_1 be the WKB solution in the region $0 < z < z_0$ and ψ_2 the corresponding solution in the region $z_0 < z < d$.

First consider the region $0 \leq z \leq z_0$. The solution ψ_1 is the sum of an upgoing and downgoing wave. The "equivalent ray" for this case is shown in Figure 5.4.1a. We wish to calculate the value of the acoustic field at an arbitrary depth z. The WKB solution at z is

$$\psi_1(z) = A_1 \beta^{-1/2} \exp\left[-ik_0 \int_{z_0}^{z} \beta \, dz \right]$$

$$+ A_1 \beta^{-1/2} \exp\left[-ik_0 \int_{z_0}^{z_u} \beta \, dz + ik_0 \int_{z_u}^{z} \beta \, dz - i\frac{\pi}{2} \right] \quad (5.4.3)$$

The first term represents the wave moving upward from the reference level z_0 to the depth z. It is represented by the "equivalent ray" AB in Figure 5.4.1a. The second term represents the wave moving downward. It is represented by

WKB SOLUTION IN A WAVEGUIDE 213

the "equivalent ray" ACD in Figure 5.4.1a. Since the "ray" has been totally reflected at the depth z_u, we have added a $-\pi/2$ phase shift as was discussed in Section 5.3.

Now consider the region $z_0 \leq z \leq d$. The geometry for this case is shown in Figure 5.4.1b. Again, the acoustic field at the depth z is the sum of an upgoing and downgoing wave. The WKB solution at z is

$$\psi_2(z) = A_2 \beta^{-1/2} \exp\left[ik_0 \int_{z_0}^{z} \beta \, dz \right]$$

$$+ A_2 \beta^{-1/2} \exp\left[ik_0 \int_{z_0}^{z_L} \beta \, dz - ik_0 \int_{z_L}^{z} \beta \, dz - i\frac{\pi}{2} \right] \quad (5.4.4)$$

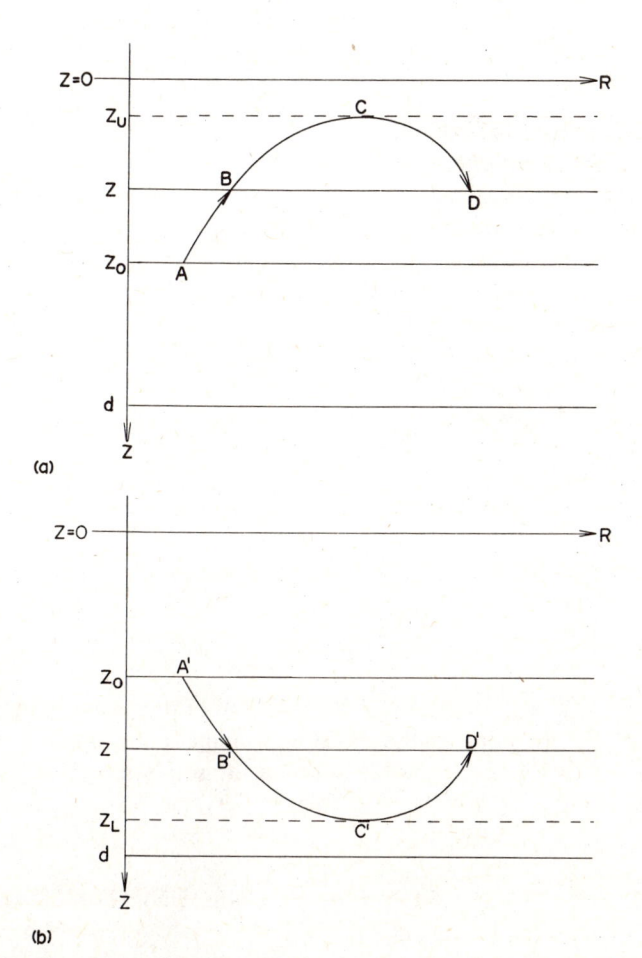

FIGURE 5.4.1. Geometry used to determine the characteristic equation in the WKB approximation.

214 APPROXIMATE SOLUTIONS OF THE WAVE EQUATION

The first term represents a wave moving downward from the reference level z_0 to the depth z. It is represented by the "equivalent ray" $A'B'$ in Figure 5.4.1b. The second term represents a wave moving upward. It is indicated by the "equivalent ray" $A'B'D'$ in Figure 5.4.1b. Since this "ray" is totally reflected at the turning point z_L, we have added a $-\pi/2$ phase shift.

Now the two solutions ψ_1 and ψ_2 must be equivalent at the arbitrary reference level z_0. Setting $z = z_0$ in Eqs. (5.4.3) and (5.4.4) and equating them gives

$$A_1\left\{1 + \exp\left(2ik_0\int_{z_u}^{z_0}\beta\,dz - i\frac{\pi}{2}\right)\right\} = A_2\left\{1 + \exp\left(2ik_0\int_{z_0}^{z_L}\beta\,dz - i\frac{\pi}{2}\right)\right\} \quad (5.4.5)$$

The derivative of ψ_1 and ψ_2 must also be continuous at z_0. This condition gives

$$A_1\left\{1 - \exp\left(2ik_0\int_{z_u}^{z_0}\beta\,dz - i\frac{\pi}{2}\right)\right\} = -A_2\left\{1 - \exp\left(2ik_0\int_{z_0}^{z_L}\beta\,dz - i\frac{\pi}{2}\right)\right\} \quad (5.4.6)$$

In deriving Eq. (5.4.6), we neglected the derivative of the amplitudes when we took the derivative of ψ_1 and ψ_2 in order to be consistent with the WKB approximation

Dividing Eq. (5.4.5) by Eq. (5.4.6), we get

$$\exp\left[2ik_0\int_{z_u}^{z_0}\beta\,dz + 2ik_0\int_{z_0}^{z_L}\beta\,dz - i\pi\right] = 1$$

This condition is satisfied if

$$2k_0\int_{z_u}^{z_L}\beta\,dz - \pi = 2\pi(n-1), \quad n = 1, 2, \ldots \quad (5.4.7)$$

Eq. (5.4.7) is the characteristic equation for the eigenvalues ξ in the WKB approximation.

In Section 6.2 we will show a comparison of the wave solution in an inhomogeneous waveguide to the WKB approximation. A more sophisticated and detailed treatment of the WKB solution in an inhomogeneous waveguide can be found in the work of Henrick [5.5].

5.5. WKB GREEN'S FUNCTION FOR AN UNBOUNDED MEDIA

In order to derive the ray approximation using the method of steepest descent in Section 5.6, we need to obtain the WKB approximation for the Green's function for the depth equation

$$\frac{d^2G}{dz^2} + k_0^2(n^2 - \xi^2)G = -\delta(z) \qquad (5.5.1)$$

We will derive the Green's function only for the specific case that we will need. We will consider an unbounded medium whose sound speed, $c(z)$, varies as shown in Figure 5.5.1a. We will further assume a point source at the origin.

Figure 5.5.1b shows an "equivalent ray," namely the ray OC, being emitted into the upper halfspace, $z > 0$. Because of the nature of the sound speed, this ray will continue on to infinity, never being refracted downward. The WKB

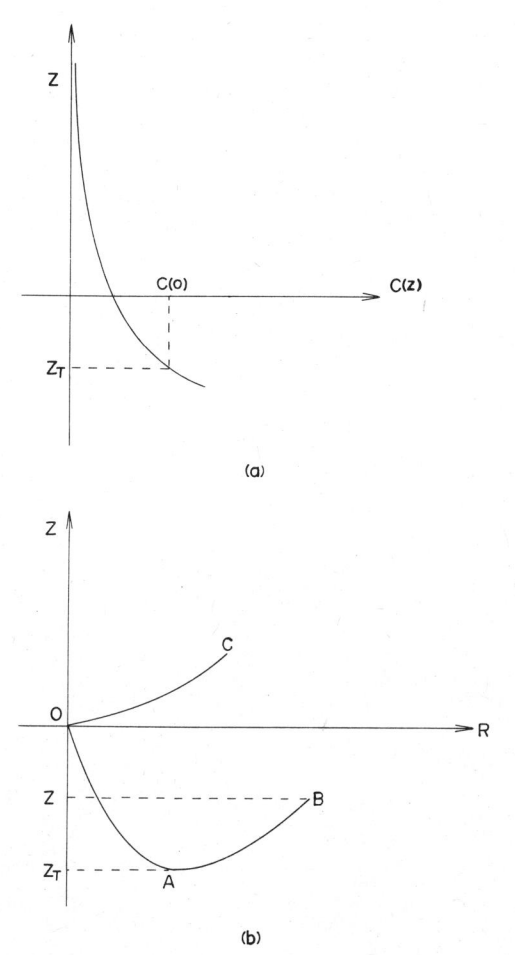

FIGURE 5.5.1. (a) Sound speed and (b) ray geometry used to derive the WKB Green's function for an unbounded medium.

216 APPROXIMATE SOLUTIONS OF THE WAVE EQUATION

solution corresponding to this "equivalent ray" is

$$G = A_1 \beta^{-1/2} \exp\left[ik_0 \int_0^z \beta \, dz\right], \quad z > 0 \tag{5.5.2}$$

where β is given by Eq. (5.4.2).

Also shown in Figure 5.5.1b is the "equivalent ray" OAB being emitted in the halfspace $z < 0$. As we saw in Section 5.3, the WKB solution corresponding to this "equivalent ray" is

$$G = A_2 \beta^{-1/2} \exp\left[-ik_0 \int_0^z \beta \, dz\right] + A_2 \beta^{-1/2} \exp\left[ik_0 \int_0^z \beta \, dz - i\frac{\pi}{2}\right], \quad z < 0$$

$$\tag{5.5.3}$$

Note that the boundary conditions at both $z \to +\infty$ and $z \to -\infty$ have been incorporated into these solutions. The condition at infinity require the wave to be outgoing.

We now have two arbitrary constants, A_1 and A_2, to evaluate. The two conditions that we must impose on this solution are given by Conditions Z.2 and Z.3 in Section 3.1. Applying these conditions at the source depth $z = 0$ gives

$$A_1 = A_2 + A_2 e^{-i\pi/2} \tag{5.5.4}$$

$$ik_0 \beta^{1/2}(0) A_1 + ik_0 \beta^{1/2}(0) A_2 - ik_0 \beta^{1/2}(0) A_2 e^{-i\pi/2} = -1 \tag{5.5.5}$$

As in Section 5.4, we have dropped the term involving the derivative of the amplitude when we differentiated G in order to be consistent with the WKB approximation.

Solving these equations for A_2 yields

$$A_2 = \frac{i}{2k_0 \beta^{1/2}(0)} \tag{5.5.6}$$

Since we will only consider the downgoing "ray" OAB, the desired Green's function is

$$G = \frac{1}{2k_0 \beta^{1/2}(0) \beta^{1/2}(z)} \exp\left[-ik_0 \int_0^z \beta \, dz + i\frac{\pi}{2}\right]$$

$$+ \frac{1}{2k_0 \beta^{1/2}(0) \beta^{1/2}(z)} \exp\left[ik_0 \int_0^z \beta \, dz\right], \quad z < 0 \tag{5.5.7}$$

where we have used the fact that $\exp(i\pi/2) = i$

Now let us rewrite the Green's function in terms of the quantity $\phi(z)$ defined by

$$\phi(z) = \int_{z_T}^{z} [n^2(z) - \xi^2]^{1/2} \, dz \tag{5.5.8}$$

where z_T is the depth of the turning point as shown in Figure 5.5.1b. Eq. (5.5.7) then becomes

$$G = \left\{ \frac{1}{2k_0(n^2 - \xi^2)^{1/4}(1 - \xi^2)^{1/4}} \right\} \exp\left[ik_0[\phi(0) - \phi(z)] + i\frac{\pi}{2} \right]$$

$$+ \left\{ \frac{1}{2k_0(n^2 - \xi^2)^{1/4}(1 - \xi^2)^{1/4}} \right\} \exp[ik_0[\phi(0) + \phi(z)]] \tag{5.5.9}$$

To obtain this final form we have used Eq. (5.4.2) and the fact that $n(0) = c_0/c(0) = c_0/c_0 = 1$, since we have been assuming c_0 to be the sound speed at the source. Eq. (5.5.9) is the final result that we will need in the next section.

5.6. RAY THEORY APPROXIMATION BY THE METHOD OF STEEPEST DESCENT

In this section we wish to investigate how ray theory arises from the application of the WKB approximation to the depth modes and the method of steepest descent. We will present the approach used by Sachs and Silbiger [5.6].

We start with the integral solution for the wave equation given by Eq. (3.B.6) in Appendix 3B.

$$p = \frac{1}{4\pi} \int_{-\infty}^{\infty} G(z, \xi) H_0^{(1)}(k_0 \xi r) k_0^2 \xi \, d\xi \tag{5.6.1}$$

where we have put $\zeta = k_0 \xi$.

We will consider only the case treated in Section 5.5. That is, we assume an unbounded medium whose sound speed $c(z)$ varies as shown in Figure 5.5.1a. As there, we also assume that the point source is located at the origin. We will only consider rays emitted downward or in the negative z direction. The Green's function for this case is given by Eq. (5.5.9).

If we use the asymptotic expansion for the Hankel function given by Eq. (3.1.32) and the Green's function given by Eq. (5.5.9), we obtain for waves which have left the source (but which have not passed through a turning point)

$$p = \frac{1}{4\pi} \int_{-\infty}^{\infty} d\xi \left[\frac{k_0 \xi}{2\pi r (n^2 - \xi^2)^{1/2} (1 - \xi^2)^{1/2}} \right]^{1/2} e^{ik_0 W + i(\pi/4)} \tag{5.6.2}$$

where

$$W(\xi, r, z) = \xi r + \phi(0) - \phi(z) \tag{5.6.3}$$

218 APPROXIMATE SOLUTIONS OF THE WAVE EQUATION

and, for the wave reflected at the turning point,

$$p = \frac{1}{4\pi}\int_{-\infty}^{\infty} d\xi \left[\frac{k_0\xi}{2\pi r(n^2-\xi^2)^{1/2}(1-\xi^2)^{1/2}}\right]^{1/2} e^{ik_0 W - i(\pi/4)} \quad (5.6.4)$$

where
$$W(\xi, r, z) = \xi r + \phi(0) + \phi(z) \quad (5.6.5)$$

Note that to obtain Eq. (5.6.2) we used only the first term in Eq. (5.5.9), and to obtain (5.6.4) we used only the second term in Eq. (5.5.9).

The branch cuts of the square roots in Eqs. (5.6.2) and (5.6.4) lie along the negative ξ axis and, for positive real values of the arguments, the positive square root is to be taken.

At this point let us examine the structure of the integral solution given by Eqs. (5.6.2) and (5.6.4). For convenience, consider only Eq. (5.6.4). The integrand has the form of a quasi-plane wave with amplitude

$$\frac{1}{4\pi}\left[\frac{k_0\xi}{2\pi r(n^2-\xi^2)^{1/2}(1-\xi^2)^{1/2}}\right]^{1/2}$$

and phase

$$\left(k_0 W - \frac{\pi}{4}\right)$$

Eq. (5.6.4) says that the total acoustic pressure at a point (r, z) is obtained by summing (integrating) these quasi-plane waves over all arrival angles $\theta = \cos^{-1}(\xi/n)$.

Now it may happen that there are values of ξ for which $\partial W/\partial \xi = 0$. Let us call the zeros of $\partial W/\partial \xi = 0$, ξ_i. The condition $\partial W/\partial \xi = 0$ says that those waves, with arrival angles ξ in some small neighborhood of ξ_i, all have the same phase. Consequently, the main contributions to the integral will come from the addition of coherent waves in each of the neighborhoods of the zeros ξ_i.

Now if $\partial W/\partial \xi$ has zeros at $\xi = \xi_j$, and k_0 is sufficiently large, we can use the method of steepest descent discussed in Section 2.8. This method informs us, as we have just noted above, that the main contributions to the integral are in the vicinity of these zeros.

However, even if we applied the method of steepest descent to the integral at this point, we would still not have the full ray theory approximation. As we shall see in Section 5.8, the solution would still account for diffraction and caustic effects at least to a first-order approximation.

In order to arrive at the ray theory approximation, we must impose one additional condition on the phase W. Let us expand W in a Taylor's series about $\xi = \xi_j$:

$$W(\xi) = W_j + (\xi - \xi_j)W_j' + \tfrac{1}{2}(\xi - \xi_j)^2 W_j'' + \tfrac{1}{6}(\xi - \xi_j)^3 W_j''' + \cdots \tag{5.6.6}$$

where

$$W_j = W(\xi_j), \qquad W' = \left[\frac{\partial W}{\partial \xi}\right]_{\xi=\xi_j}, \qquad W'' = \left[\frac{\partial^2 W}{\partial \xi^2}\right]_{\xi=\xi_j}, \qquad \text{and so on.}$$

By definition,

$$W_j' = 0 \tag{5.6.7}$$

We now assume the amplitudes in the integrands of Eqs. (5.6.2) and (5.6.4) are slowly varying in the neighborhood of ξ_j, and we neglect third- and higher-order terms in the Taylor expansion of Eq. (5.6.6). Thus, to obtain the ray theory approximation, we must assume the phase W is a quadratic function of ξ.

Now provided $W_j'' \neq 0$, we get, by applying the method of steepest descent for the wave incident upon the turning point,

$$p_{\text{inc}} = \frac{1}{4\pi} \sum_j \left[\frac{\xi_j}{r(1-\xi_j^2)^{1/2}(n^2-\xi_j^2)^{1/2}[\phi''(0) - \phi''(z)]}\right]^{1/2}$$

$$\times \exp\{ik_0[\xi_j r + \phi(0) - \phi(z)] + i\pi/2\} \tag{5.6.8}$$

and, for the reflected wave,

$$p_{\text{REFl}} = \frac{1}{4\pi} \sum_j \left[\frac{\xi_j}{r(1-\xi_j^2)^{1/2}(n^2-\xi_j^2)^{1/2}[\phi''(0) + \phi''(z)]}\right]^{1/2}$$

$$\times \exp\{ik_0[\xi_j r + \phi(0) + \phi(z)]\} \tag{5.6.9}$$

In writing Eqs. (5.6.8) and (5.6.9), we have assumed that W_j'' is positive, and we confine ourselves to rays with $\xi_j < n$ and $\xi_j < 1$. Furthermore, with the assumption that $c(z)$ monotonically decreases with increasing height, it can be shown that $\xi_j \geq 0$.

Eqs. (5.6.8) and (5.6.9) are the Green's function in the ray theory approximation. Notice in Section 5.1 we only derived an expression for the intensity along a ray while here we have both the amplitude and phase. Moreover, the ray theory Green's function gives us the total pressure at a fixed receiver location for a point source at the origin. According to Eq. (5.2.19), each term in the sum in Eqs. (5.6.8) and (5.6.9) correspond to a ray leaving the source at an angle $\theta_j = \cos^{-1}\xi_j$. Thus, these equations describe all rays that leave the source and arrive at the receiver; that is, they describe the multipath structure between source and receiver.

5.7. PHYSICAL DESCRIPTION OF THE FORMATION OF A SMOOTH CAUSTIC

Let us digress for a moment and discuss the physics of caustic formation.

We will see in Section 6.2 that under certain oceanographic conditions there will exist range intervals in which the signal strength will be greater than would be present due to cylindrical spreading only. The range intervals at which this enhancement occurs are called convergence zones.

The dominant feature of convergence zones is the focusing of energy and the consequent formation of caustics. A caustic is a focal surface. The most prevalent type of caustic in convergence zones is the smooth caustic shown in Figure 5.7.1. The envelope of the rays after a single refraction at depth is the caustic. Figure 5.7.2 shows the sound velocity used to generate this caustic. The source is situated at a 100-meter depth.

Now as we saw in Section 5.1, it is incorrect to speak about intensity or power per unit area along a single ray. Therefore, associated with each ray is

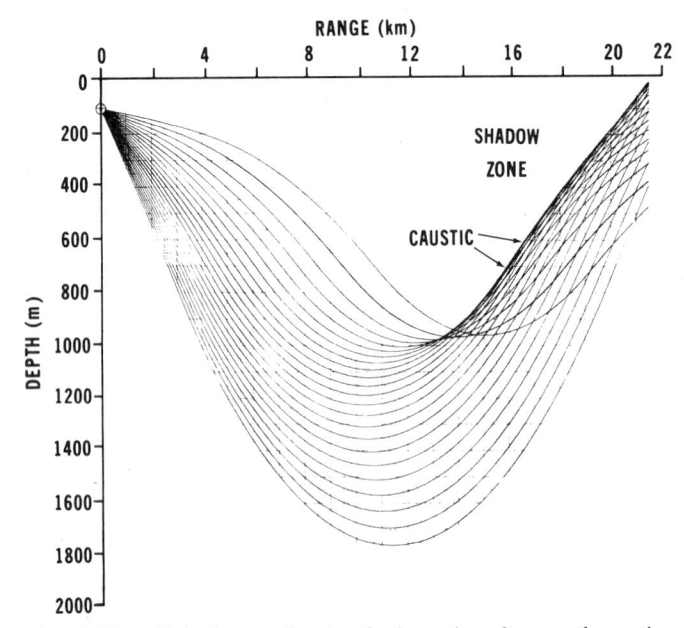

FIGURE 5.7.1. Ray diagram showing the formation of a smooth caustic.

PHYSICAL DESCRIPTION OF THE FORMATION OF A SMOOTH CAUSTIC 221

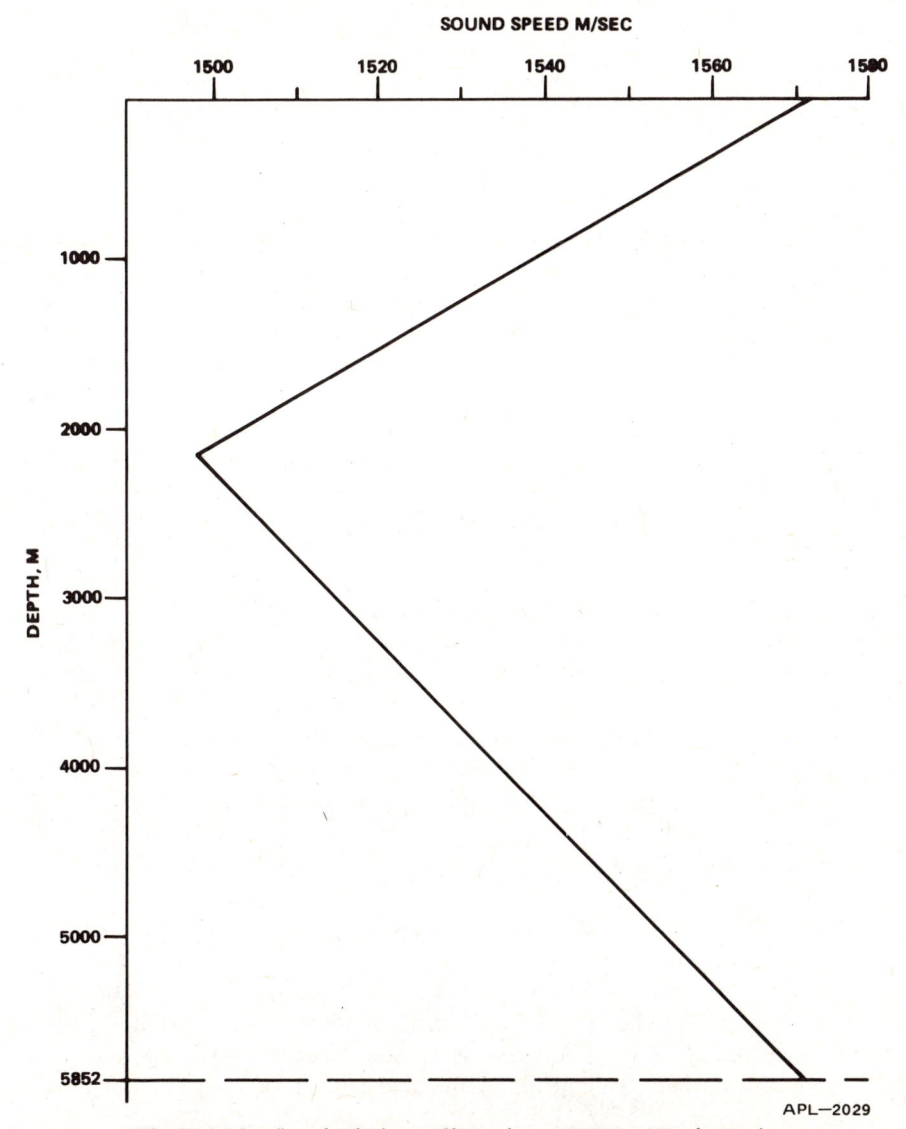

FIGURE 5.7.2. Sound velocity profile used to generate a smooth caustic.

an infinitesimal area of the wave front about the ray and perpendicular to it. The collection of all rays that go through this small area of wave front is called the infinitesimal ray bundle associated with the original ray. This ray bundle then describes how a small but finite area of the wave front is refracted. In the ray theory approximation, the energy is confined to a ray bundle. Leakage of energy out of the ray bundle is called diffraction.

Figure 5.7.3 shows two infinitesimal ray bundles. The diameter of the bundles have been greatly exaggerated for clarity. Each bundle is represented

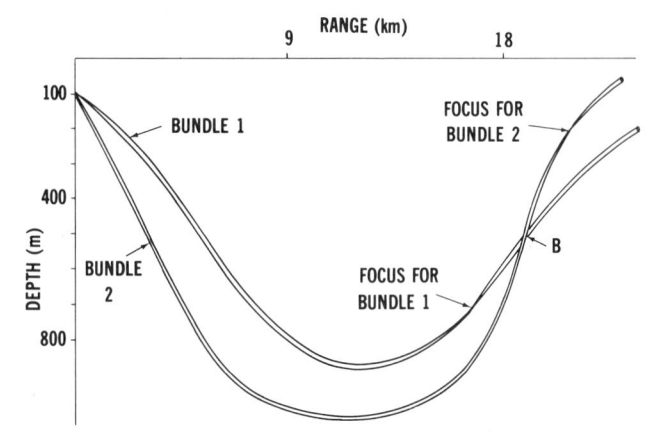

222 APPROXIMATE SOLUTIONS OF THE WAVE EQUATION

FIGURE 5.7.3. Two of the infinitely many ray bundles that form the caustic.

as a single ray on a normal ray plot such as Figure 5.7.1. Bundle 1 focuses soon after its lower vertex. Bundle 2, leaving the source at a steeper grazing angle, focuses at a shallower depth. Ray bundles leaving the source at angles between those of bundles 1 and 2 will have focal points between the focal points of bundles 1 and 2. The locus of all these focal points is the caustic. In the ray theory approximation, the intensity at the caustic is infinite. This result follows directly from Eq. (5.1.38). The important thing to note is that the entire wave front is not focused. Small areas of the wave front are individually focused at different points, giving rise to high intensity regions in the convergence zones. Also, the small portion of the wave front that passes through the focal point undergoes a 90-degree phase retardation. This result is derived in Appendix 5A.

In Figure 5.7.1, note that to the left of the caustic is a shadow zone, that is, a region of zero intensity in the ray theory approximation. This is a shadow zone only for the totality of ray bundles that form the caustic. In general, there will be other ray bundles which get into the shadow zone, but none of these bundles will have a focal point on the caustic. Next, consider the multipath structure behind the caustic in the illuminated region. This is shown in Figure 5.7.4.

Consider a receiver being towed along at a depth of 400 meters. The ray bundle that leaves the source at 5.5 degrees focuses at 400 meters and forms the caustic there. To the left of this point, at a 400-meter depth, is the shadow zone. To the right of the caustic, two ray bundles go through each point. A ray bundle is represented by a single ray in Figure 5.7.4 and in all ray diagrams that follow. One such range point, namely, Point B, is illustrated in Figure 5.7.4. One of the bundles leaves the source at a grazing angle (here 3.9 degrees) less than that of the caustic-forming bundle, and the other bundle always leaves the source at a grazing angle (here 7.4 degrees) greater than that of the caustic-forming bundle. Point B here corresponds to Point B on Figure 5.7.3. It

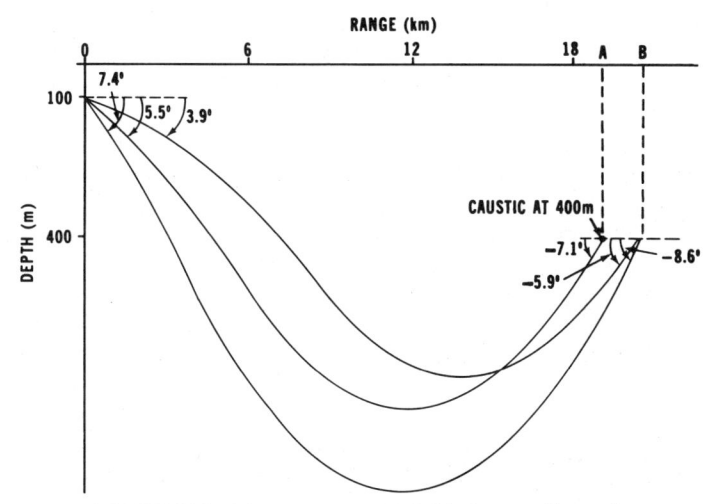

PHYSICAL DESCRIPTION OF THE FORMATION OF A SMOOTH CAUSTIC   223

FIGURE 5.7.4. The two-ray system behind a smooth caustic.

is seen that, for the receiver at Point B, one ray bundle has already focused before arriving at B and, so in addition to a phase difference due to the relative travel times, there will be an additional 90-degree phase shift of bundle 1 relative to bundle 2.

An extremely useful plot is that of arrival angle versus range. This is shown in Figure 5.7.5 for the case under consideration. A negative arrival angle means the ray is traveling upward when it arrives at the receiver and a positive angle means it is traveling downward when it arrives at the receiver. The vertex of

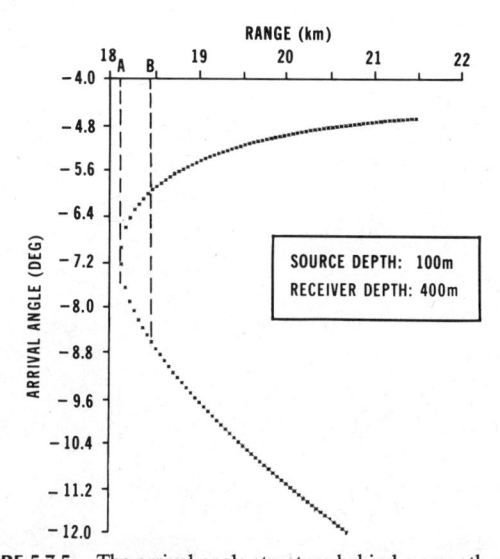

FIGURE 5.7.5. The arrival angle structure behind a smooth caustic.

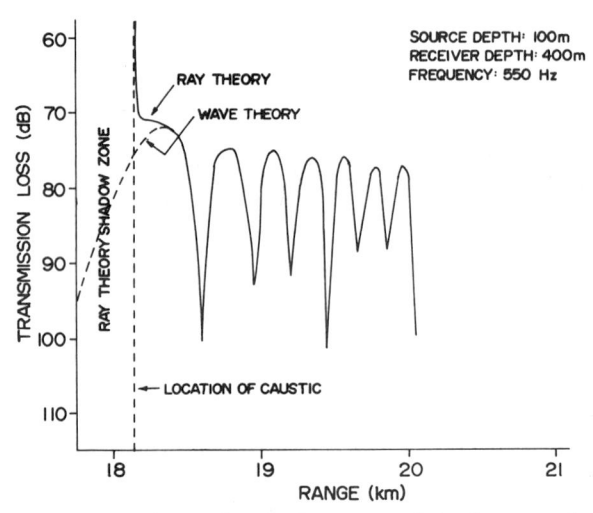

224 APPROXIMATE SOLUTIONS OF THE WAVE EQUATION

FIGURE 5.7.6. Ray theory and normal mode theory transmission loss around a smooth caustic.

the curve, Point A, corresponds to the arrival angle and range of the ray bundle forming the caustic. The two arrivals at Point B in Figure 5.7.4 are also shown in Figure 5.7.5. In general, we see from the arrival angle plot that there are two rays through each range point behind the caustic.

Figure 5.7.6 shows TL as a function of range for a range interval about the caustic. This plot of TL is for a frequency of 550 Hz. The vertical dotted line is

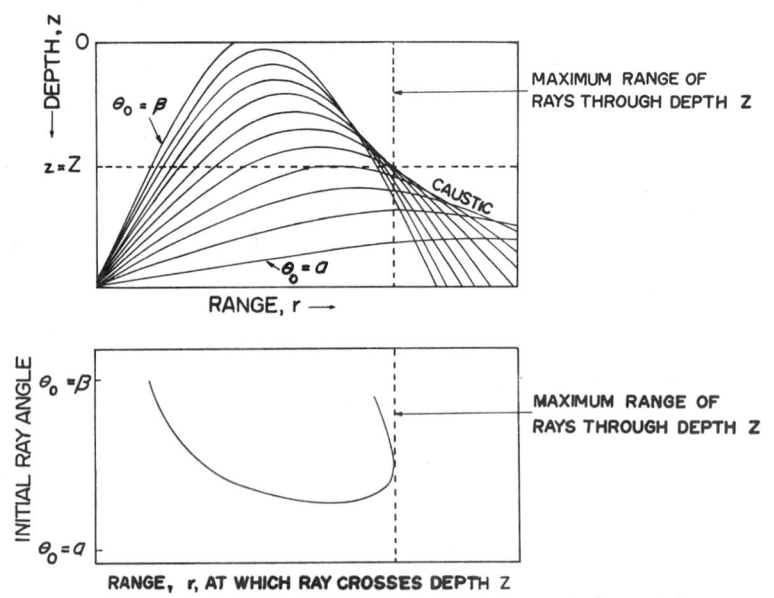

FIGURE 5.7.7. Smooth caustic where arrival angle structure exhibits a maximum in range instead of the usual minimum.

the location of the caustic. The TL shown by the solid curve is calculated from ray theory. This predicts zero intensity in the shadow zone and infinite intensity on the caustic. The lobing structure behind the caustic is due to the interference of the two rays beating in and out of phase. Now to remove the infinite intensity at the caustic, which is of course due to the ray theory approximation, wave theory could be used. The TL due to wave theory is shown by the dotted curve. Note that the maximum intensity is behind the caustic and not on it. This is due to the 90-degree phase shift of one ray relative to the other. Also note that wave theory predicts energy diffracted into the shadow zone.

Figure 5.7.7 shows a smooth caustic that is the reverse of the one just examined. Here the two-ray system is to the left of the caustic and the shadow

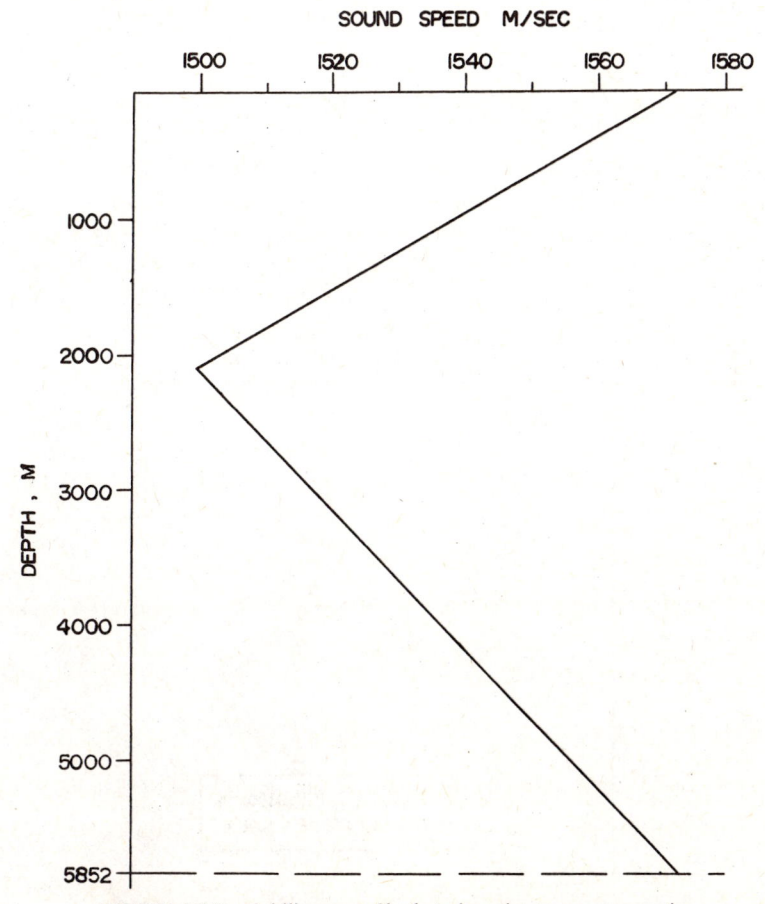

FIGURE 5.7.8. A bilinear profile that gives rise to a cusp caustic.

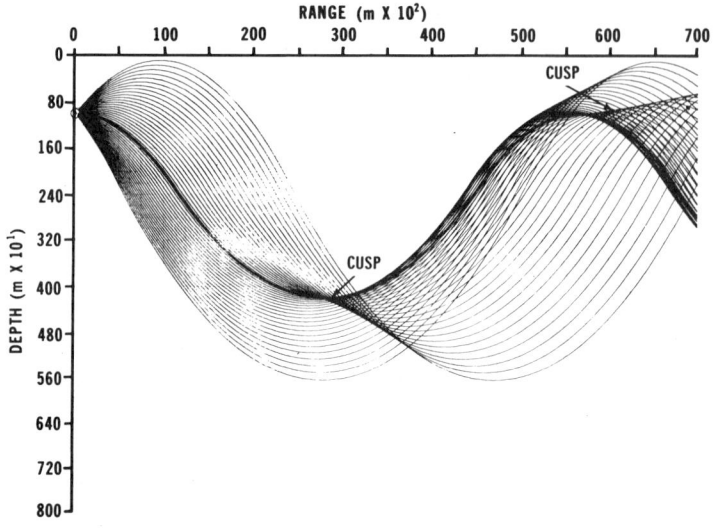

226 APPROXIMATE SOLUTIONS OF THE WAVE EQUATION

FIGURE 5.7.9. Ray plot showing a smooth and cusp caustic.

zone is to the right. Otherwise, everything said for the other caustic is true for this one.

A second type of caustic that can be formed is the cusp caustic. Consider a source at a depth of 1000 meters in a ocean which has a sound velocity profile as shown in Figure 5.7.8. A ray plot for this case is shown in Figure 5.7.9. In this figure the caustic that is formed at a depth of 1000 meters and a range of

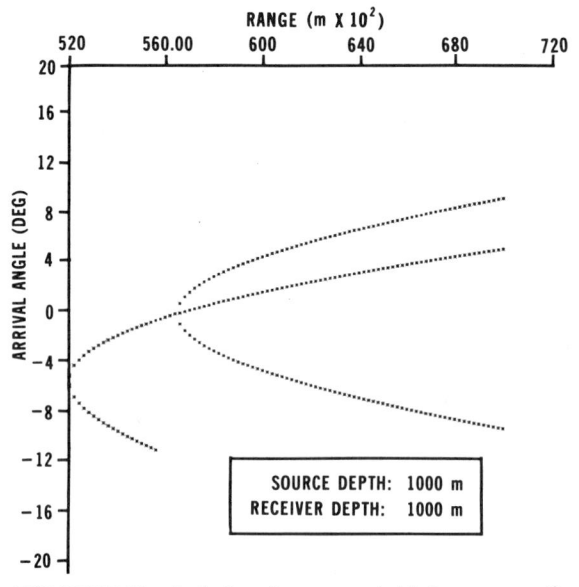

FIGURE 5.7.10. Arrival angle structure behind a cusp caustic.

56 kilometers is a cusp caustic. Note in the figures that one cusp is formed at source depth and a second cusp is formed at a reciprocal source depth at a range of approximately 27 kilometers. Reciprocal source depth is that depth where the speed of sound equals that at the source depth. A cusp is actually the intersection of two smooth caustics. The arrival angle structure behind the cusp is different than that for a smooth caustic. Figure 5.7.10 shows the arrival angle structure versus range for a receiver at a depth of 1000 meters. Note that a smooth caustic is formed at a range of 52 kilometers and one of the rays coming off the smooth caustic goes through the cusp at a range of 56 kilometers. We see that a three-ray system exists behind the cusp in contrast to a two-ray system behind a smooth caustic.

5.8. CORRECTION TO RAY THEORY FOR SMOOTH CAUSTICS

Returning now to our ray solution of Eqs. (5.6.8) and (5.6.9), we see that they cannot be used at caustics.

Mathematically, a caustic will arise when the denominator in Eqs. (5.6.8) or (5.6.9) vanishes because W_j'' goes to zero. In the range under consideration (in which $n > \xi_j$), $\phi''(z)$ is an increasing function of decreasing z. Hence, the denominator in the expression for p_{inc} will have no zeros other than at the origin. However, for increasing z, $\phi''(z)$ decreases and there may exist a point z at which $\phi''(z) = -\phi''(0)$. The location of the caustic is then found from the solution of this equation, which is independent of r, and from Eq. (5.6.7) or $r = \phi'(z) - \phi'(0)$, where ϕ is defined in Eq. (5.5.8). In differentiation with respect to ξ, it should be kept in mind that the lower limit of integration in Eq. (5.5.8) is a function of ξ through the relation $n(z_T) = \xi$. The order of integration and differentiation should not be reversed unless this dependence is taken into account. We conclude that the rays will not encounter a caustic between the source and the first turning point, but may pass through at most one caustic after passing the turning point. In other words, only the ray reflected at the turning point can pass through a caustic.

Consider some point on the caustic (r_c, z_c). The ray going through this point is given by $\xi = \xi_c$, the solution of $W'(\xi_c, r_c, z_c) = 0$. By definition, we also have for a point on the caustic $W''(\xi_c, z_c) = 0$. W'' is independent of r [see Eq. (5.6.5)]. Hence, for some point at the same depth but not on the caustic $(r \neq r_c, z = z_c)$, we have $W'(\xi_j, r, z_c) = 0$ but $W''(\xi_j, z_c) \neq 0$, while at this same point $W'(\xi_c, r, z_c) \neq 0$ but $W''(\xi_c, z_c) = 0$.

For points not on the caustic, we expand $W(\xi)$ around ξ_c:

$$W(\xi, r, z_c) = W(\xi_c, r, z_c) + W'(\xi_c, r, z_c)(\xi - \xi_c)$$
$$+ \frac{W''(\xi_c, r, z_c)}{2}(\xi - \xi_c)^2$$
$$+ \frac{W'''(\xi_c, r, z_c)}{6}(\xi - \xi_c)^3 \quad (5.8.1)$$

228   APPROXIMATE SOLUTIONS OF THE WAVE EQUATION

neglecting other terms. Now letting

$$W_c(r) = W(\xi_c, r, z_c) \qquad (5.8.2)$$

$$W_c'(r) = W'(\xi_c, r, z_c) \qquad (5.8.3)$$

$$W_c'''(r) = W'''(\xi_c, z_c) \qquad (5.8.4)$$

$$\Delta\xi = \xi - \xi_c \qquad (5.8.5)$$

$$\Delta r = r - r_c \qquad (5.8.6)$$

and noting

$$W'''(\xi_c, r, z_c) = W'''(\xi_c, z_c) = 0 \qquad (5.8.7)$$

since W''' is independent of r.

Eq. (5.8.1) then becomes

$$W(\xi, r, z_c) = W_c(r) + W_c'(r)\Delta\xi + \tfrac{1}{6}W_c'''(\Delta\xi)^3 \qquad (5.8.8)$$

Now

$$W_c'(r) - W_c'(r_c) = [r + \phi'(0) + \phi'(z_c)] - [r_c + \phi'(0) + \phi'(z_c)] = \Delta r$$

and since $W_c'(r_c) = 0$, we have

$$W_c'(r) = \Delta r \qquad (5.8.9)$$

Define the quantity s to be

$$s^3 = \frac{k_0}{2}(\Delta\xi)^3 W_c''' \qquad (5.8.10)$$

so that

$$s = \pm k_0^{1/3} \Delta\xi \left|\frac{W_c'''}{2}\right|^{1/3} \qquad (5.8.11)$$

where the $+$ sign is chosen if $W_c''' > 0$ and the $-$ sign is chosen if $W_c''' < 0$. Then we have

$$ds = \pm k_0^{1/3} \left|\frac{W_c'''}{2}\right|^{1/3} d\xi \qquad (5.8.12)$$

and

$$k_0 W = k_0 W_c + k_0 \Delta r \Delta \xi + \frac{k_0}{6} \frac{W_c'''}{6} (\Delta \xi)^3$$

$$= k_0 W_c \pm k_0^{2/3} (\Delta r) \left| \frac{W_c'''}{2} \right|^{-1/3} s + \frac{s^3}{3}$$

$$= k_0 W_c \pm \rho s + \frac{s^3}{3}$$

where for convenience we have put

$$\rho = k^{2/3} \Delta r \, \zeta \qquad (5.8.13)$$

$$\zeta = \left| \frac{W_c'''}{2} \right|^{-1/3} \qquad (5.8.14)$$

Assuming the amplitude in Eq. (5.6.4) is slowly varying, we get

$$p(r, z_c) = \frac{1}{4\pi} \left[\frac{k_0 \xi_c}{2\pi r [n^2(z_c) - \xi_c^2]^{1/2} (1 - \xi_c^2)^{1/2}} \right]^{1/2} \exp\left(ik_0 W_c - i\frac{\pi}{4} \right)$$

$$\times k_0^{-1/3} \zeta \int_{-\infty}^{\infty} \exp\left(\pm i\rho s + i\frac{s^3}{3} \right) ds \qquad (5.8.15)$$

Now it can be shown that an integral representation for the Airy function $Ai(\rho)$ is

$$Ai(\pm \rho) = \frac{1}{2\pi} \int_{-\infty}^{\infty} \exp\left(\pm i\rho s + i\frac{s^3}{3} \right) ds$$

Consequently, Eq. (5.8.15) can be written in the form

$$p(r, z_c) = \frac{1}{4\pi} k_0^{1/6} \zeta \left[\frac{2\pi \xi_c}{r [n^2(z_c) + \xi_c^2]^{1/2} (1 - \xi_c^2)^{1/2}} \right]^{1/2}$$

$$\times Ai(\pm \rho) \exp\left(ik_0 W_c - i\frac{\pi}{4} \right) \qquad (5.8.16)$$

This solution expresses the field at an arbitrary height z_c as a function of Δr, the horizontal distance from the caustic. The sign of W_c''' is negative if the shadow zone lies between the z-axis and the caustic boundary; it is positive if

230   APPROXIMATE SOLUTIONS OF THE WAVE EQUATION

the illuminated zone covers the same region. The applicability of Eq. (5.8.16) is restricted not only to large frequencies, but also to the vicinity of the caustic. It is, however, valid on both sides of the caustic ($\Delta r > 0$, $\Delta r < 0$) as well as on the caustic itself. On the caustic, the pressure is seen to increase with the one-sixth power of frequency.

Next, let us show that the definition of a smooth caustic given here, namely, $W'' = 0$, coincides with the definition which arises from Eq. (5.1.43) which is $\partial r/\partial \theta = 0$.

Differentiating Eq. (5.6.5) with respect to ξ yields

$$\frac{\partial W}{\partial \xi} = r + \frac{\partial \phi(0)}{\partial \xi} + \frac{\partial \phi(z)}{\partial \xi} \qquad (5.8.17)$$

and a second differentiation yields

$$\frac{\partial^2 W}{\partial \xi^2} = \frac{\partial}{\partial \xi}\left[\frac{\partial \phi(0)}{\partial \xi} + \frac{\partial \phi(z)}{\partial \xi}\right] \qquad (5.8.18)$$

Let (r_c, z_c) be a point on the caustic and $\xi = \xi_c$ be the ray going through this point. On this ray

$$\frac{\partial W}{\partial \xi} = 0$$

and so, from Eq. (5.8.17),

$$r_c = -\left[\frac{\partial \phi(0)}{\partial \xi} + \frac{\partial \phi(z_c)}{\partial \xi}\right]_{\xi=\xi_c} \qquad (5.8.19)$$

Substituting this expression for r_c into Eq. (5.8.18) we have, as long as we are on the ray going through the caustic point (r_c, z_c),

$$\frac{\partial^2 W}{\partial \xi^2} = -\frac{\partial r_c}{\partial \xi} \qquad (5.8.20)$$

Since $\xi = \cos \theta_s$ from Eq. (5.2.19), this last equation is the desired result. Thus, we see that the two conditions are identical.

Even though we will not discuss the cusp caustic, we can easily show how it arises.

Recall that the conditions for the smooth caustic are

$$W'(\xi_j, r, z) = 0 \qquad (5.8.21)$$

$$W''(\xi_j, z) = 0 \qquad (5.8.22)$$

$$W'''(\xi_j, z) \neq 0 \qquad (5.8.23)$$

If the parameter ξ_j is eliminated between Eqs. (5.8.21) and (5.8.22), we arrive at the equation for the caustic.

A cusp caustic is the intersection of two smooth caustics. The conditions for a cusp caustic are

$$W'(\xi_j, r, z) = 0 \qquad (5.8.24)$$

$$W''(\xi_j, z) = 0 \qquad (5.8.25)$$

$$W'''(\xi_j, z) = 0 \qquad (5.8.26)$$

$$W''''(\xi_j, z) \neq 0 \qquad (5.8.27)$$

If the parameter ξ_j is eliminated among Eqs. (5.8.24) through (5.8.26), we obtain the equation for the point at which the cusp caustic exists.

Finally, let us examine the accuracy of the caustic correction given by Eq. (5.8.16). To do this we will present an example from the work of Blatstein [5.7]. Blatstein incorporated Sachs and Silbiger's solution, Eq. (5.8.16), into the CONGRATS ray trace program written by Weinberg [5.8].

We will consider an oceanic waveguide which has a sound speed variation with depth as shown in Figure 5.8.1. The source, which is radiating at a frequency of 100 Hz, is located at a depth of 305 meters, and a receiver is located at a depth of 500 meters. A ray trace is shown in Figure 5.8.2. The caustic that we shall investigate occurs at a range of 56.3 km for the receiver depth of 500 meters. The two-ray system associated with this caustic is shown in Figure 5.8.3.

A comparison of the transmission loss calculated by normal mode theory and ray theory is shown in Figure 5.8.4. As we have already seen in Section 5.7, ray theory does not predict any sound intensity in the shadow zone to the left of the caustic. On the caustic, the ray theory intensity is infinite (zero transmission loss). However, we see that as we move deeper into the illuminated region, ray theory agrees quite well with the normal mode result.

In Figure 5.8.5 a comparison of the transmission loss calculated by normal mode theory and the modified ray theory given by Eq. (5.8.16) is shown. We see that the agreement is excellent in the shadow zone and in the illuminated region up to the first peak behind the caustic. However, from this range on, the two solutions diverge.

Consequently, we use the modified ray theory [Eq. (5.8.16)] only in the interval $|\rho| \leq 1.5$ where ρ is defined by Eq. (5.8.13). Ordinary ray theory is used in the region $\rho > 1.5$ (compare Figure 2.10.1). Unfortunately, the modified ray solution does not do this well for all caustics. Depending on the sound speed variation, the caustic structure can be so complicated that Eq. (5.8.16) is

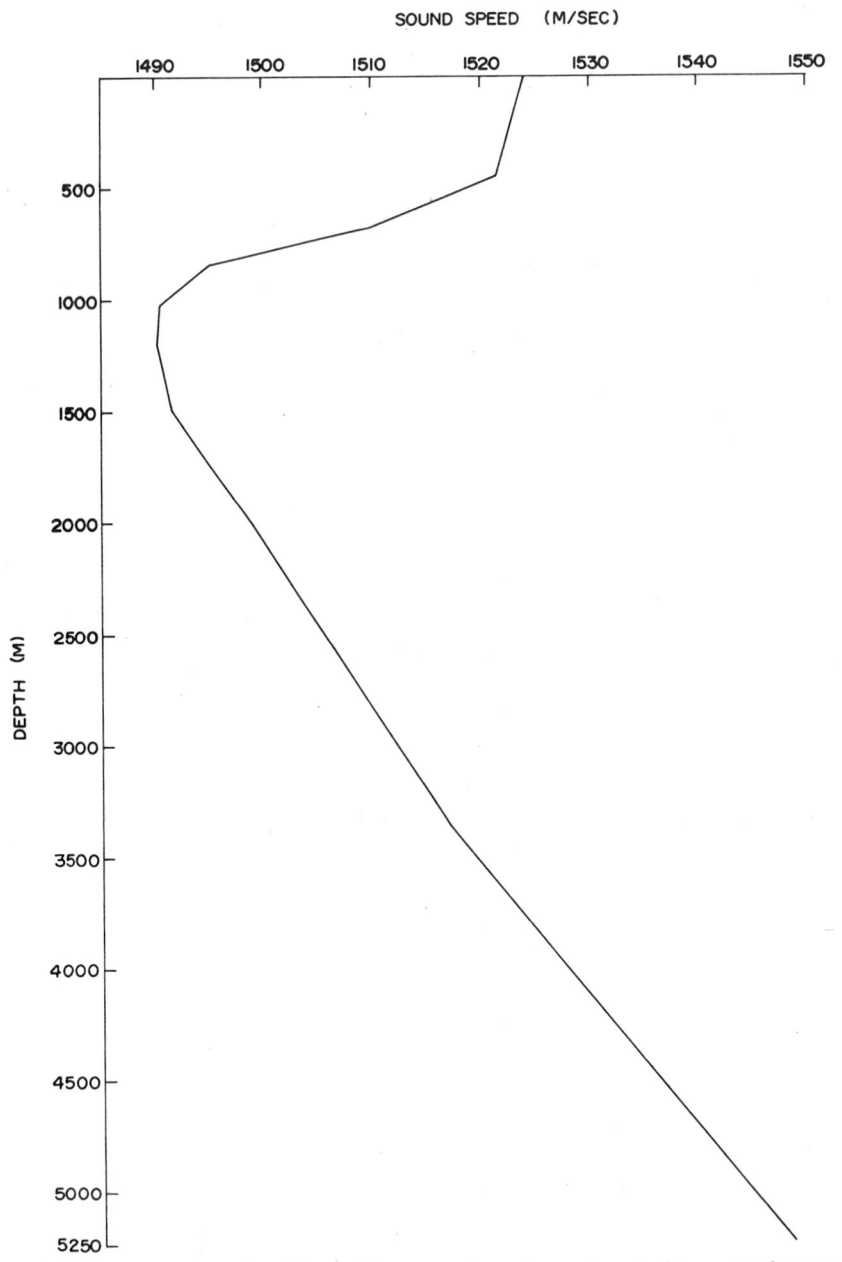

FIGURE 5.8.1. Sound speed profile used for comparison of normal mode theory, ray theory and modified ray theory.

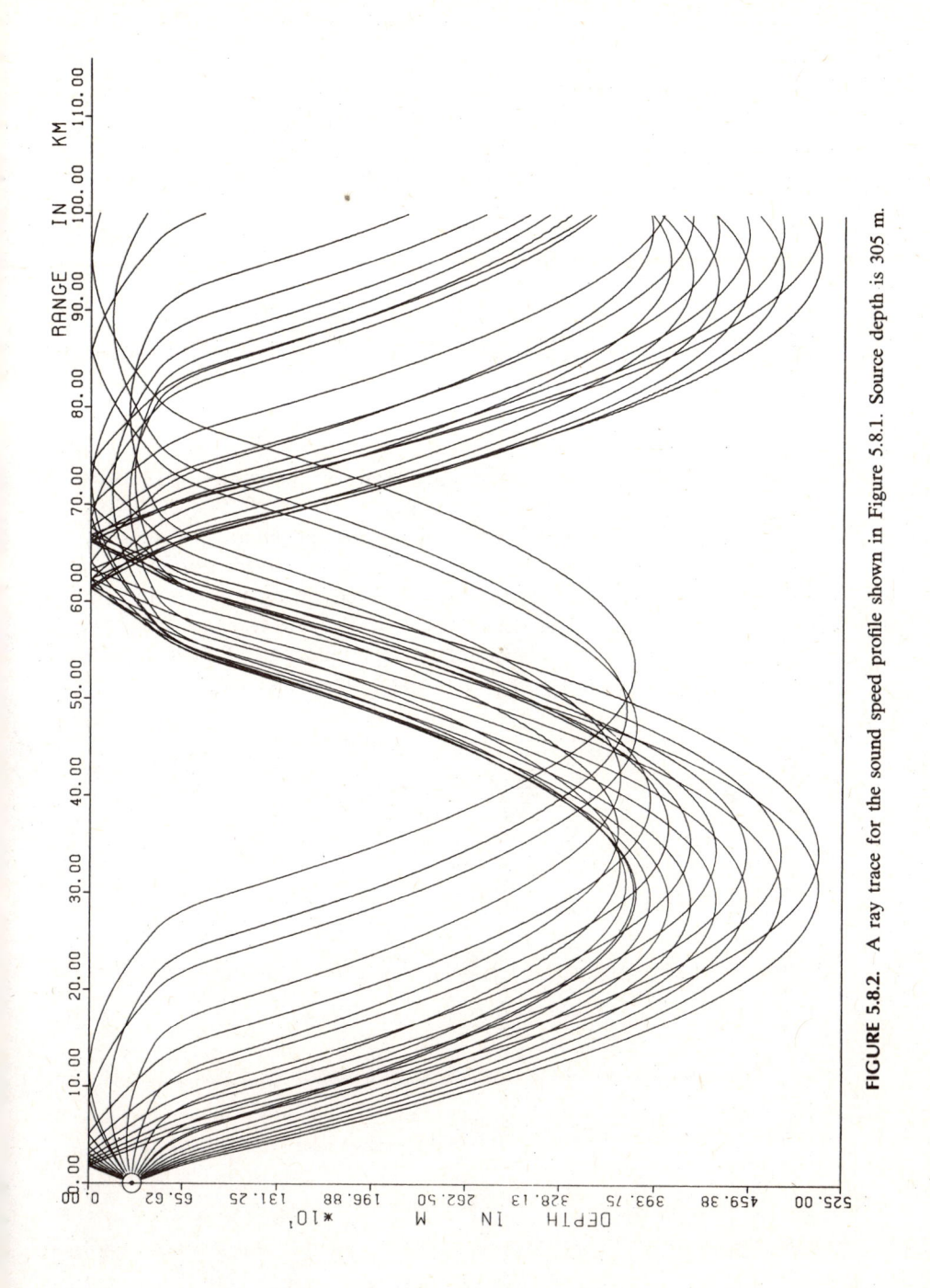

FIGURE 5.8.2. A ray trace for the sound speed profile shown in Figure 5.8.1. Source depth is 305 m.

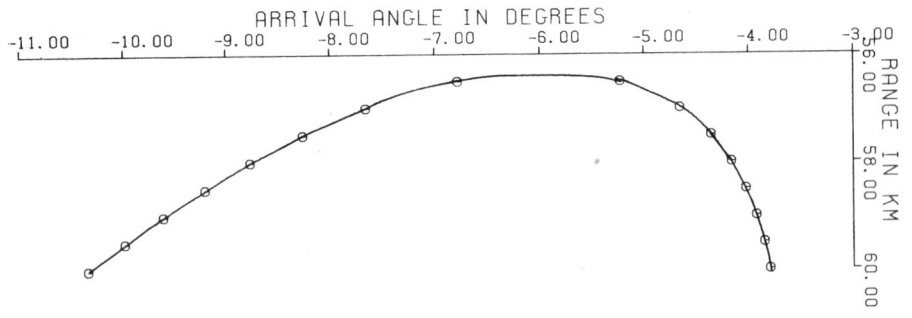

234 APPROXIMATE SOLUTIONS OF THE WAVE EQUATION

FIGURE 5.8.3. Arrival angle structure for a receiver at a depth of 500 m at a range about 56.3 km.

not an accurate solution. Other examples are shown in the report by Blatstein where the agreement is not as good as shown here.

The CONGRATS ray program with corrections for smooth caustics has been compared with experimental transmission loss data by Boyles and Joice [5.9] [5.10]. The transmission loss data were measured during a sea test in April 1976 in the Pacific Ocean. The data consist of 1/2 Hertz bandwidth, two second samples taken continuously from 5 nautical miles to 100 nautical miles. The agreement was extremely good even out to 100 nautical miles.

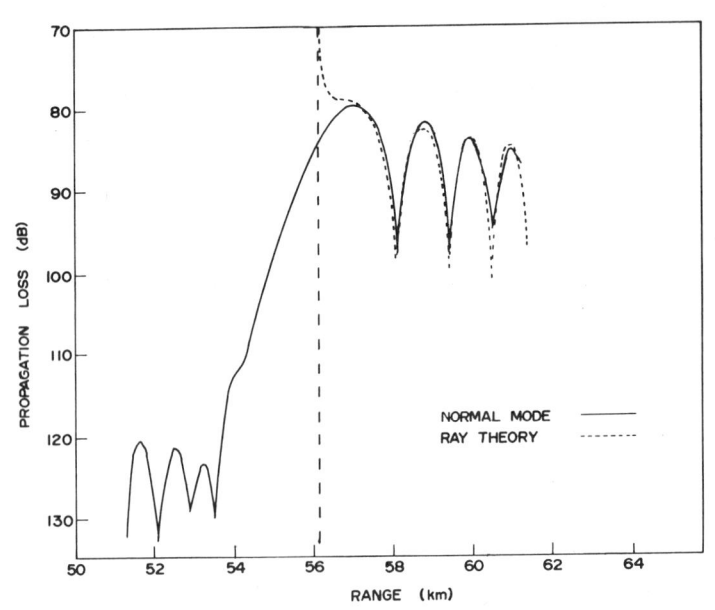

FIGURE 5.8.4. Comparison of transmission loss calculated by ray theory and normal mode theory.

APPENDIX 5.A. – 90-DEGREE PHASE SHIFT DUE TO A FOCAL POINT 235

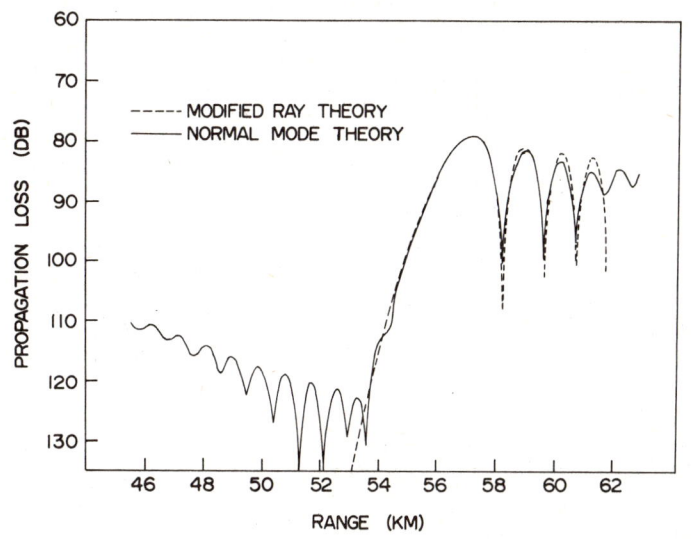

FIGURE 5.8.5. Comparison of transmission loss calculated by modified ray theory and normal mode theory.

APPENDIX 5.A. DERIVATION OF THE −90-DEGREE PHASE SHIFT AS A WAVE FRONT PASSES THROUGH A FOCAL POINT

We will use the expression given by Eq. (3.A.24) to calculate the phase shift experienced by a wave front as it goes through a focal point. That result was

$$\psi(\mathbf{r}) = \left(\frac{1}{8\pi^3 k}\right)^{1/2} e^{-i(\pi/4)} \int \left\{ \frac{e^{ikR}}{\sqrt{R}} \nabla_0 \psi - \psi \nabla_0 \left[\frac{e^{ikR}}{\sqrt{R}}\right] \right\} \cdot \hat{n}_0 \, d\xi_0$$

(5.A.1)

Since we are interested only in the portion of a wave front subtended by an infinitesimal bundle of rays, we need only consider a small area of the wave front. Suppose the center of the wave front is located at the origin as shown in Figure 5.A.1.

Assuming the wave front is the arc of a circle centered at $x = a$, the equation for the circle is

$$(x - a)^2 + y^2 = a^2$$

or

$$2ax - x^2 = y^2$$

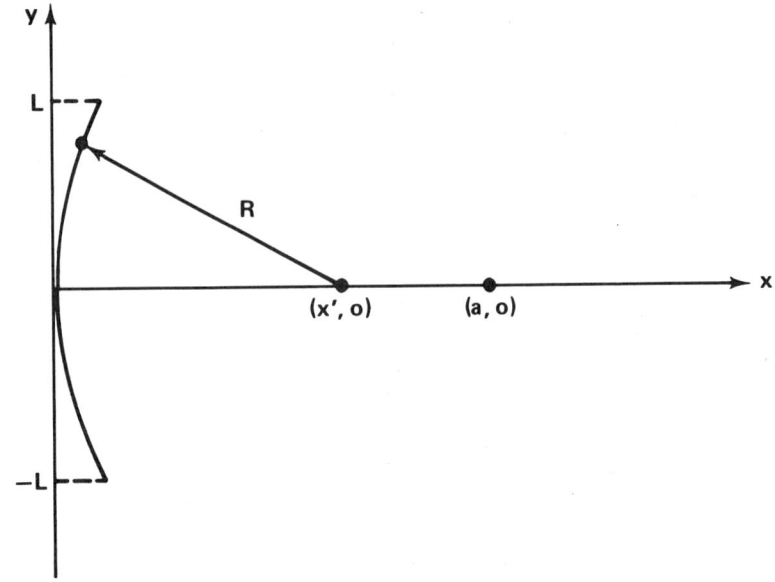

236   APPROXIMATE SOLUTIONS OF THE WAVE EQUATION

FIGURE 5.A.1. A portion of a cylindrical wave front collapsing on the focal point located at $x = a$.

We will assume that we can neglect x^2 and take for the equation describing the small portion of wave front centered about the origin

$$x = \frac{y^2}{2a} \tag{5.A.2}$$

Now, let $(x', 0)$ be the point we wish to evaluate the field:

$$R = \left[(x - x')^2 + y^2\right]^{1/2} = \left[x^2 - 2xx' + x'^2 + y^2\right]^{1/2}$$

$$= x'\left[1 - \frac{2x}{x'} + \frac{x^2}{x'^2} + \frac{y^2}{x'^2}\right]^{1/2}$$

$$= x'\left[1 - \frac{2}{x'}\frac{y^2}{2a} + \frac{x^2}{x'^2} + \frac{y^2}{x'^2}\right]^{1/2}$$

$$\approx x'\left[1 - \frac{y^2}{2ax'} + \frac{x^2}{2x'^2} + \frac{y^2}{2x'^2}\right]$$

$$\approx x' - \frac{y^2}{2}\left[\frac{1}{a} - \frac{1}{x'}\right] \tag{5.A.3}$$

APPENDIX 5.A. – 90-DEGREE PHASE SHIFT DUE TO A FOCAL POINT

We will make the further assumption that

$$\hat{n}_0 \approx -\hat{e}_x$$

$$\hat{n}_0 \cdot \nabla_0 \approx -\frac{\partial}{\partial x}$$

$$\hat{n}_0 \cdot \nabla_0 \psi \approx -\frac{\partial \psi}{\partial x} \approx -ik\psi$$

$$\hat{n}_0 \cdot \nabla_0 \left(\frac{e^{ikR}}{\sqrt{R}}\right) \approx -\frac{\partial}{\partial x}\left(\frac{e^{ikx'}}{\sqrt{x'}}\right) = 0$$

Hence, (5.A.1) becomes

$$\psi(x') = -ik\left(\frac{1}{8\pi^3 k}\right)^{1/2} e^{-i(\pi/4)} \frac{e^{ikx'}}{\sqrt{x'}} \int e^{-ik(y^2/2)[(1/a)-(1/x')]} \psi \, dy \quad (5.A.4)$$

The field ψ can be considered constant on the wave surface. Hence, (5.A.4) can be written

$$\psi(x') = A \int_{-\infty}^{\infty} e^{-ik(y^2/2)[(1/a)-(1/x')]} \, dy \quad (5.A.5)$$

where we have put

$$A = -ik\psi\left(\frac{1}{8\pi^3 k}\right)^{1/2} e^{-i\pi/4} \frac{e^{ikx'}}{\sqrt{x'}} = |A|e^{i\phi} \quad (5.A.6)$$

We have also extended the range of integration to infinity since the field amplitude is assumed zero outside the limits $-L < y < L$.

To evaluate the integral occurring in (5.A.5) we need the following result:

$$\int_{-\infty}^{\infty} e^{\pm i\zeta^2} \, d\zeta = \int_{-\infty}^{\infty} \cos \zeta^2 \, d\zeta \pm i \int_{-\infty}^{\infty} \sin \zeta^2 \, d\zeta$$

$$= \frac{\sqrt{\pi}}{2}(1 \pm i) \quad (5.A.7)$$

Now for $x' < a$, let

$$\alpha = \left[-\frac{k}{2}\left(\frac{1}{a} - \frac{1}{x'}\right)\right]^{1/2} = \left[\left|\frac{k}{2}\left(\frac{1}{x'} - \frac{1}{a}\right)\right|\right]^{1/2}$$

and

$$\zeta = \alpha y$$

238 APPROXIMATE SOLUTIONS OF THE WAVE EQUATION

so that

$$\int_{-\infty}^{\infty} e^{i\alpha^2 y^2} \, dy = \frac{1}{\alpha} \int_{-\infty}^{\infty} e^{i\zeta^2} \, d\zeta = \frac{1}{\alpha} \sqrt{\frac{\pi}{2}} \, (1 + i)$$

and, consequently,

$$\psi(x') = \frac{|A|}{\alpha} \sqrt{\pi} \, e^{i\phi} e^{i\pi/4} \qquad (5.A.8)$$

For $x' > a$, let

$$\beta = \left[\frac{k}{2} \left(\frac{1}{a} - \frac{1}{x'} \right) \right]^{1/2} = \left| \frac{k}{2} \left(\frac{1}{a} - \frac{1}{x'} \right) \right|^{1/2}$$

and

$$\zeta = \beta y$$

so that

$$\int_{-\infty}^{\infty} e^{-i\beta^2 y^2} \, dy = \frac{1}{\beta} \int_{-\infty}^{\infty} e^{-i\zeta^2} \, d\zeta$$

$$= \frac{1}{\beta} \sqrt{\frac{\pi}{2}} \, (1 - i)$$

and, consequently,

$$\psi(x') = \frac{|A|}{\beta} \sqrt{\pi} \, e^{i\phi} e^{-i\pi/4} \qquad (5.A.9)$$

Comparing the phase of the wave after it has passed through the focal point [Eq. (5.A.9)] to the phase before it passes through the focal point [Eq. (5.A.8)], we see that there is an additional phase shift of -90 degrees that the wave undergoes.

REFERENCES

5.1. M. A. Pedersen, "Acoustic Intensity Anomalies Introduced by Constant Velocity Gradients," *J. Acoust. Soc. Am.* **33**, 465–474 (1961).

5.2. M. A. Pedersen and D. F. Gordon, "Comparison of Curvilinear and Linear Profile Approximation in the Calculation of Underwater Sound Intensities by Ray Theory," *J. Acoust. Soc. Am.* **41**, 419–438 (1967).

5.3. H. Bremmer, "The WKB Approximation as the First Term of a Geometric-Optical Series," *Comm. Pure and Applied Math.*, vol IV, no. 1 (1951), 105–115.

5.4. L. M. Brekhovskikh, *Waves in Layered Media* (Academic Press, New York, 1960), pp. 206–215.

5.5. R. F. Hendrick, "The Uniform WKB Approach to Pulsed and Broadband Propagation," *J. Acoust. Soc. Am.*, forthcoming.

5.6. D. A. Sachs and A. Silbiger, "Focusing and Refraction of Harmonic Sound and Transient Pulses in Stratified Media," *J. Acoust. Soc. Am.* **49**, 824–840 (1971).

5.7. I. M. Blatstein, "Comparison of Normal Mode Theory, Ray Theory, and Modified Ray Theory for Arbitrary Sound Velocity Profiles Resulting in Convergence Zones," NOLTR 74-95, 29 August 1974, Naval Surface Weapons Center, White Oak, Md.

5.8. H. Weinberg, "CONGRATS I: Ray Plotting and Eigenray Generation," NUSC Report No. 1052, 30 October 1969, Naval Underwater System Center, New London, Conn.

5.9. C. A. Boyles and G. W. Joice, "A Comparison of Three Propagation Models with Detailed Convergence Zone Transmission Loss Data," *J. Acoust. Soc. Am.* **64**, Supplement No. 1, S74 (1978).

5.10. C. A. Boyles and G. W. Joice, "Comparison of Three Acoustic Transmission Loss Models with Experimental Data," Johns Hopkins APL Technical Digest, vol. 3, no. 1, 1982. (Reprints available on request to authors.)

6 APPLICATION TO CONVERGENCE ZONE AND SURFACE DUCT PROPAGATION

6.0. INTRODUCTION

In this chapter we discuss the application of ray theory and normal mode theory to convergence zone and surface duct propagation. This is not meant to be an exhaustive treatment of the subject, but rather it is designed to point out some basic features of these types of propagation.

The basic ray theory model used to produce the results shown in this chapter is that due to H. Weinberg [6.1]. The normal mode model used is the one described in Chapter 4.

6.1. SPEED OF SOUND IN THE OCEAN

There are just two parameters in the wave equation (1.7.50) that describe the physical characteristics of the ocean and that affect the propagation of acoustic waves. These are the density ρ_0 and the sound speed $c(x, y, z)$.

The variation of density from the ocean's surface to the ocean's bottom is so slight that it is usually neglected. Typically, a single value of the density is used throughout the water column. It is only in the sediment layers of the bottom that density variations are usually taken into account.

However, the variation of sound speed from the ocean's surface to the ocean's bottom cannot be neglected. It has been found that the speed of sound in the ocean is a function of temperature, salinity, and pressure. The nature of the variation of these three quantities with depth, of course, depends on the geographic location.

There have been many equations proposed for determining the speed of sound as a function of temperature, salinity, and pressure. We do not intend to discuss the relative merits of these many equations; we will simply present one of the most recent determinations given by Mackenzie [6.2] and [6.3].

$$c = 1448.96 + 4.591T - 5.304 \times 10^{-2}T^2$$
$$+ 2.374 \times 10^{-4}T^3 + 1.340(S - 35)$$
$$+ 1.630 \times 10^{-2}D + 1.675 \times 10^{-7}D^2$$
$$- 1.025 \times 10^{-2}T(S - 35) - 7.139 \times 10^{-13}TD^3$$

Here T is the temperature in degrees Celsius, S is the salinity in parts per thousand (‰), and D is the depth in meters. One of the conveniences of this equation is that the more easily calculated quantity of depth is used instead of the quantity of pressure. The range of validity of this equation encompasses temperature from -2 to 30 degrees C, salinity at 25 to 40 ‰, and depth at 0 to 8000 meters. The variation of sound speed with depth is called the sound velocity profile.

One example of the variation of temperature, salinity, and, consequently, sound speed with depth is shown in Figure 6.1.1. These measurements were taken in the North Pacific at 35° 38.3′N latitude, 128° 1.2′W longitude during April.

We see here that the temperature decreases rapidly with increasing depth from the sea surface to a depth of about 500 meters. This region is called the thermocline. We also see in the deep ocean, say below 2500 meters, that the temperature is fairly constant. Consequently, in the thermocline region the temperature has the greatest effect on the speed of sound, while in the deep ocean, pressure is the dominant factor. Except in a few ocean areas, the variation of salinity with depth has little effect on the speed of sound.

The horizontal variation of the sound speed are usually quite weak and for many applications can be neglected. This greatly simplifies the theory and allows us to use the techniques of separation of variables. However, for most long-range propagation, particularly in a surface duct environment (which we will discuss in Section 6.4), horizontal variation in the sound speed cannot be neglected. The mathematical solution of the wave equation for this case is quite complex and will be presented in Chapter 7.

6.2. CONVERGENCE ZONE PROPAGATION—A SINGLE CHANNEL NORTH PACIFIC PROFILE

A North Pacific sound velocity profile for June is shown in Figure 6.2.1. Table 6.2.1 gives the values of sound speed versus depth. We see that the sound speed decreases steadily from its value of 1507.2 m/sec at the surface to a minimum of 1478.0 m/sec at a depth of 686 meters. Then the sound speed increases until

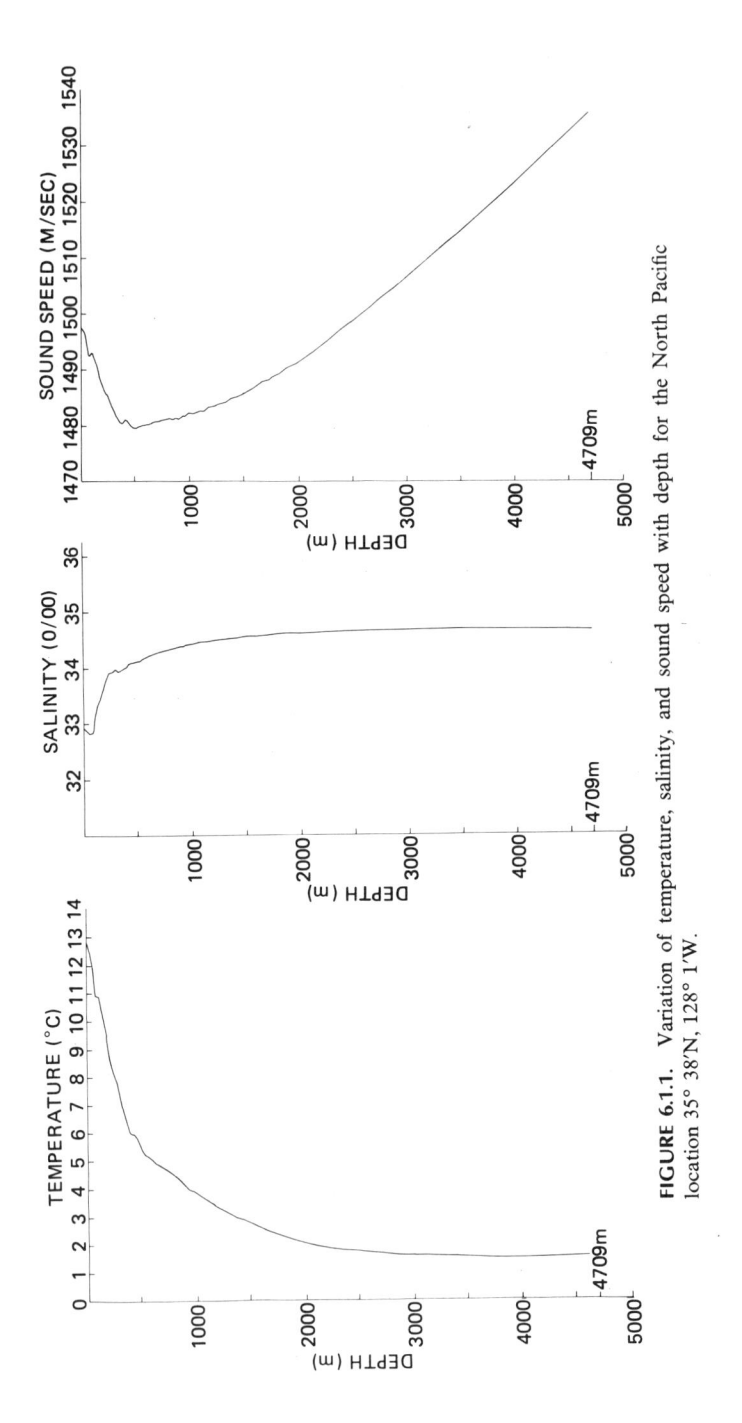

FIGURE 6.1.1. Variation of temperature, salinity, and sound speed with depth for the North Pacific location 35° 38'N, 128° 1'W.

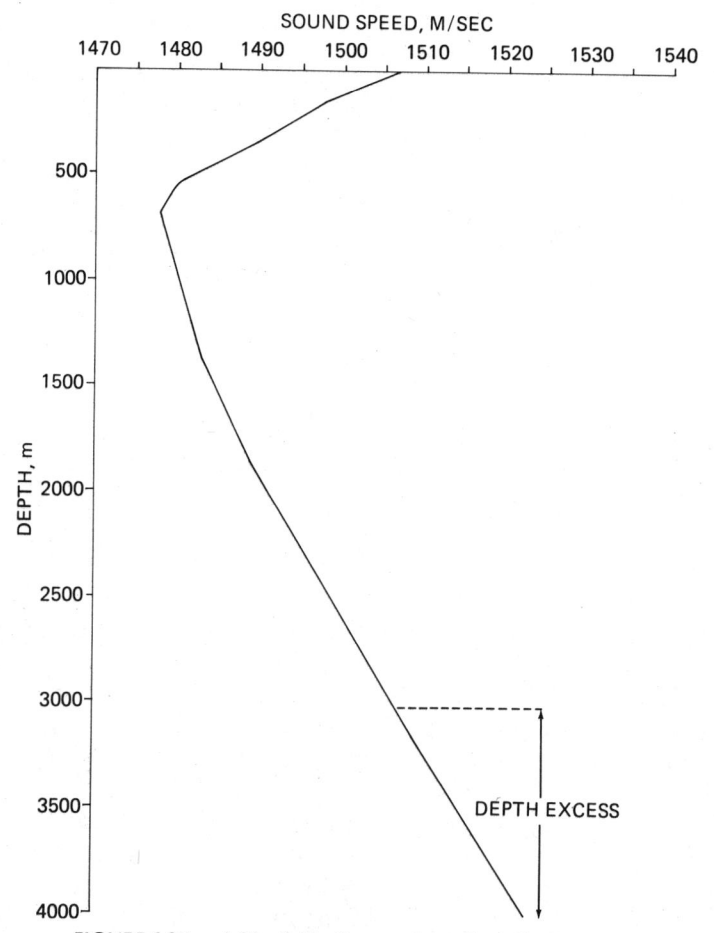

FIGURE 6.2.1. A North Pacific sound speed profile for June.

TABLE 6.2.1. Values of Sound Velocity Versus Depth for the North Pacific Profile

Depth (meters)	Sound Velocity (meters/second)
0.0	1507.2
150.0	1498.1
305.0	1491.7
533.0	1480.7
610.0	1478.9
686.0	1478.0
762.0	1478.6
1372.0	1483.2
1829.0	1488.6
3048.0	1507.5
4000.0	1523.0

the bottom is reached. We will call a profile with a single minimum a single channel profile. We also see from Figure 6.1.1 that at a depth of 3020 meters the sound speed increases to a value that equals the surface sound speed. The depth increment from this point, namely 3020 meters, to the bottom is called the depth excess. We will see shortly that a necessary condition for convergence zone propagation is that a sound velocity profile have depth excess.

The dominant feature of convergence zones is the focusing of energy and the consequent formation of caustics. The most prevalent type of caustic is the smooth caustic (discussed in Sections 5.7 and 5.8), since this can be formed at any depth. As we saw in Section 5.7, the cusp caustic, at least under the ideal conditions of the bilinear profile, is only formed at source depth and reciprocal source depth. The departure of a real profile from the ideal bilinear profile destroys some of this symmetry.

Figure 6.2.2 shows a ray trace for this profile for a source depth of 25 meters. Only those rays that are refracted in the water columns are shown. The rays that reflect off the bottom (the bottom bounce rays) are omitted from the ray trace. Consequently, the shadow zones seen in this figure are only shadow zones for the rays that are trapped in the water column. These shadow zones would be filled in by rays that reflect off the bottom.

The first convergence zone (CZ) is formed after the rays leaving the source are refracted once at depth. This deep refraction takes place at an approximate range of 15 nautical miles. We would say that the first CZ extends approximately from this first deep refraction at 15 nautical miles to the second deep refraction at approximately 45 nautical miles. We see that at different depths the CZ has a different structure. Below a depth of 200 meters, the first CZ splits into two distinct portions separated by a shadow zone.

Figure 6.2.3 shows the arrival angle structure versus range in the first CZ for a receiver at a depth of 200 meters. At this depth the CZ starts at 27.5 nautical

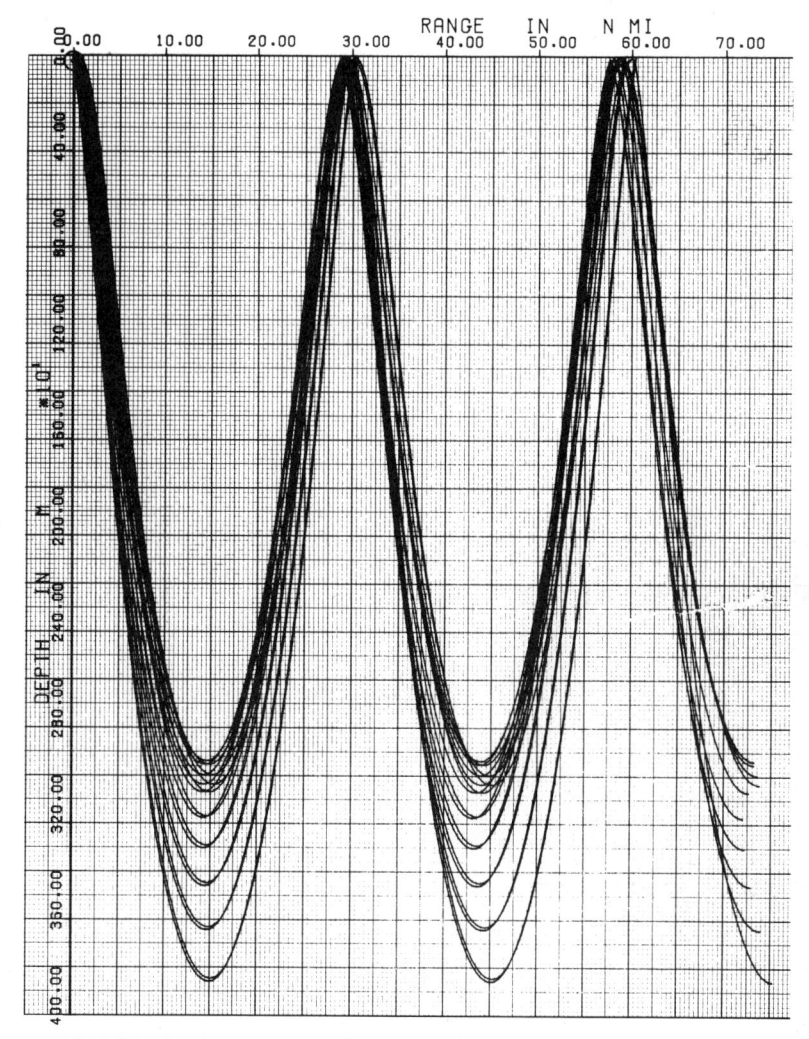

FIGURE 6.2.2. A ray trace for the North Pacific profile. Source depth is 25 m.

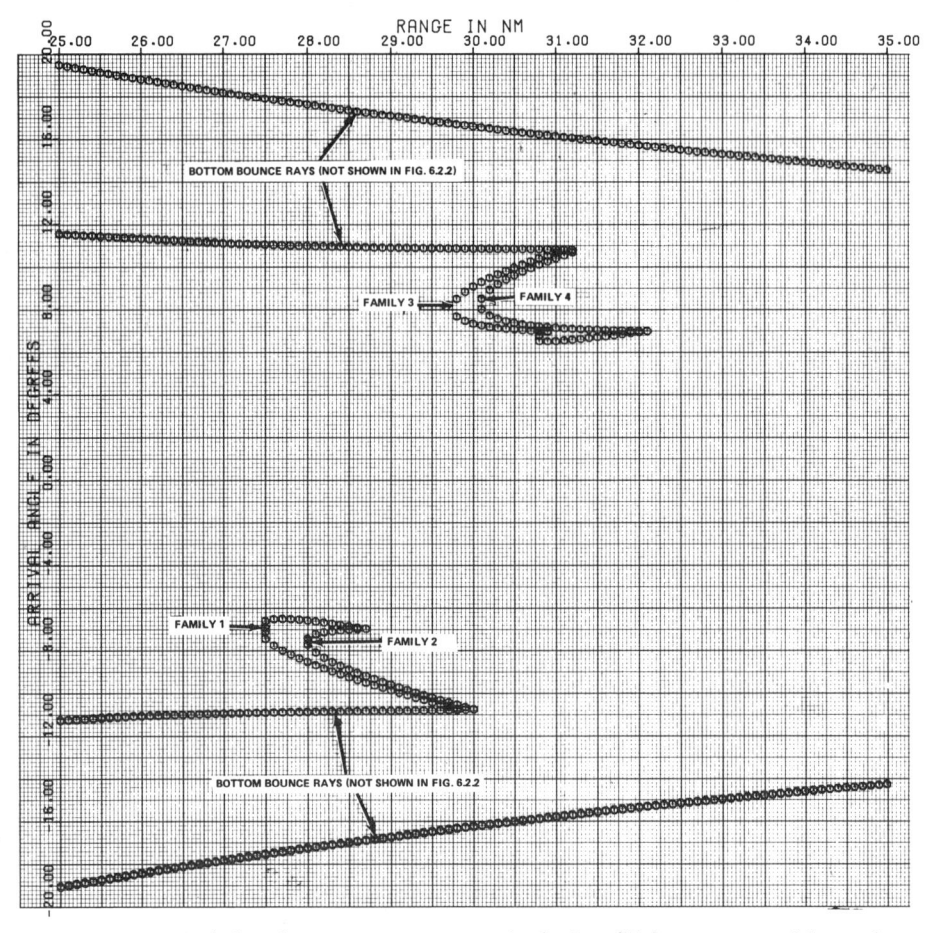

FIGURE 6.2.3. Arrival angle structure versus range in the first CZ for a source at 25 m and a receiver at 200 m.

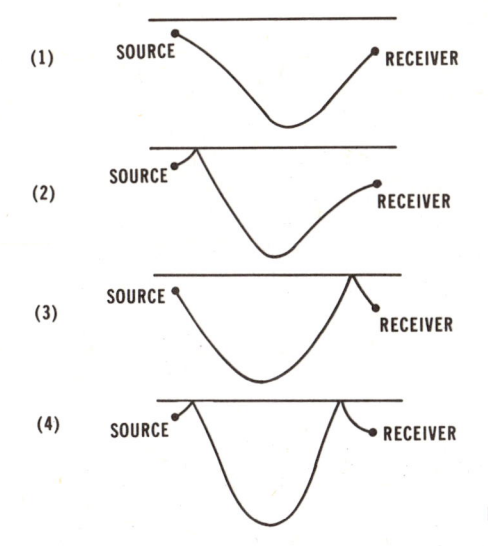

FIGURE 6.2.4. Ray families in the first CZ.

miles with family 1. A typical ray from each of the families is shown in Figure 6.2.4. We see that family 1 is a two-ray system behind the smooth caustic, which occurs at a range of 27.5 nautical miles at a receiver depth of 200 meters. We see that each family of rays forms its own smooth caustic.

Figures 6.2.5 through 6.2.7 illustrate the effect of changing the source depth on propagation. In Figure 6.2.5 the source is at a depth of 300 meters, in Figure 6.2.6 the source is at a depth of 3000 meters, and in Figure 6.2.7 this source is located at the sound velocity minimum at a depth of 686 meters. It is seen that the source depth has a profound effect on the propagation characteristics. Consequently, what we have said about a source at 25 meters will not be true at other depths. The arrival angle structure, for example, may be considerably different for other source depths.

The statement that we made above concerning shadow zones was not strictly true. We said that the only energy that gets into the shadow zones shown in Figure 6.2.2 was due to rays that have reflected off the bottom. This statement is only true in the ray theory approximation. For low frequencies where ray theory breaks down, diffracted energy also gets into the shadow zones. However, we need normal mode theory to correctly account for this energy. So let us examine the normal mode solution for this profile.

Table 6.2.1 gives the values of sound speed versus depth that were used in the normal mode solution. The density of the seawater was taken to be 1000 kg/m^3. For a frequency of 15 Hz, there were 14 modes trapped in the water column. Recall that the phase velocity for the nth mode c_n is related to the eigenvalue κ_n by the relation $c_n = \omega/\kappa_n$. Table 6.2.2 gives the eigenvalues and corresponding phase velocities for these 14 modes. The phase velocities of the modes trapped in the water column lie between the minimum sound speed and the maximum sound speed in the water column. The eigenvalues κ_n are solutions of the characteristic equation (4.1.37). We saw in Section 4.1 that the

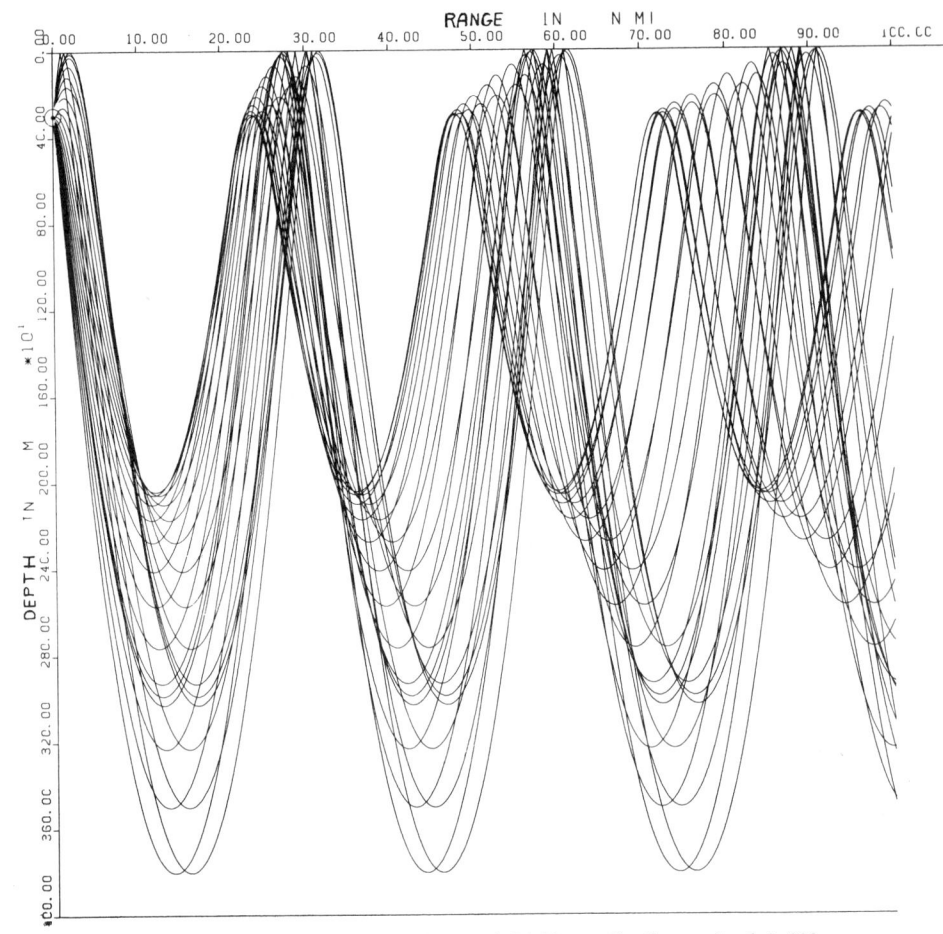

FIGURE 6.2.5. A ray trace for the North Pacific profile. Source depth is 300 m.

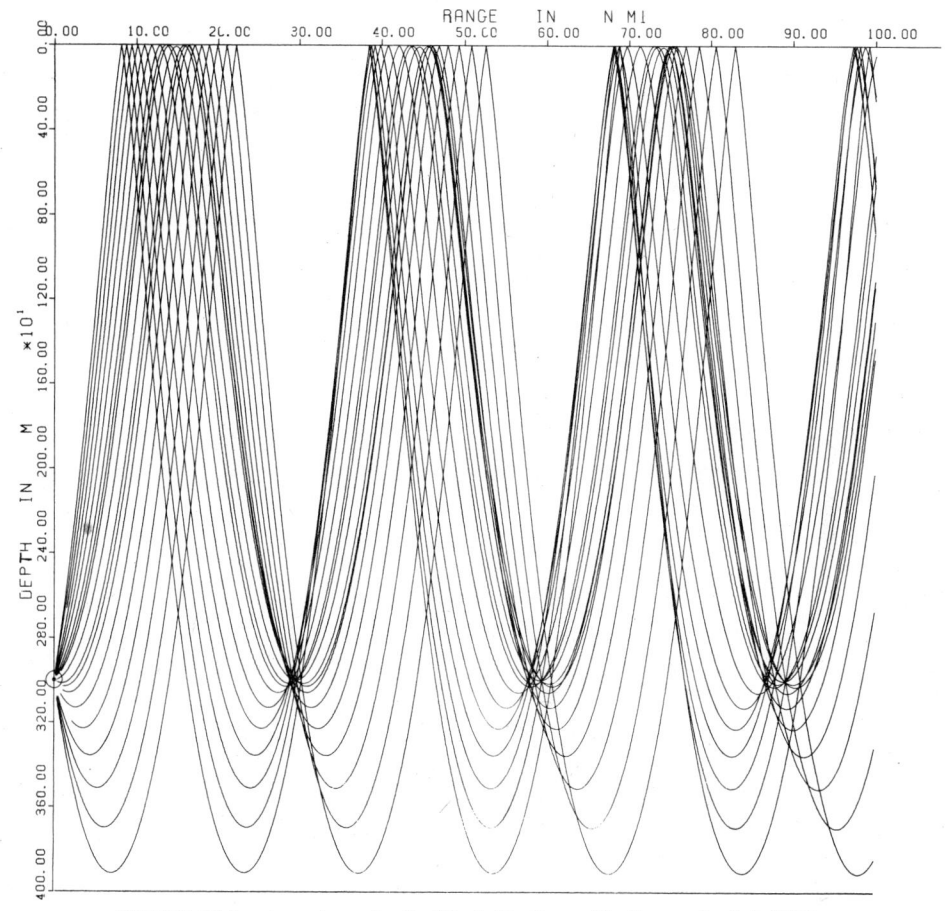

FIGURE 6.2.6. A ray trace for the North Pacific profile. Source depth is 3000 m.

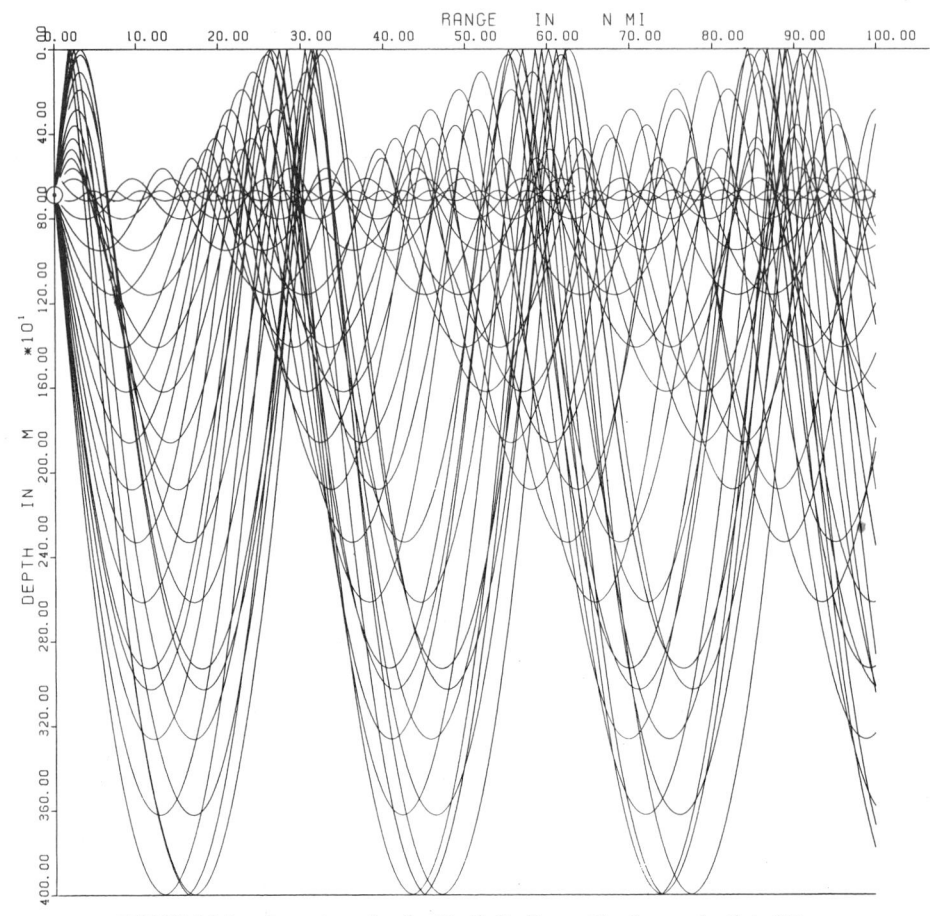

FIGURE 6.2.7. A ray trace for the North Pacific profile. Source depth is 686 m.

CONVERGENCE ZONE PROPAGATION   251

TABLE 6.2.2. Eigenvalues and Phase Velocities for Modes Trapped in the Water Column at a Frequency of 15 Hz

Mode No.	Eigenvalues (1/meters)	Phase Velocity (meters/second)
1	0.6363781932426366D − 01	0.1481002658614974D + 04
2	0.6347346140104315D − 01	0.1484837560885642D + 04
3	0.6331965981718260D − 01	0.1488444187473643D + 04
4	0.6317475405052120D − 01	0.1491858275100262D + 04
5	0.6303540011497052D − 01	0.1495156363500429D + 04
6	0.6290321516131378D − 01	0.1498298288346592D + 04
7	0.6277552466854680D − 01	0.1501345948207040D + 04
8	0.6265102758083363D − 01	0.1504329350162588D + 04
9	0.6252885249331866D − 01	0.1507268658380776D + 04
10	0.6240857845202955D − 01	0.1510173472067427D + 04
11	0.6228999439031204D − 01	0.1513048452326594D + 04
12	0.6217376778540618D − 01	0.1515876919877072D + 04
13	0.6206220954032581D − 01	0.1518601743408040D + 04
14	0.6195863505537230D − 01	0.1521140346675888D + 04

eigenvalues are determined by the frequency, the nature of the sound velocity profile (and hence water depth), and the boundary conditions at the surface and bottom.

The mode depth functions $\psi_n(z)$ are also determined by the same quantities that determine the eigenvalues. Figure 6.2.8 shows the normalized depth modes $\psi_n(z)$ given by Eq. (4.1.16) and normalized according to Eq. (4.1.38). Recall from Section 5.2 how we associated a ray with a mode. We saw there that the modal phase velocity c_n was the velocity at which the associated ray became horizontal. Also recall that the associated ray was only a heuristic concept and was not fully equivalent to the mode since the ray did not account for any diffraction effects. The depth at which the associated ray becomes horizontal was called the turning point for the ray. In Section 5.3 we examined the wave solution in the vicinity of a turning point. We saw there that on the side of the turning point which contained the incident and totally reflected wave, the depth function was oscillatory in nature because of the interference of these two waves. On the other side of the turning point, the depth function was seen to be exponentially damped, indicating that in this region there is a nonpropagating diffraction field.

Now let us examine mode 1 in detail. The phase velocity for the first mode is $c_1 = 1481.0$ m/sec. Now all modes have a zero at the surface because of the pressure-release condition. Except for the zero at the surface, mode 1 has no other zeroes. Applying the results discussed in the last paragraph to mode 1, we see that mode 1 has two turning points. That is, the modal phase velocity $c_1 = 1481.0$ m/sec is realized twice in the water column. These two turning points are at depths of 520 and 1080 meters. From Figure 6.2.8 we see that for

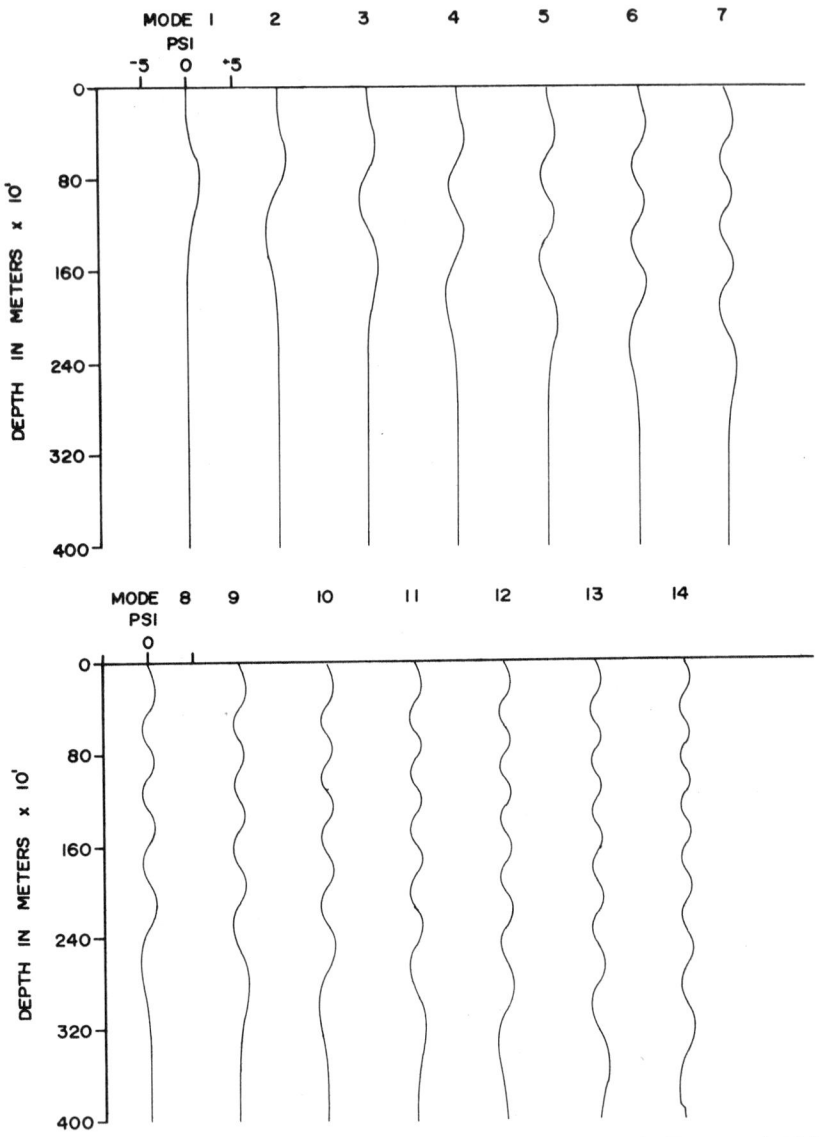

FIGURE 6.2.8. Normalized depth eigenfunctions $\psi_n(z)$ for the Pacific profile. Frequency is 15 Hz.

$0 \leq z < 520$ m and for 1080 m $< z \leq 4000$ m the depth function is exponentially damped and that between the turning points in the interval 520 m $< z < 1080$ m the depth function is oscillatory. Because the mode has two turning points, the associated ray will propagate down the oceanic waveguide by refracting up and down between these turning points as illustrated in Figure 6.2.9. Consequently, this ray never reflects off the surface or bottom. Modes with two turning points are called RR modes since the associated ray is refracted at the two turning points.

Now each succeeding mode has one more zero than the previous mode. The analysis that we applied to mode 1 can be applied to all the modes. We see from Figure 6.2.8 that modes 1 through 8 are RR modes, that is, they have two turning points. However, modes 9 through 14 have only one turning point since $c_9 = 1507.27$ m/sec and this is greater than the sound speed at the surface. Because modes 9 through 14 only have one deep turning point, the

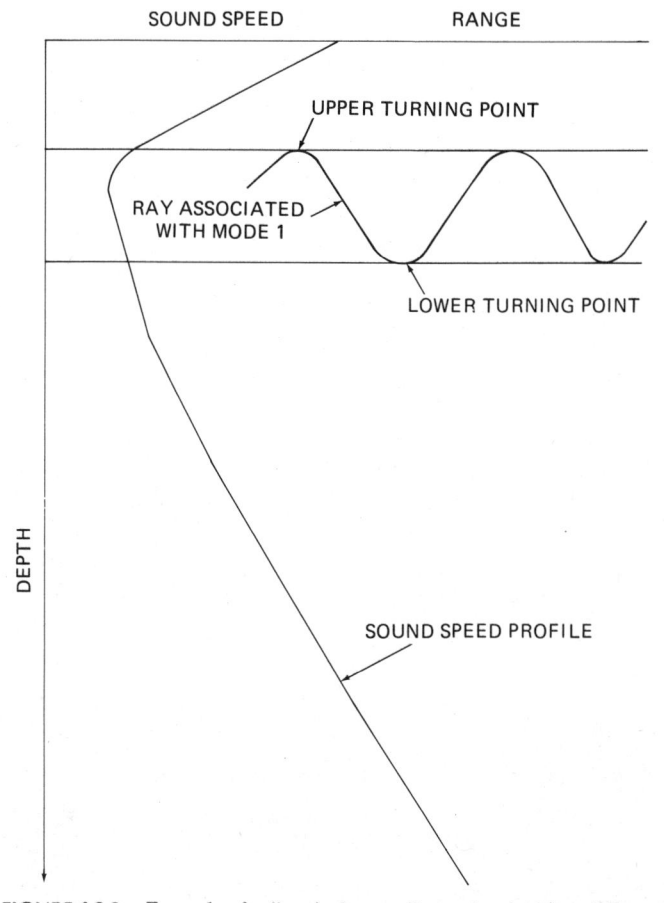

FIGURE 6.2.9. Example of a "equivalent ray" associated with an RR mode.

254 APPLICATION TO CONVERGENCE ZONE AND SURFACE DUCT PROPAGATION

associated ray after being refracted at depth must reflect off the surface in order to continue propagating down the oceanic wave guide. Consequently, these modes are called Refracted/Surface-Reflected modes (RSR).

Now let us examine transmission loss as a function of range. The transmission loss is computed from Eqs. (4.1.55) and (4.1.65). Note from Eq. (4.1.55) that the depth functions enter into the transmission loss computation as the product $\psi_n(z_s)\psi_n(z)$ where the one depth function is evaluated at source depth and the other depth function is evaluated at receiver depth. Consider mode 1 again. Our source depth of 25 meters is well above the turning point of mode 1. Looking at Figure 6.2.8, we see that, for this source depth, mode 1 will be very weakly excited. In order for mode 1 to be strongly excited, the source would have to be placed between its turning points where the mode peaks up.

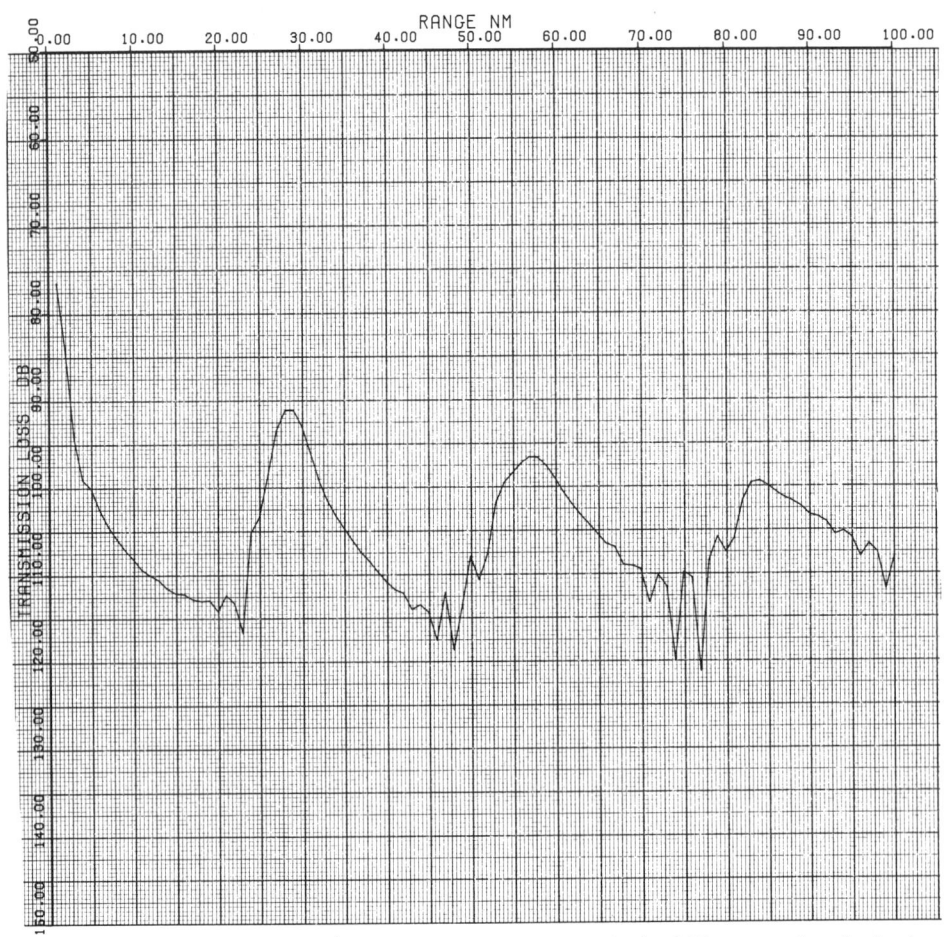

FIGURE 6.2.10. Transmission loss versus range for a source depth of 25 m, a receiver depth of 200 m, and a frequency of 15 Hz.

Figure 6.2.10 shows transmission loss versus range for a source depth of 25 meters, a receiver depth of 200 meters, and a frequency of 15 Hz. We see that the transmission loss is smallest or intensity greatest at the ranges we have said were convergence zones. This increase of energy is of course due to the focusing of energy. Also notice that considerable energy gets diffracted into the ray theory shadow zones between the CZs. Figure 6.2.11 shows the transmission loss versus range for modes 9 through 14 only. We see that these are the modes that form the CZs. The modal phase velocities c_n for modes 9 through 14 are all greater than the surface sound speed. Thus, if there were no depth excess, there would be no CZ formation.

Figure 6.2.12 shows transmission loss versus range for modes 1 through 8 only. These modes basically fill in between CZs. For these source and receiver

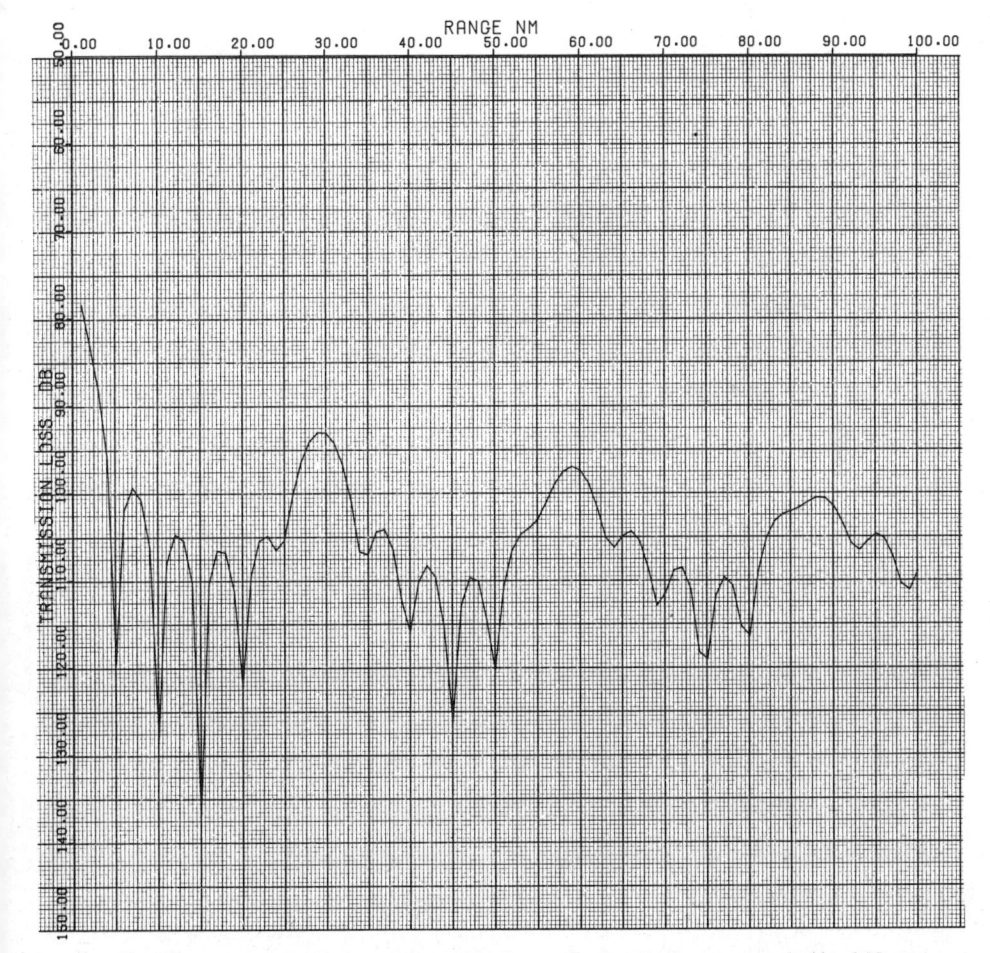

FIGURE 6.2.11. Transmission loss versus range for modes 9 to 14 for a source depth of 25 m, a receiver depth of 200 m, and a frequency of 15 Hz.

depths, these modes also exhibit another effect. Even though these modes are RR modes and the source depth is above the upper turning point, the receiver depth is such that some of these modes are weakly coupled (between source and receiver).

Figures 6.2.13 through 6.2.17 show examples of transmission loss for other frequencies and source and receiver depths. Figures 6.2.10, 6.2.13, and 6.2.14 should be compared. They are for the same source and receiver depths, but with the frequency increasing from 15 to 100 to 300 Hz, respectively.

Figures 6.2.13, 6.2.15, and 6.2.16 should be compared. These are all for the same frequency of 100 Hz and receiver depth of 200 meters. The source depth varies from 25 meters in Figure 6.2.13, to 300 meters in Figure 6.2.15, to 3000

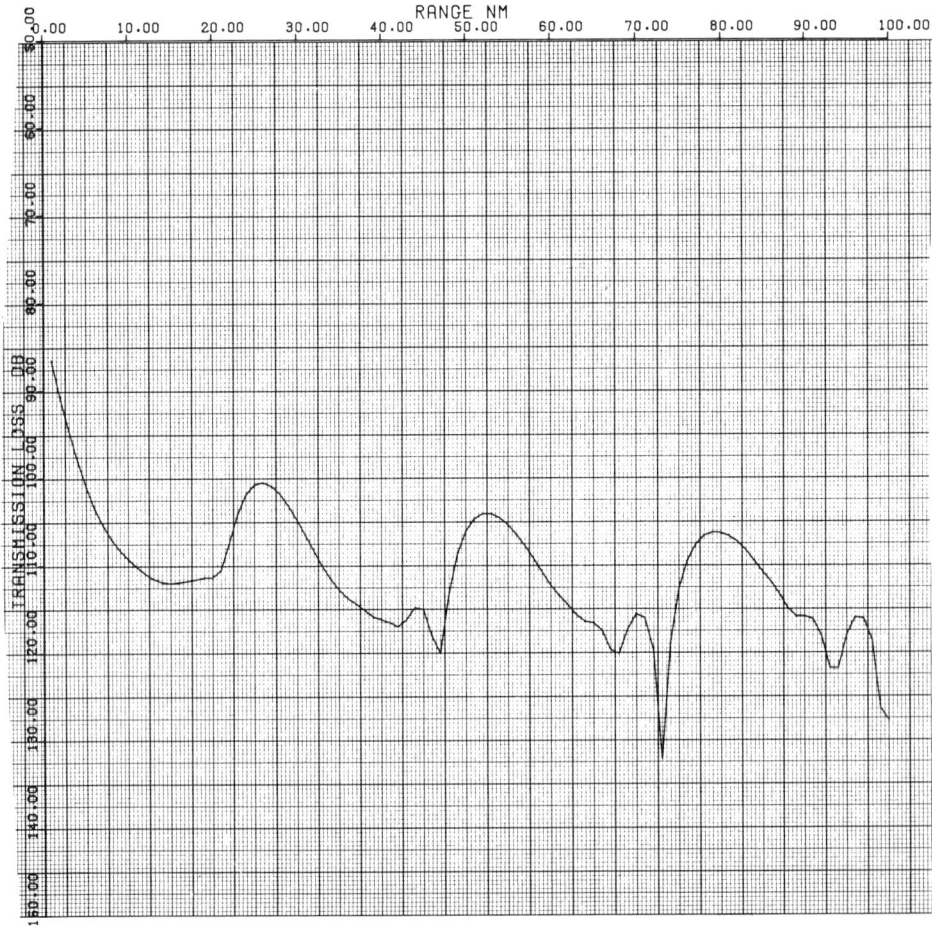

FIGURE 6.2.12. Transmission loss versus range for modes 1 to 8 for a source depth of 25 m, a receiver depth of 200 m, and a frequency of 15 Hz.

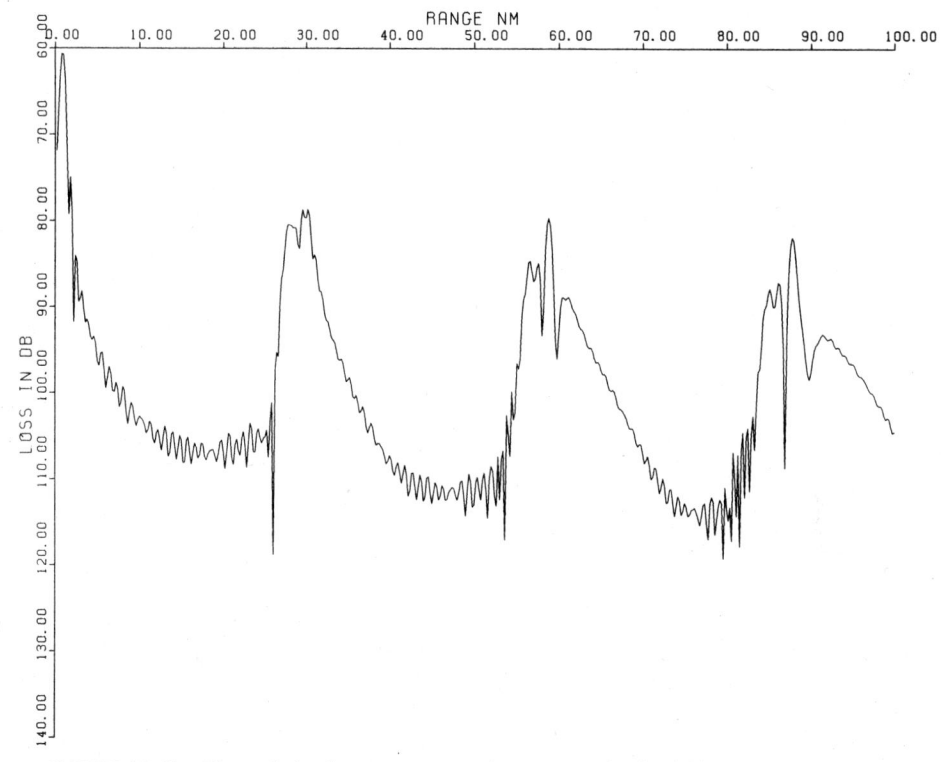

FIGURE 6.2.13. Transmission loss versus range for a source depth of 25 m, a receiver depth of 200 m, and a frequency of 100 Hz.

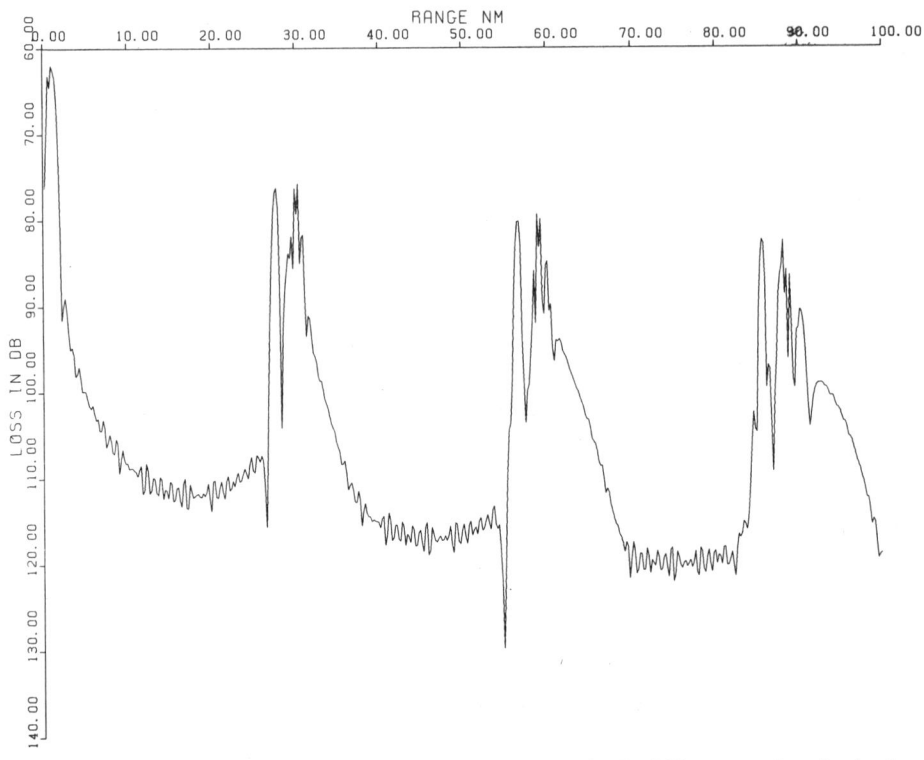

FIGURE 6.2.14. Transmission loss versus range for a source depth of 25 m, a receiver depth of 200 m, and a frequency of 300 Hz.

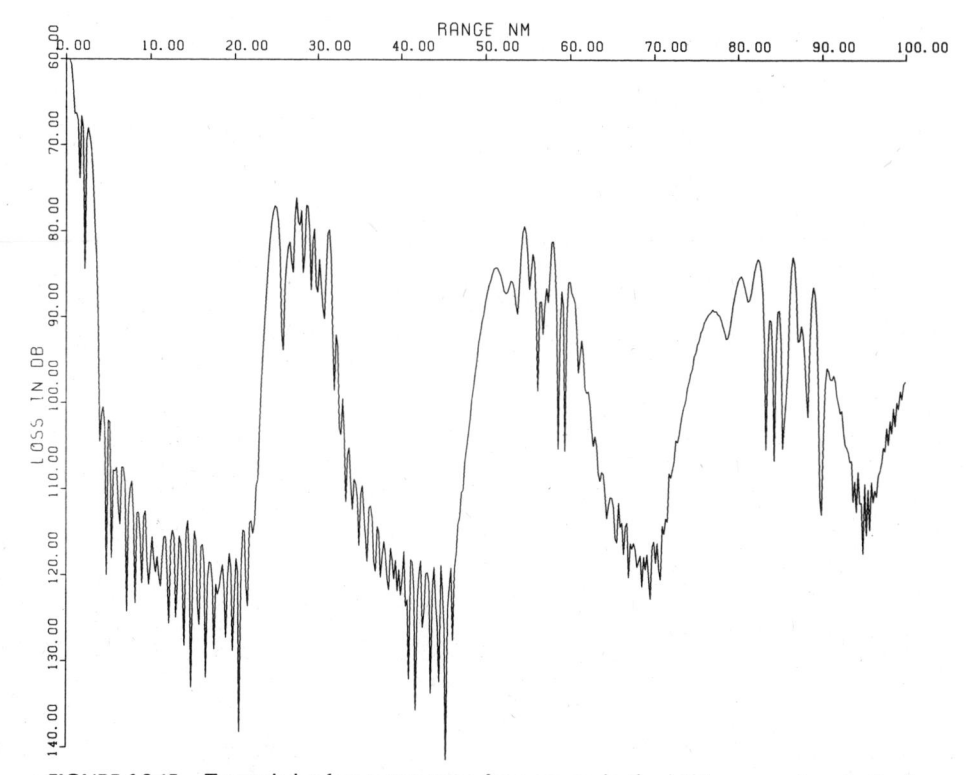

FIGURE 6.2.15. Transmission loss versus range for a source depth of 300 m, a receiver depth of 200 m, and a frequency of 100 Hz.

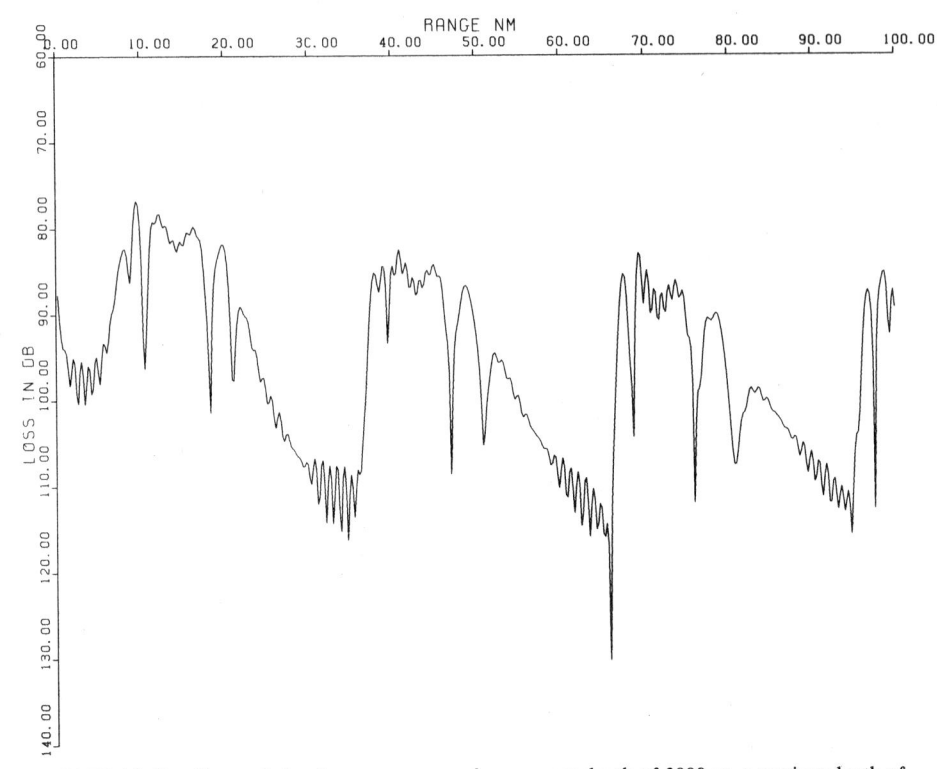

FIGURE 6.2.16. Transmission loss versus range for a source depth of 3000 m, a receiver depth of 200 m, and a frequency of 100 Hz.

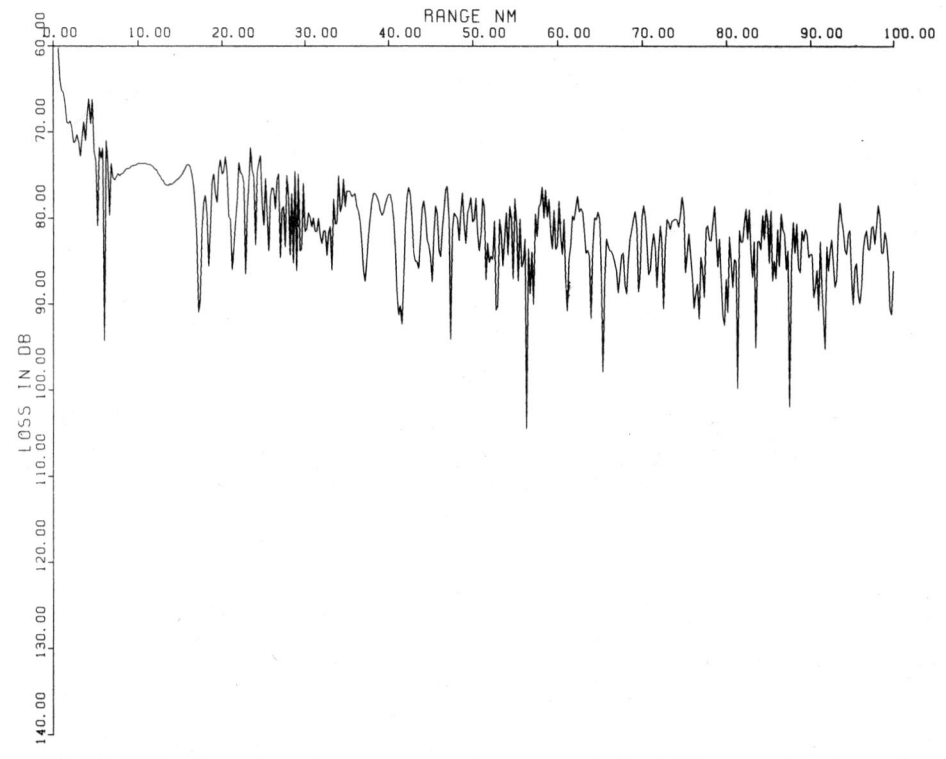

FIGURE 6.2.17. Transmission loss versus range for a source depth of 686 m, a receiver depth of 686 m, and a frequency of 100 Hz.

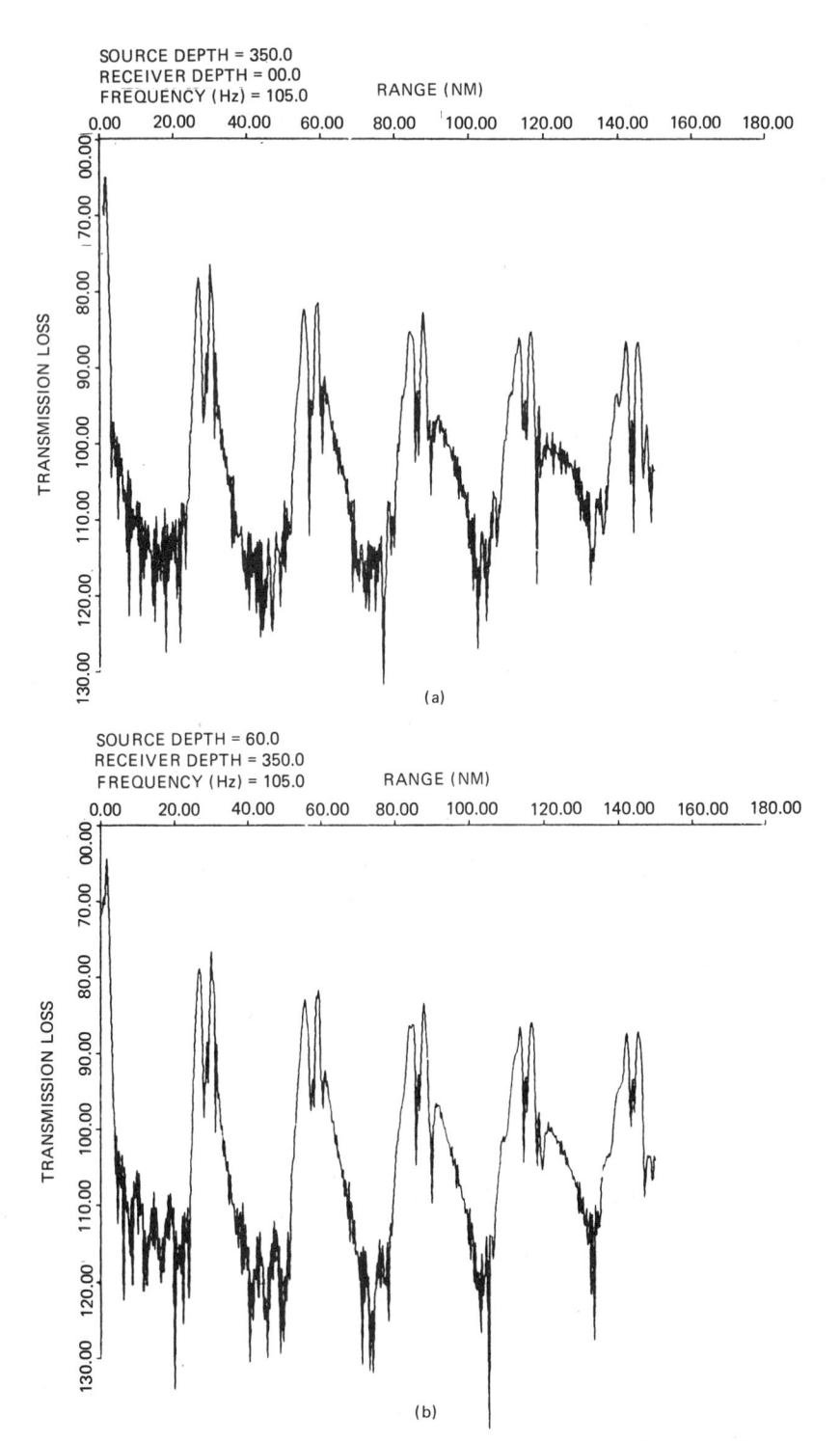

FIGURE 6.2.18. Comparison of (a) WKB with (b) normal mode transmission loss for a source depth of 60 m, a receiver depth of 350 m, and a frequency of 105 Hz.

meters in Figure 6.2.16. Notice how the CZs broaden and move into a smaller range.

Finally, Figure 6.2.17 is for a source and receiver at the sound velocity minimum of 686 meters and for a frequency of 100 Hz. Notice how the transmission loss is filled in at all ranges. If we refer to the ray trace of Figure 6.2.7, we see this fill-in is due to RR rays that cycle down the minimum axis.

Finally, let us show a comparison between the WKB solution for the Pacific profile and the normal mode solution. Figure 6.2.18a shows the WKB solution and Figure 6.2.18b shows the normal mode solution for a frequency of 105 Hz, a source depth of 60 meters, and a receiver depth of 350 meters. This comparison is due to Henrick [5.5]. It is seen that the WKB approximation is extremely good, even at the low frequency of 105 Hz.

6.3. CONVERGENCE ZONE PROPAGATION—A DOUBLE CHANNEL NORTH ATLANTIC PROFILE

A sound velocity profile common to many areas of the North Atlantic in summer is the double channel profile. One variation of the double channel profile is shown in Figure 6.3.1, and the values are tabulated in Table 6.3.1. Note that this profile has two relative minima. The first minimum occurs at a depth of 100 meters and the second minimum at a depth of 1300 meters.

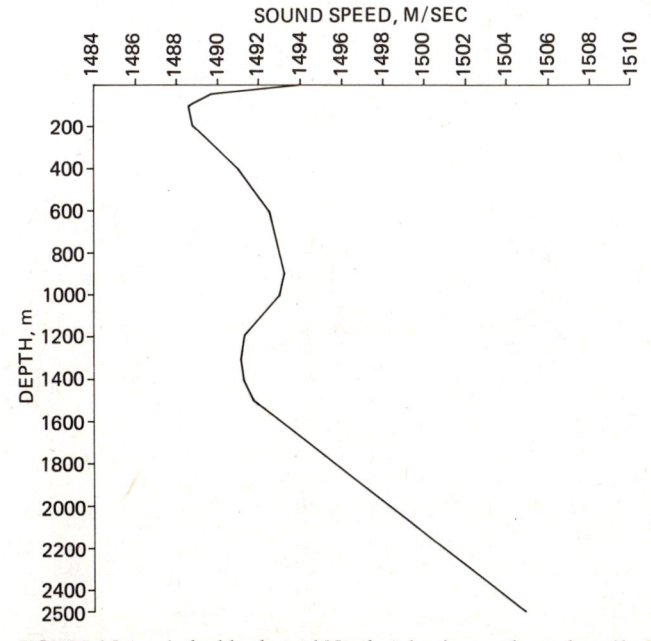

FIGURE 6.3.1. A double channel North Atlantic sound speed profile.

TABLE 6.3.1. Values of Sound Velocity versus Depth for the Double Channel North Atlantic Profile

Depth (meters)	Sound Velocity (meters/second)
0	1495.0
50	1489.8
100	1488.6
200	1488.8
400	1491.0
600	1492.5
900	1493.3
1000	1493.0
1200	1491.3
1300	1491.2
1400	1491.3
1500	1491.7
2500	1505.0

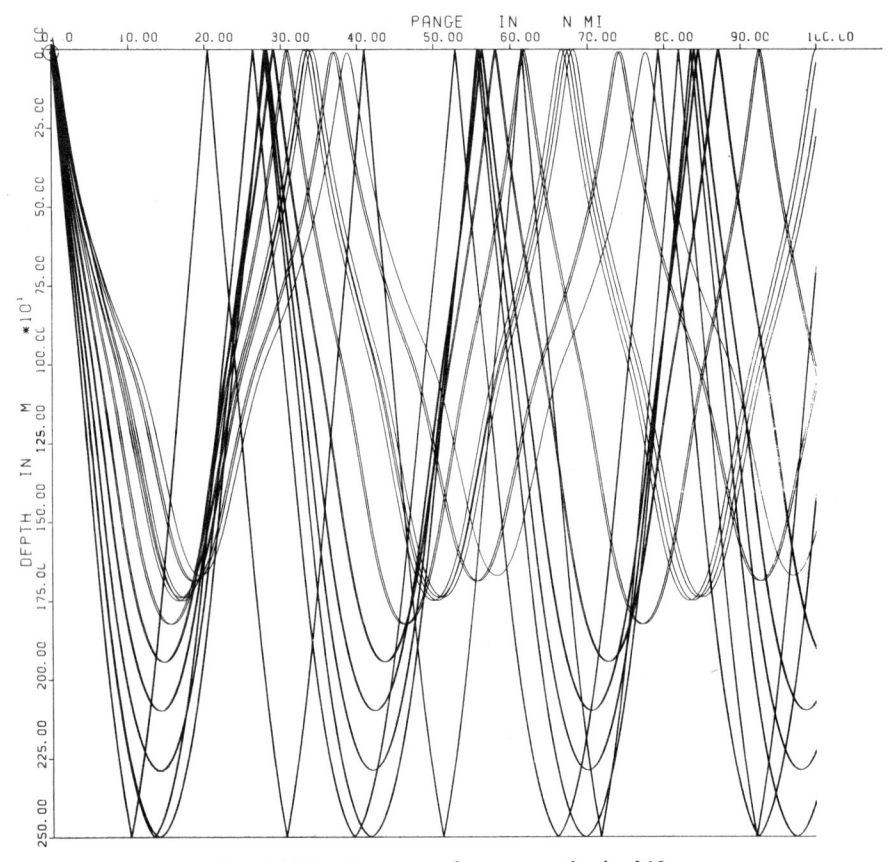

FIGURE 6.3.2. A ray trace for a source depth of 10 m.

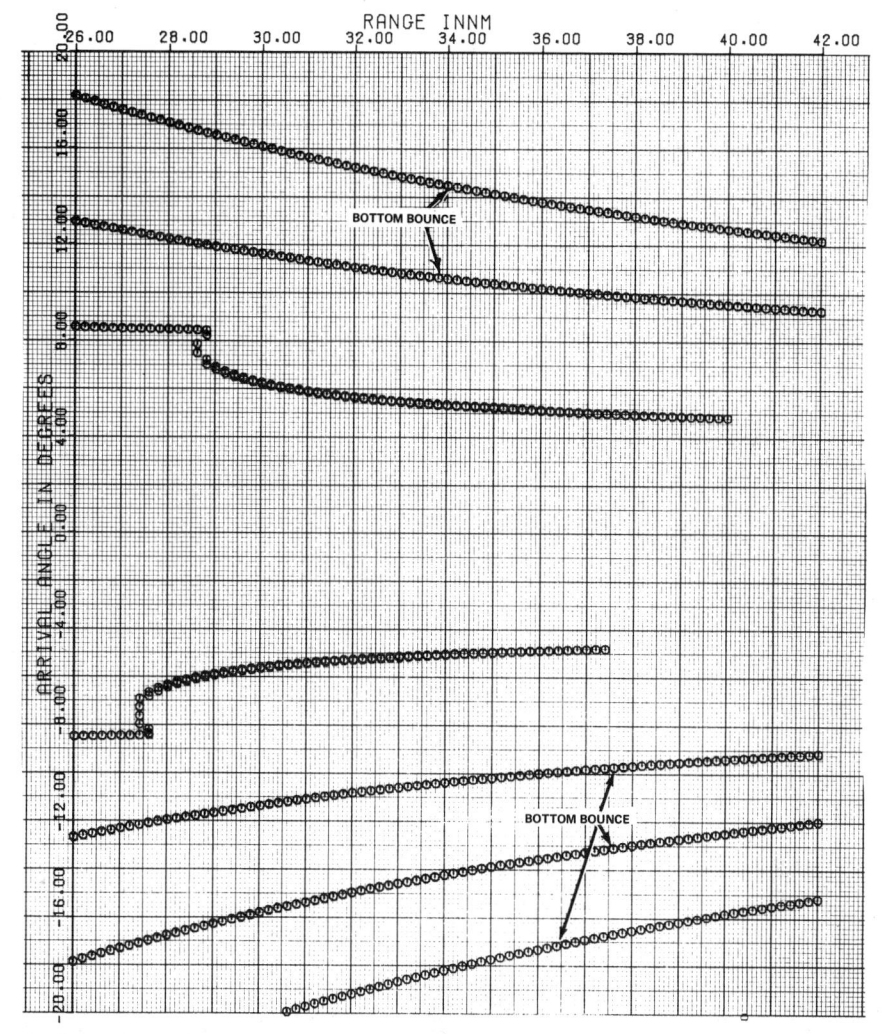

FIGURE 6.3.3. Arrival angle structure versus range in the first CZ for a source depth of 10 m and a receiver depth of 150 m.

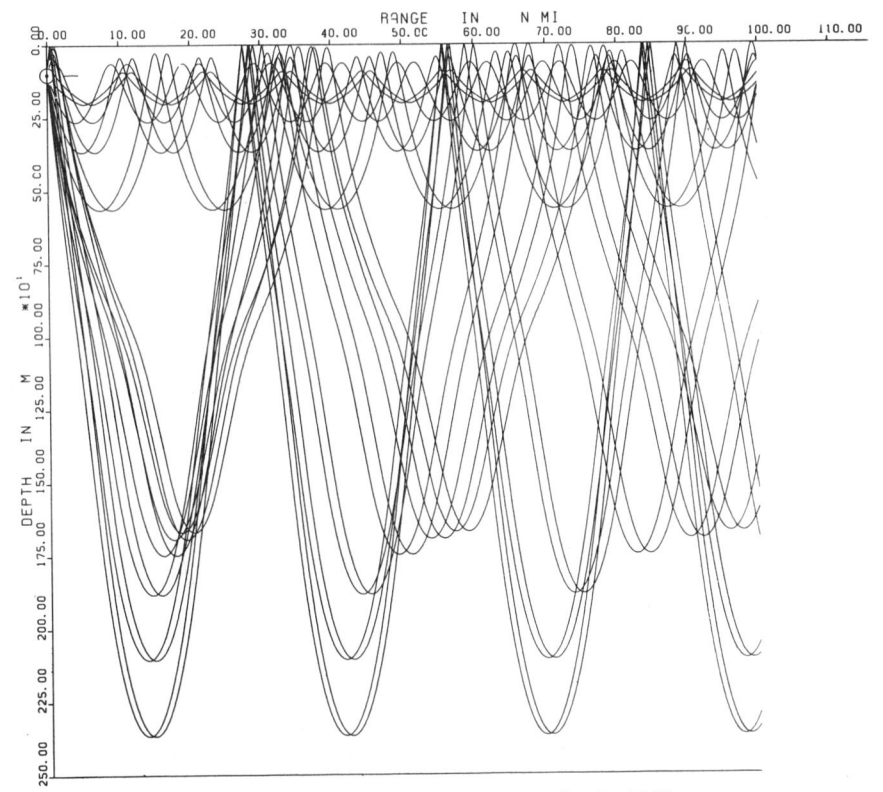

FIGURE 6.3.4. A ray trace for a source depth of 100 m.

Because of the two minima, we speak of this profile as a double channel profile. Note for this profile there is also depth excess, giving rise to CZ formation.

Figure 6.3.2 shows a ray trace for a source depth of 10 meters. Comparing this ray trace with that of Figure 6.2.2 for the North Pacific profile, we see that, because of the double channel nature of the North Atlantic profile, the CZ structure is quite different. Notice in particular how much wider the second CZ is in the double channel case. Figure 6.3.3 shows the arrival angle structure of the first CZ for a receiver depth of 150 meters.

If the source is placed at the upper minimum in the sound speed profile, namely at 100 meters, the acoustic energy couples strongly into the upper channel. Figure 6.3.4 shows a ray trace for a source at 100 meters. Note that the upper channel is filled in with RR rays between CZs.

Table 6.3.1 gives the values of the sound speed versus depth that were used for the normal mode calculations. Again, water density was taken as 1000 kg/m^3. For a frequency of 60 Hz, there are 23 modes trapped in the water column. The eigenvalues and corresponding phase velocities for these modes

TABLE 6.3.2. Eigenvalues and Phase Velocities for Modes Trapped in the Water Column at a Frequency of 60 Hz

Mode No.	Eigenvalues (1/meters)	Phase Velocity (meters/second)
1	0.2530604417110248D + 00	0.1489727576075559D + 04
2	0.2527600863323288D + 00	0.1491497811701953D + 04
3	0.2527339450573399D + 00	0.1491652094241809D + 04
4	0.2525751207745474D + 00	0.1492590074884231D + 04
5	0.2525530658560275D + 00	0.1492720419579805D + 04
6	0.2524466522518557D + 00	0.1493349644639640D + 04
7	0.2523894294230716D + 00	0.1493688223363894D + 04
8	0.2522917797762370D + 00	0.1494266356062559D + 04
9	0.2521979876739435D + 00	0.1494822071769153D + 04
10	0.2520930521736507D + 00	0.1495444301936137D + 04
11	0.2519846013526383D + 00	0.1496087921274195D + 04
12	0.2518700222014145D + 00	0.1496768512329365D + 04
13	0.2517547714238338D + 00	0.1497453717753392D + 04
14	0.2516370737477298D + 00	0.1498154118612566D + 04
15	0.2515178443202289D + 00	0.1498864303046410D + 04
16	0.2513972157423760D + 00	0.1499583506991200D + 04
17	0.2512764732762919D + 00	0.1500304081457930D + 04
18	0.2511556188749992D + 00	0.1501026017731281D + 04
19	0.2510336189825935D + 00	0.1501755501747818D + 04
20	0.2509133237154431D + 00	0.1502475487743788D + 04
21	0.2507934882699608D + 00	0.1503193408374990D + 04
22	0.2506785278494146D + 00	0.1503882768360751D + 04
23	0.2505665120212196D + 00	0.1504555079566456D + 04

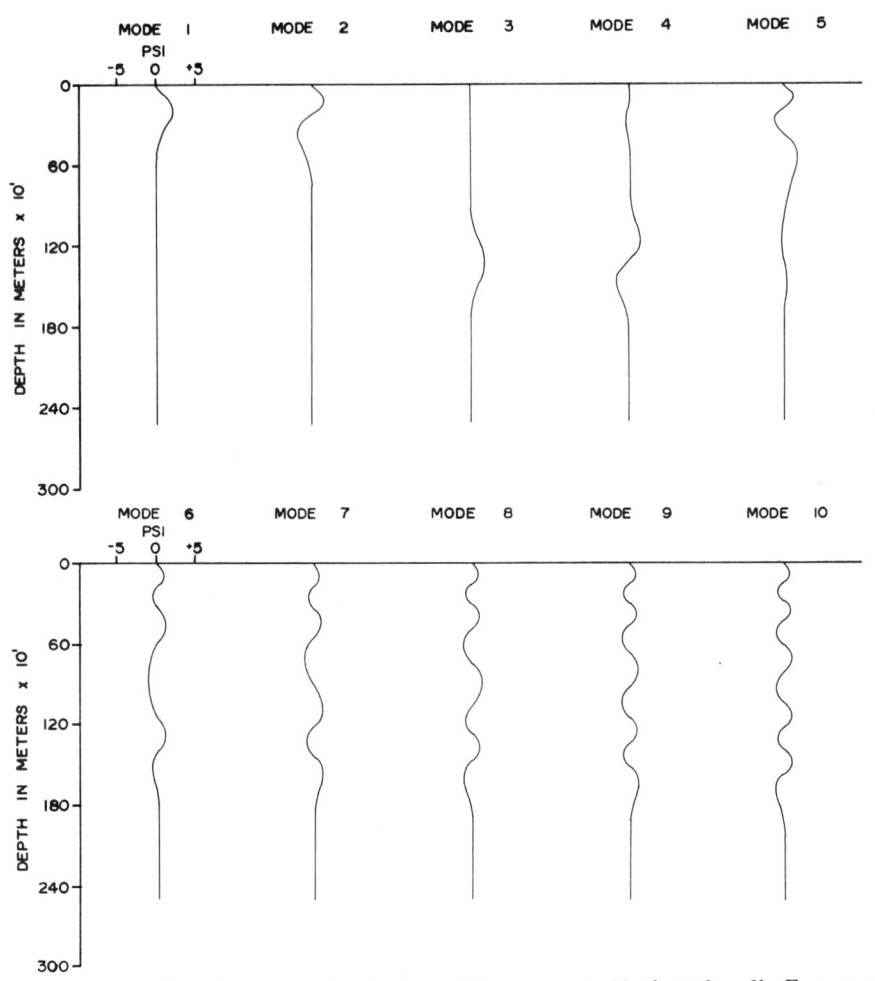

FIGURE 6.3.5. Normalized depth eigenfunctions $\psi_n(z)$ for the double channel profile. Frequency is 60 Hz.

are shown in Table 6.3.2. Examples of the mode shapes are shown in Figure 6.3.5.

The phase velocities of modes 1 and 2 lie in region I, as shown in Figure 6.3.6. These modes are trapped in the upper channel only. Note that these modes have two turning points and so are RR modes. Modes 3, 4, and 5 fall in region II, as shown in Figure 6.3.6. Mode 3 is trapped entirely in the lower channel. Mode 4 is strongly trapped in the lower channel and very weakly trapped in the upper channel. In contrast, mode 5 is strongly trapped in the upper channel and weakly trapped in the lower channel. Modes 3 through 5 are also RR modes. Modes 6 through 9 lie in region III, as shown in Figure 6.3.6. These modes are also RR modes. Finally, modes 10 through 23 lie in region IV, as shown in Figure 6.3.6. Modes 10 through 23 are RSR modes.

Figure 6.3.7 shows transmission loss versus range for a source depth of 10 meters, a receiver depth of 150 meters, and a frequency of 60 Hz. Figure 6.3.8 shows the transmission loss due to modes 1 through 9, and Figure 6.3.9 shows the transmission loss due to the CZ forming modes 10 through 23.

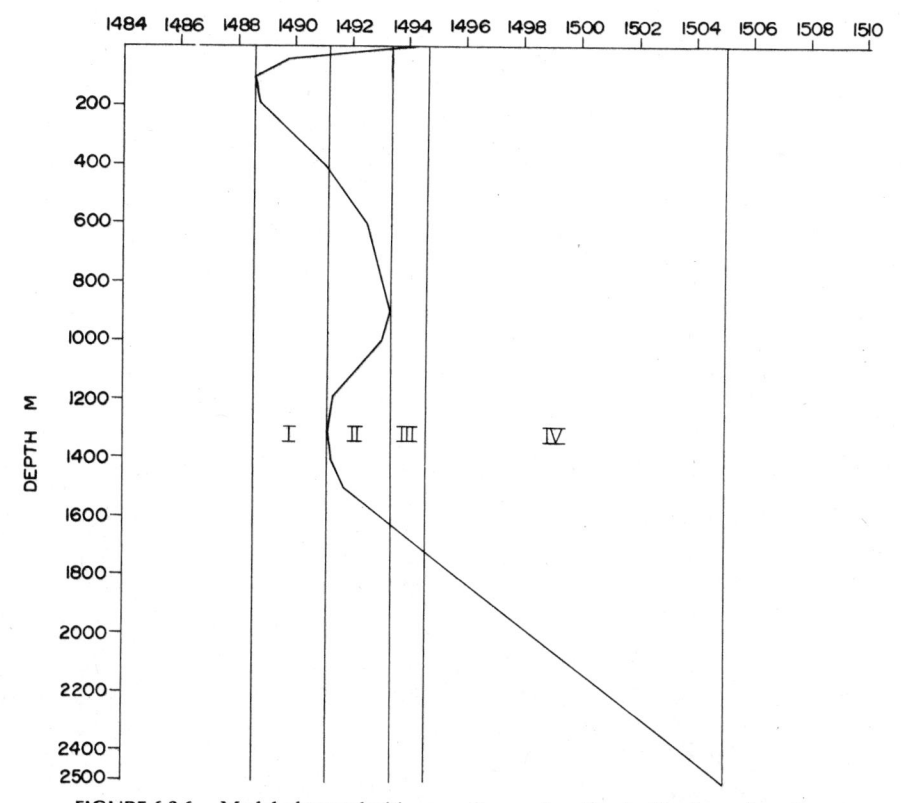

FIGURE 6.3.6. Modal phase velocities superimposed on the double channel profile.

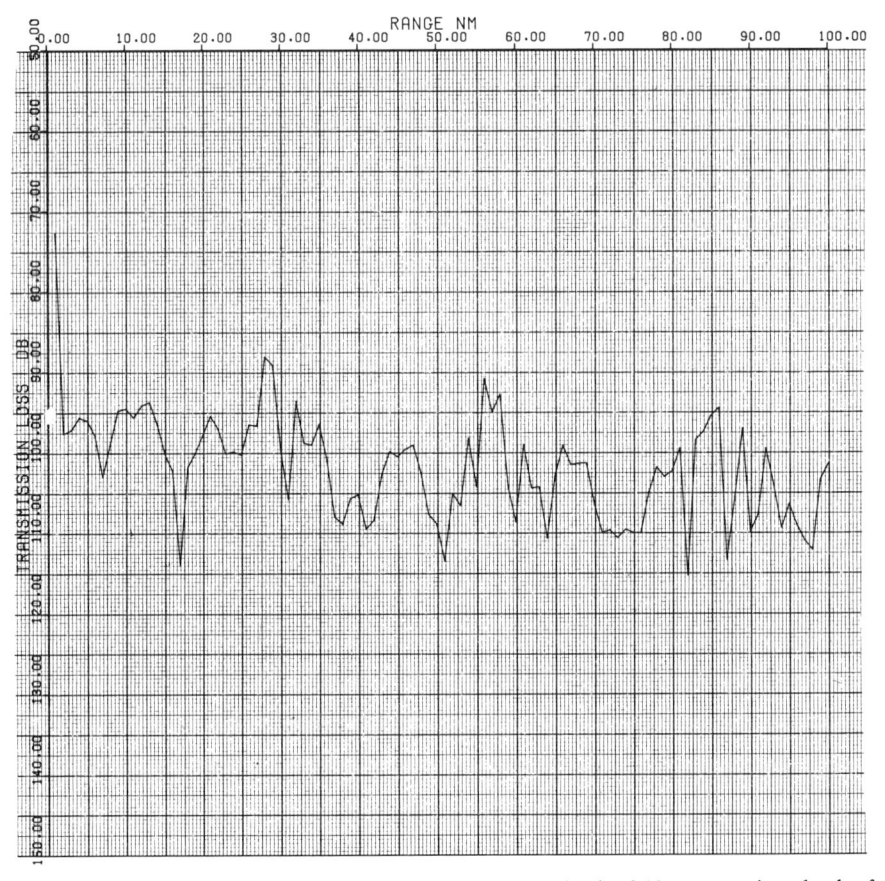

FIGURE 6.3.7. Transmission loss versus range for a source depth of 10 m, a receiver depth of 150 m, and a frequency of 60 Hz.

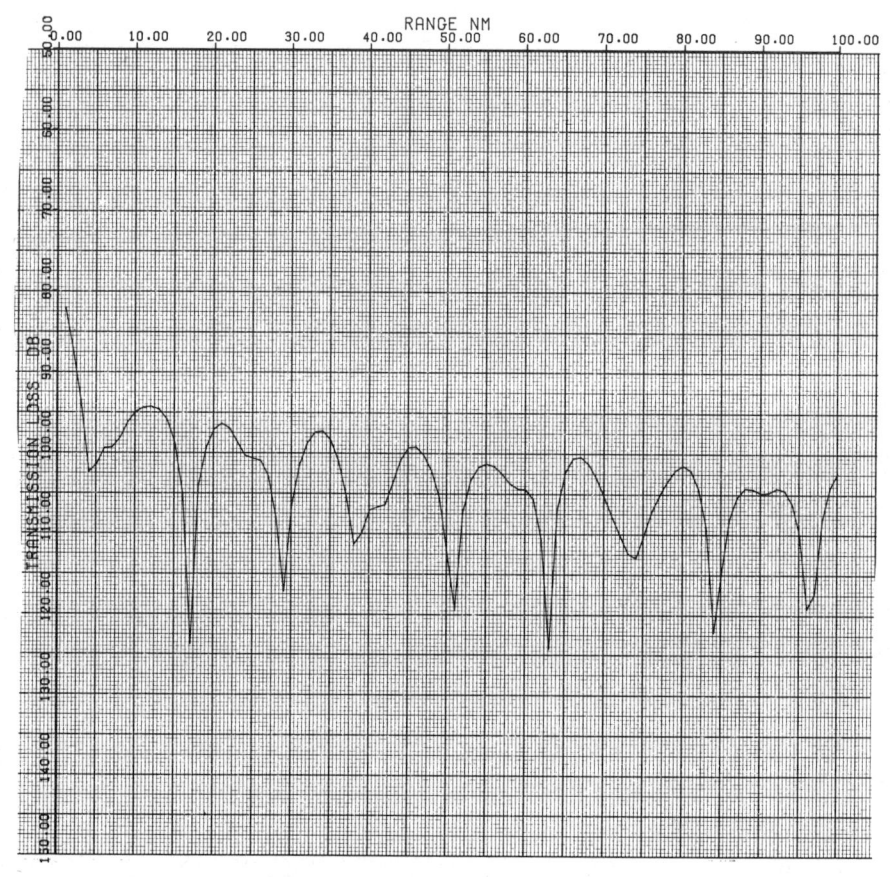

FIGURE 6.3.8. Transmission loss versus range for modes 1 to 9 for a source depth of 10 m, a receiver depth of 150 m, and a frequency of 60 Hz.

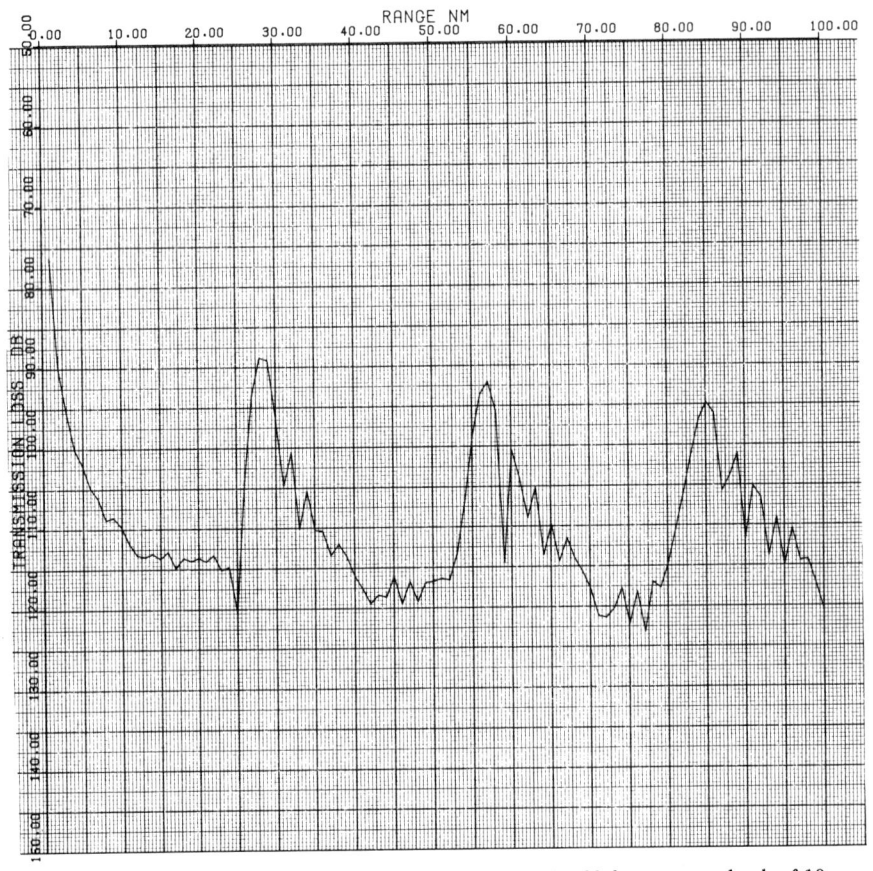

FIGURE 6.3.9. Transmission loss versus range for modes 10 to 23 for a source depth of 10 m, a receiver depth of 150 m, and a frequency of 60 Hz.

6.4. SURFACE DUCT PROPAGATION—A NORTH ATLANTIC PROFILE

In certain areas of the world's oceans, the temperature profile shows the presence of an isothermal layer beneath the sea surface. This layer of isothermal water is created and maintained by wind mixing. Being isothermal, this layer is simply called a mixed layer. The sound velocity in this layer increases with depth because of the pressure effect. In some areas this positive sound velocity gradient can extend to many hundreds of meters because of the existence of other oceanographic conditions that prevail below the wind-mixing layer and that also create a positive sound velocity gradient. With exceptions in some northern waters, the positive sound velocity gradient layer is followed by a layer with a negative sound speed gradient. Such a sound speed profile is illustrated in Figure 6.4.1.

If an acoustic source is placed in the layer with the positive sound speed gradient, a cone of rays will be completely trapped in this layer because of upward refractions. Some typical rays are shown in Figure 6.4.1. Because sound is trapped in this upper layer, it is acoustically referred to as a surface duct. In the ray theory approximation, all the energy that is put into the cone of rays trapped in the duct will remain in this cone as it propagates down the duct.

However, a full wave solution shows that energy can "leak" out of the duct due to diffraction effects as a wave propagates down the duct. The amount of leakage depends mainly on the frequency, duct thickness, and the sound speed gradient in and below the duct.

Now let us examine propagation in a surface duct in much more detail. Figure 6.4.2 shows a North Atlantic profile in March with a surface duct

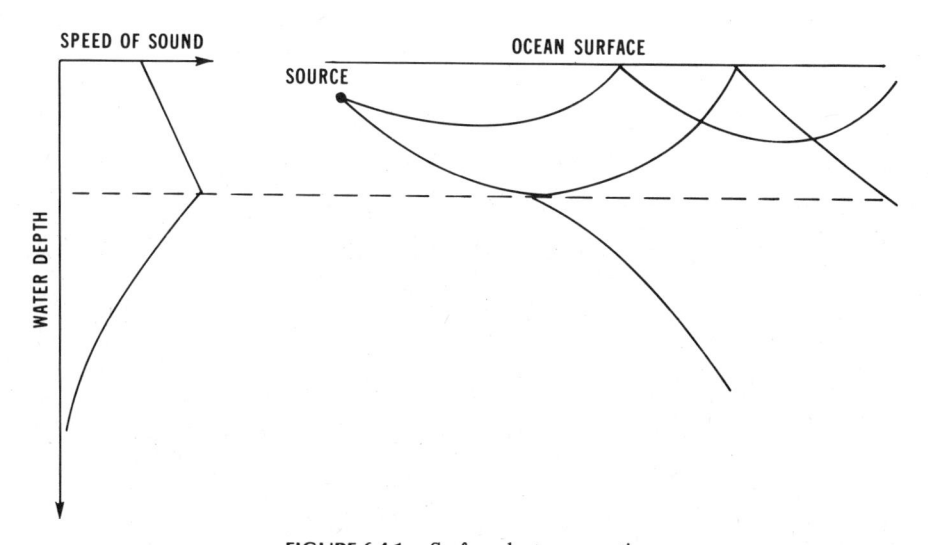

FIGURE 6.4.1. Surface duct propagation.

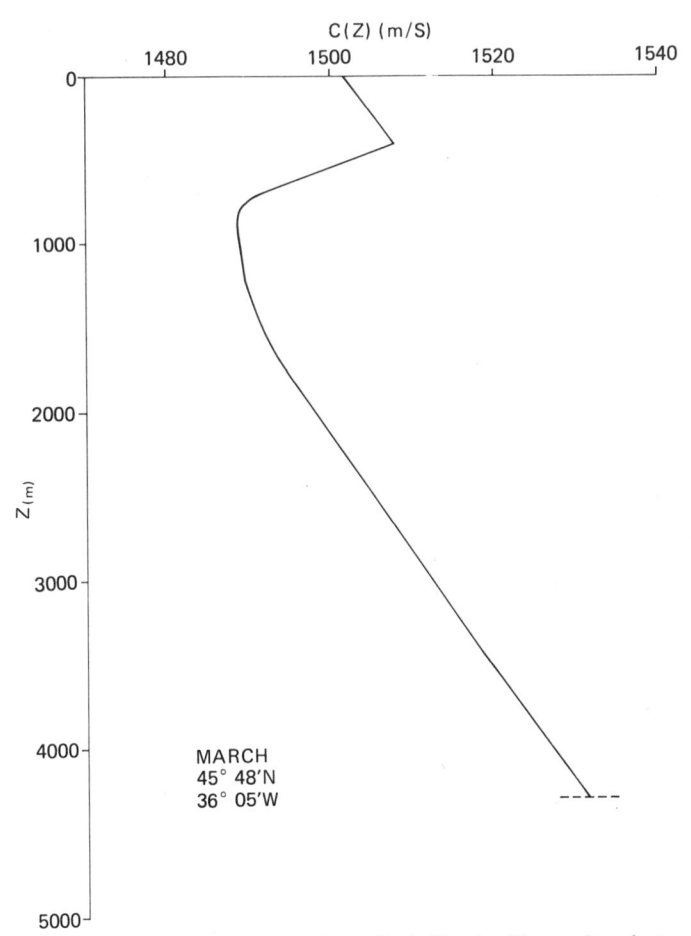

FIGURE 6.4.2. A North Atlantic profile in March with a surface duct.

SURFACE DUCT PROPAGATION 275

extending to a depth of 400 meters. Note that this sound speed profile also has considerable depth excess.

Figure 6.4.3 shows a ray trace for this profile for a source located at a depth of 100 meters. We see some of the rays trapped in the duct and those exiting the source at larger grazing angles forming CZs. The cone of angles trapped in the duct can easily be calculated from Snell's law.

Figure 6.4.4 shows the arrival angle structure versus range for a source depth of 100 meters and a receiver depth of 100 meters. At a range of about 30 nautical miles, we can see the four caustics appearing in the first CZ. We also see that the entire length of the duct is filled with caustics. In fact, there is a series of a smooth caustic followed by a cusp caustic. Note the three-ray systems behind each cusp as was described in Section 5.7 and illustrated in Figure 5.7.10.

Figure 6.4.5 shows the arrival angle structure versus range for a different source and receiver depth. The source depth is 50 meters and the receiver

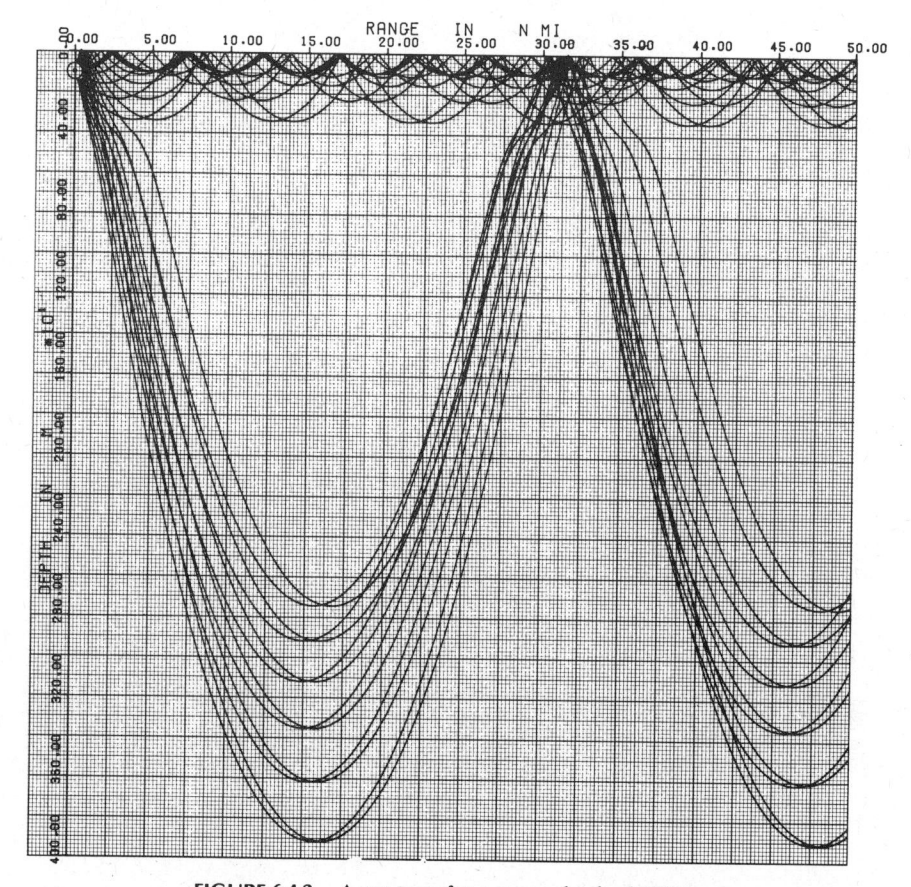

FIGURE 6.4.3. A ray trace for a source depth of 100 m.

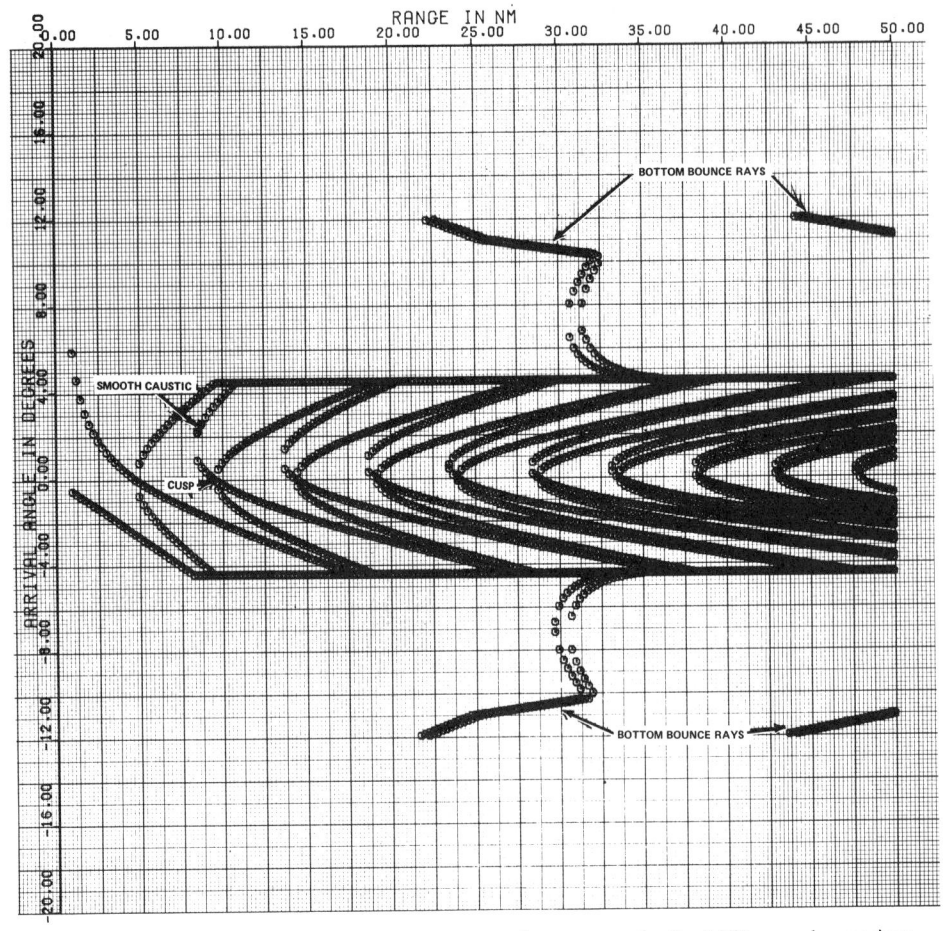

FIGURE 6.4.4. Arrival angle structure versus range for a source depth of 100 m and a receiver depth of 100 m.

276

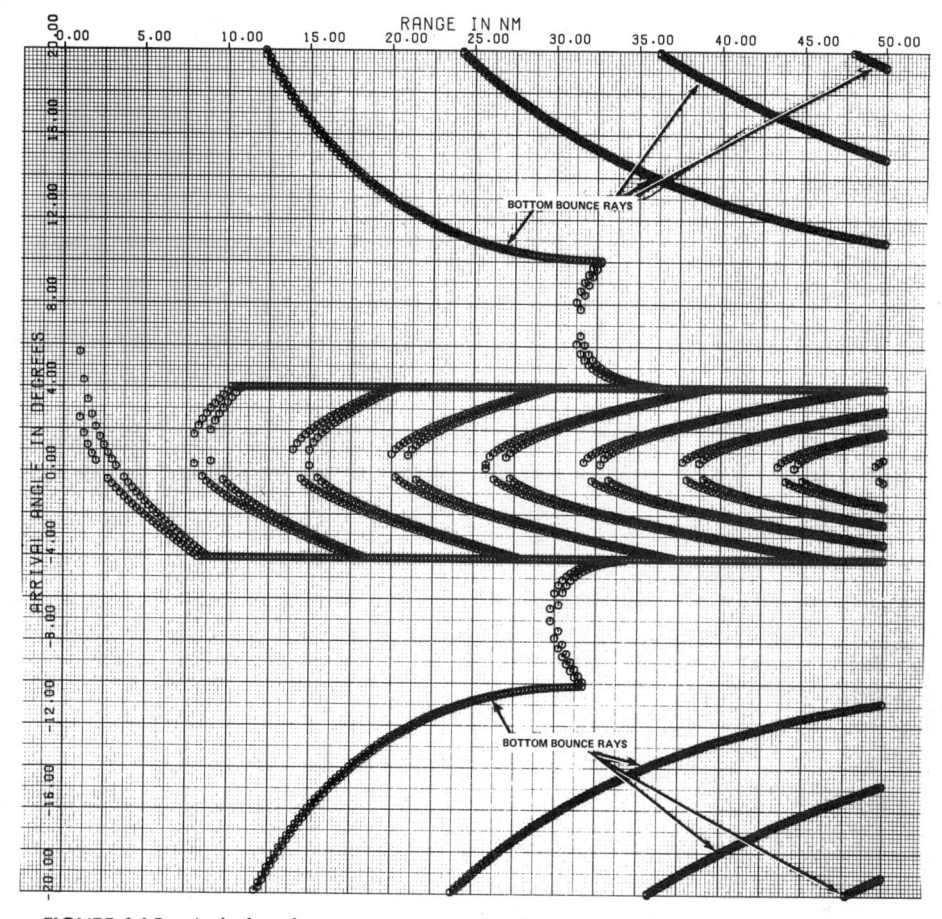

FIGURE 6.4.5. Arrival angle structure versus range for a source depth of 50 m and a receiver depth of 100 m.

depth is 150 meters. We again see the CZ caustics appearing at about 30 nautical miles, but now the duct is filled with a series of smooth caustics instead of a smooth/cusp combination. This is because the receiver is no longer at source depth. In each case, one important thing to note is that the surface duct is filled with caustics and ray theory does not treat caustics. Thus, one should always use ray theory with caution in a surface duct, even at high frequencies when diffraction leakage is negligible.

Now let us discuss the normal mode solution for this sound speed profile. The actual values for the profile used in the normal mode model are shown in Table 6.4.1.

First, let us consider a frequency of 60 Hz. For this case, there are 53 modes trapped in the water column, that is, there are 53 modes with phase velocities that lie between the minimum sound speed and maximum sound speed in the water column. The eigenvalues and corresponding phase velocities for these modes are shown in Table 6.4.2. and examples of the mode shapes are shown in Figure 6.4.6.

Modes 1 through 14 have phase velocities that lie between the minimum sound speed in the water column and the sound speed at the surface. These modes are RR modes with two turning points. Modes 24 through 53 have phase velocities between the maximum sound speed at the bottom of the duct and the sound speed at the bottom. These modes are RSR modes and, as before, form the CZs.

Modes 15 through 23 have phase velocities that lie between the surface sound speed and the maximum sound speed at the bottom of the duct at 400 meters. That is, these modes have phase velocities that are realized in the duct. Note that these modes have three turning points. We see that modes 14, 16, 17, and 18 would only be very weakly excited by a source in the duct and then it would have to be near the bottom of the duct.

TABLE 6.4.1. Values of Sound Velocity Versus Depth for the Surface Duct Profile

Depth (meters)	Sound Speed (meters/second)
0	1501.9
400	1508.0
700	1491.2
800	1488.9
1200	1489.8
1400	1491.0
1750	1494.9
2500	1505.6
3000	1512.9
4000	1527.9

TABLE 6.4.2. Eigenvalues and Phase Velocities for Modes Trapped in the Water Column at a Frequency of 60 Hz

Mode No.	Eigenvalues (1/meters)	Phase Velocity (meters/second)
1	0.2530872223749054D + 00	0.1489569939142670D + 04
2	0.2529518039221444D + 00	0.1490367384558399D + 04
3	0.2527976636496818D + 00	0.1491276117777720D + 04
4	0.2526373654149685D + 00	0.1492222331449467D + 04
5	0.2524700171884530D + 00	9.1493211442012044D + 04
6	0.2523088435096699D + 00	0.1494165298317524D + 04
7	0.2521505309081397D + 00	0.1495103409352392D + 04
8	0.2519926999164388D + 00	0.1496039839867529D + 04
9	0.2518334338158807D + 00	0.1496985974890051D + 04
10	0.2516750469753666D + 00	0.1497928074163325D + 04
11	0.2515212738744355D + 00	0.1498843865664329D + 04
12	0.2513709713517190D + 00	0.1499740071022454D + 04
13	0.2512239291750374D + 00	0.1500617873738099D + 04
14	0.2510803283256449D + 00	0.1501476124970759D + 04
15	0.2509375942338396D + 00	0.1502330169306840D + 04
16	0.2507972385616943D + 00	0.1503170930400967D + 04
17	0.2506590928595086D + 00	0.1503999372734004D + 04
18	0.2505239860703684D + 00	0.1504810474813713D + 04
19	0.2504624199105920D + 00	0.1505180372230495D + 04
20	0.2503913044716717D + 00	0.1505607869355649D + 04
21	0.2502626820260006D + 00	0.1506381676160605D + 04
22	0.2501404716410420D + 00	0.1507117644567998D + 04
23	0.2500536208606932D + 00	0.1507641109667434D + 04
24	0.2499819656811210D + 00	0.1508073262019502D + 04
25	0.2498698320263335D + 00	0.1508750037463885D + 04
26	0.2497545963888822D + 00	0.1509446168065960D + 04
27	0.2496455927408376D + 00	0.1510105242763640D + 04
28	0.2495425624823249D + 00	0.1510728735777114D + 04
29	0.2494387023393361D + 00	0.1511357760024278D + 04
30	0.2493298481880299D + 00	0.1512017599058059D + 04
31	0.2492179176334767D + 00	0.1512696626645613D + 04
32	0.2491053313813593D + 00	0.1513380369421454D + 04
33	0.2489935718569137D + 00	0.1514059642661845D + 04
34	0.2488831695807590D + 00	0.1514731265540425D + 04
35	0.2487737717839339D + 00	0.1515397365756875D + 04
36	0.2486642919522010D + 00	0.1516064552216616D + 04
37	0.2485543042932204D + 00	0.1516735425293772D + 04
39	0.2483329506493753D + 00	0.1518087380047501D + 04
40	0.2482221951088873D + 00	0.1518764743279308D + 04
41	0.2481119602076655D + 00	0.1519439522847831D + 04
42	0.2480022942501162D + 00	0.1520111414979778D + 04
43	0.2478931975480889D + 00	0.1520780409303365D + 04
44	0.2477845935724744D + 00	0.1521446967281881D + 04
45	0.2476762967222315D + 00	0.1522112222364057D + 04
46	0.2475682589178251D + 00	0.1522776466089336D + 04
47	0.2474604127676265D + 00	0.1523440109932926D + 04
48	0.2473528185361005D + 00	0.1524101780238805D + 04
49	0.2472456271694635D + 00	0.1524763542824495D + 04
50	0.2471391690340593D + 00	0.1525420352849130D + 04
51	0.2470343446891354D + 00	0.1526067636081840D + 04
52	0.2469326116955442D + 00	0.1526696355909388D + 04
53	0.2468332120130599D + 00	0.1527311156210327D + 04

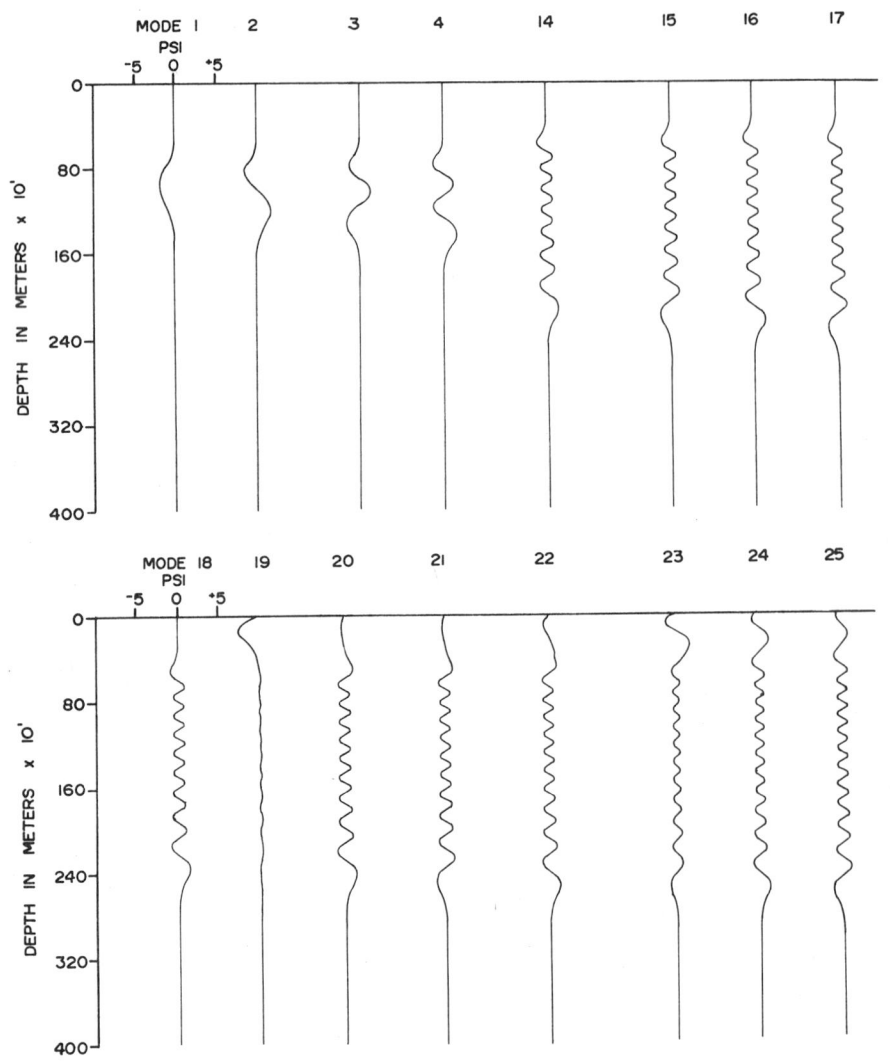

FIGURE 6.4.6. Normalized depth eigenfunctions $\psi_n(z)$ for the surface duct profile for a frequency of 60 Hz.

SURFACE DUCT PROPAGATION

Mode 19 is the first strong mode in the duct. It has turning points at depths of 200, 450, and 2450 meters. In the intervals $0 < z < 200m$ and $450m < z < 2450m$, the mode is oscillatory, and in the intervals $200m < z < 450m$ and $2450m < z < 4000m$, the mode is exponentially decaying. The second strong mode in the duct does not occur until mode 23. While modes 20, 21, and 22 are stronger in the duct than modes 15, 16, 17, and 18, most of their strength still resides in the deep ocean channel.

The two strong modes in the duct, namely, modes 19 and 23, are called "virtual modes" and they can be predicted by approximate theories. An excellent paper on this subject is that written by F. Labianca [6.4]. These modes behave like modes of order 1 and 2 if only the duct depth is considered. For example, mode 19 has a zero at the ocean surface due to the pressure-release condition, but then has no other zeros in the duct. Mode 23, in addition to the zero at the surface, has one zero in the duct. So let us refer to these two modes as virtual mode 1 and virtual mode 2.

We also see that virtual mode 2 has more energy in the deep channel than virtual mode 1. We might be tempted to say that virtual mode 2 is more "leaky" than virtual mode 1. While this concept of "diffraction leakage" is associated with the modal partition of energy, we must be very careful in the interpretation of the normal mode theory.

First, let us discuss leakage based on the investigation in Section 5.3. Consider some virtual mode trapped in the duct. Suppose the ray shown in Figure 6.4.7 is the ray associated with this virtual mode. In Section 5.3 we saw that above the upper turning point there was an incident and reflected wave; this is represented in the figure by the ray. We also saw in Section 5.3 that between the upper and lower turning points there existed a nonpropagating

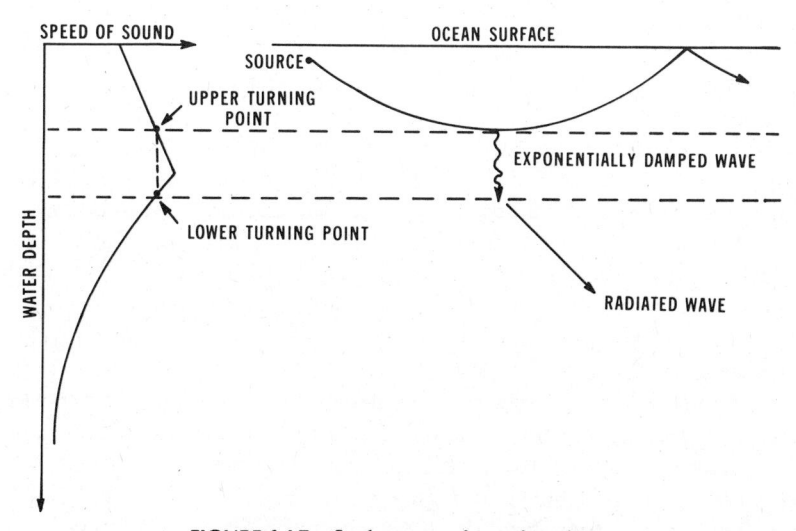

FIGURE 6.4.7. Leakage out of a surface duct.

wave with an amplitude that exponentially decreased with depth. Below the lower turning point there again was a radiated wave. We interpreted this as energy leaking off by diffraction between the turning points and then being carried off to infinity by the radiation field below the lower turning point.

While this description of leakage from a duct is physically appealing, it cannot be applied to our normal mode solution. The modal partition of energy with depth as exhibited by the modes in Figure 6.4.6 is range-independent. For example, mode 19 (virtual mode 1) has the same depth partition of energy at any range. As mode 19 propagates down the duct, its shape does not change. Energy that started out in the duct does not shift below the duct as the mode propagates. The phenomenon of "diffraction leakage" is naturally taken into account by modal interference.

Figure 6.4.8 shows transmission loss versus range for a source depth of 100 meters, a receiver depth of 100 meters, and a frequency of 60 Hz. In contrast to CZ propagation, we see that there are no shadow zones between the CZs. In fact, at 60 Hz, since there are nine modes trapped in the duct, the transmission loss is so little between CZs that it is difficult to tell when a CZ comes in.

Next, let us consider propagation in a duct at 30 Hz. Here there are only 27 modes trapped in the water column and only five with phase velocities in the duct. Table 6.4.3 shows the eigenvalues and corresponding phase velocities for the modes trapped in the water column. Figure 6.4.9 shows the mode shapes. Modes 8, 9, 10, 11, and 12 have a phase velocity that is realized in the upper 400 meters of water. Of these five modes, only one is a virtual mode; this is mode 11. Thus, as we decrease the frequency, the number of modes trapped in the duct decreases and, as we would expect, the transmission loss increases. The transmission loss for a source depth of 100 meters, a receiver depth of 100 meters, and a frequency of 30 Hz is shown in Figure 6.4.10. We see that since there is only one virtual mode trapped in the duct instead of two as at 60 Hz, the transmission loss is greater before the first CZ. However, between the first and second CZ the transmission loss decreases but not quite to the value it had at 60 Hz. This pattern of increased and decreased transmission loss then repeats itself with range.

If we add modes 1 through 7, we find that they do not contribute at all. The transmission loss due to modes 8, 9, 10, and 12 are shown in Figure 6.4.11. We see that these modes alone do not make up the full contribution observed in the duct. If we sum modes 8 through 12, as shown in Figure 6.4.12, we see that this accounts for all the energy in the duct between CZs. In fact, as shown in Figure 6.4.13, modes 11 and 12 alone account for the energy between CZs. Figure 6.4.14 shows the transmission loss due to modes 13 through 27. As before, these modes form the CZs.

Now let us consider propagation in a duct at 10 Hz. The source and receiver depths are still 100 meters. Now there are only nine modes trapped in the water column. The eigenvalues and corresponding phase velocities for these modes are shown in Table 6.4.4. Figure 6.4.15 shows the mode shapes. Modes 1 and 2 are RR modes with phase velocities less than the surface velocity.

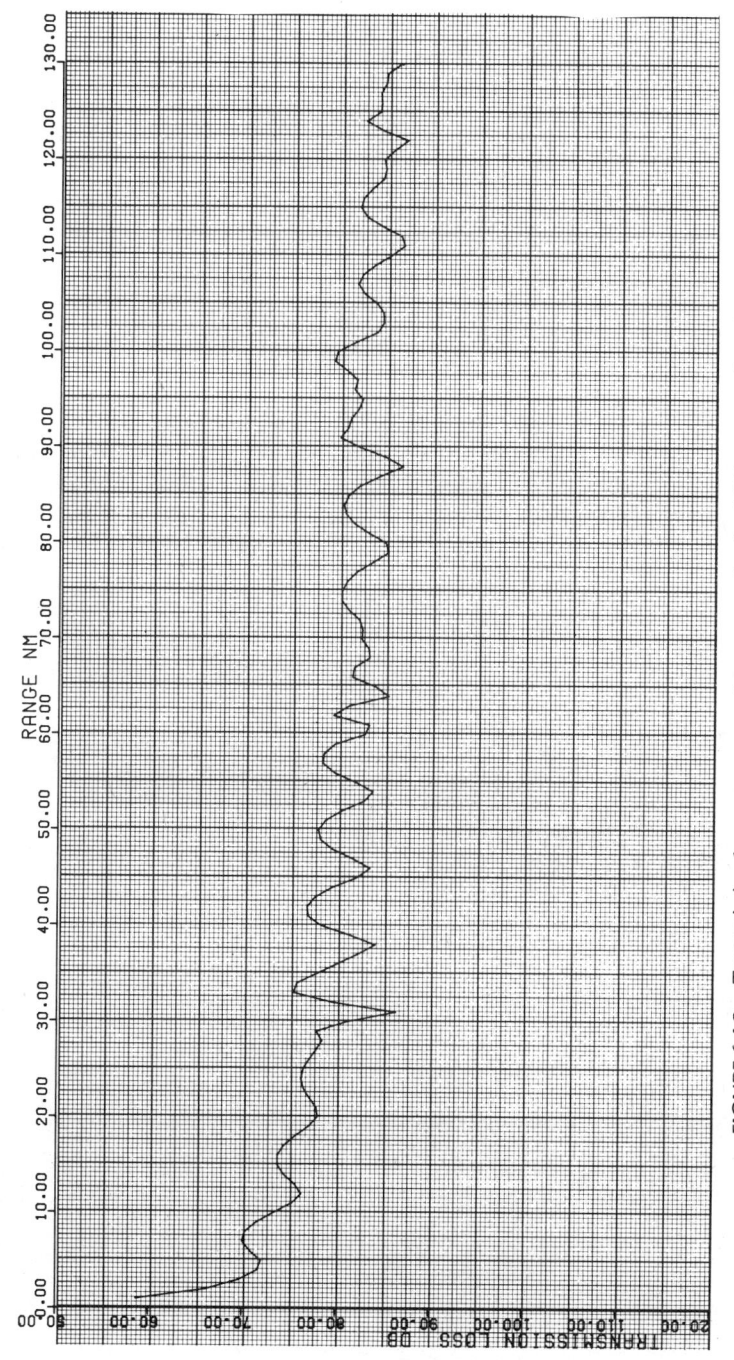

FIGURE 6.4.8. Transmission loss versus range for a source depth of 100 m, a receiver depth of 100 m, and a frequency of 60 Hz.

TABLE 6.4.3. Eigenvalues and Phase Velocities for Modes Trapped in the Water Column at a Frequency of 30 Hz

Mode No.	Eigenvalues (1/meters)	Phase Velocity (meters/second)
1	0.1265032668772352D + 00	0.1490044991472930D + 04
2	0.1263604650949184D + 00	0.1491728912787676D + 04
3	0.1261976439289900D + 00	0.1493653552846454D + 04
4	0.1260356653449797D + 00	0.1495573167321002D + 04
5	0.1258793776963772D + 00	0.1497430021222711D + 04
6	0.1257239137648899D + 00	0.1499281668624188D + 04
7	0.1255765696838617D + 00	0.1501040836598134D + 04
8	0.1254343595918685D + 00	0.1502742628325320D + 04
9	0.1252974309692705D + 00	0.1504384868526289D + 04
10	0.1251699651758227D + 00	0.1505916846350587D + 04
11	0.1250790230344906D + 00	0.1507011764581899D + 04
12	0.1250127992931862D + 00	0.1507810082496581D + 04
13	0.1249000129759865D + 00	0.1509171654382680D + 04
14	0.1247833391353175D + 00	0.1510582747036280D + 04
15	0.1246703023168581D + 00	0.1511952371273741D + 04
16	0.1245620876423320D + 00	0.1513265896414923D + 04
17	0.1244556846440002D + 00	0.1514559658360087D + 04
18	0.1243475115968726D + 00	0.1515877212134967D + 04
19	0.1242370645535736D + 00	0.1517224830550503D + 04
20	0.1241258491232626D + 00	0.1518584247735563D + 04
21	0.1240151383469639D + 00	0.1519939918044710D + 04
22	0.1239053608941428D + 00	0.1521286551729000D + 04
23	0.1237967008450306D + 00	0.1522621830216198D + 04
24	0.1236895060643634D + 00	0.1523941401441943D + 04
25	0.1235852246095625D + 00	0.1525227306184000D + 04
26	0.1234860076255119D + 00	0.1526452776633819D + 04
27	0.1233869172638791D + 00	0.1527678650178651D + 04

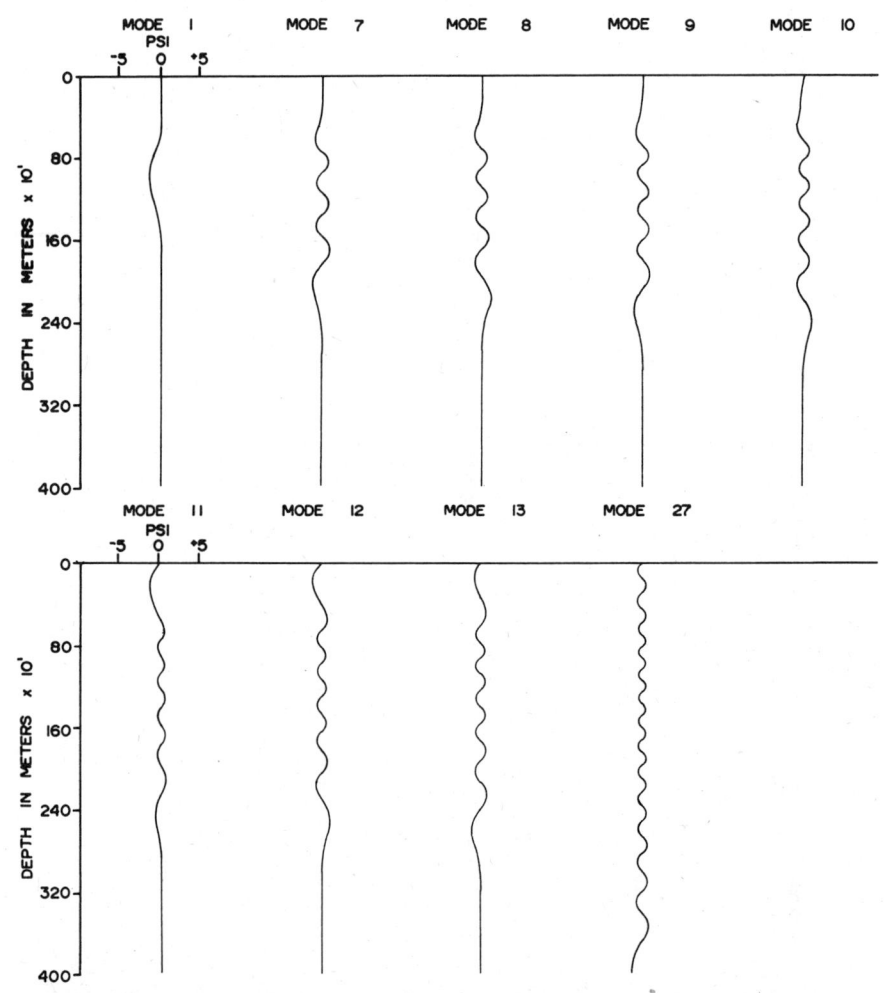

FIGURE 6.4.9. Normalized depth eigenfunctions $\psi_n(z)$ for the surface duct profile for a frequency of 30 Hz.

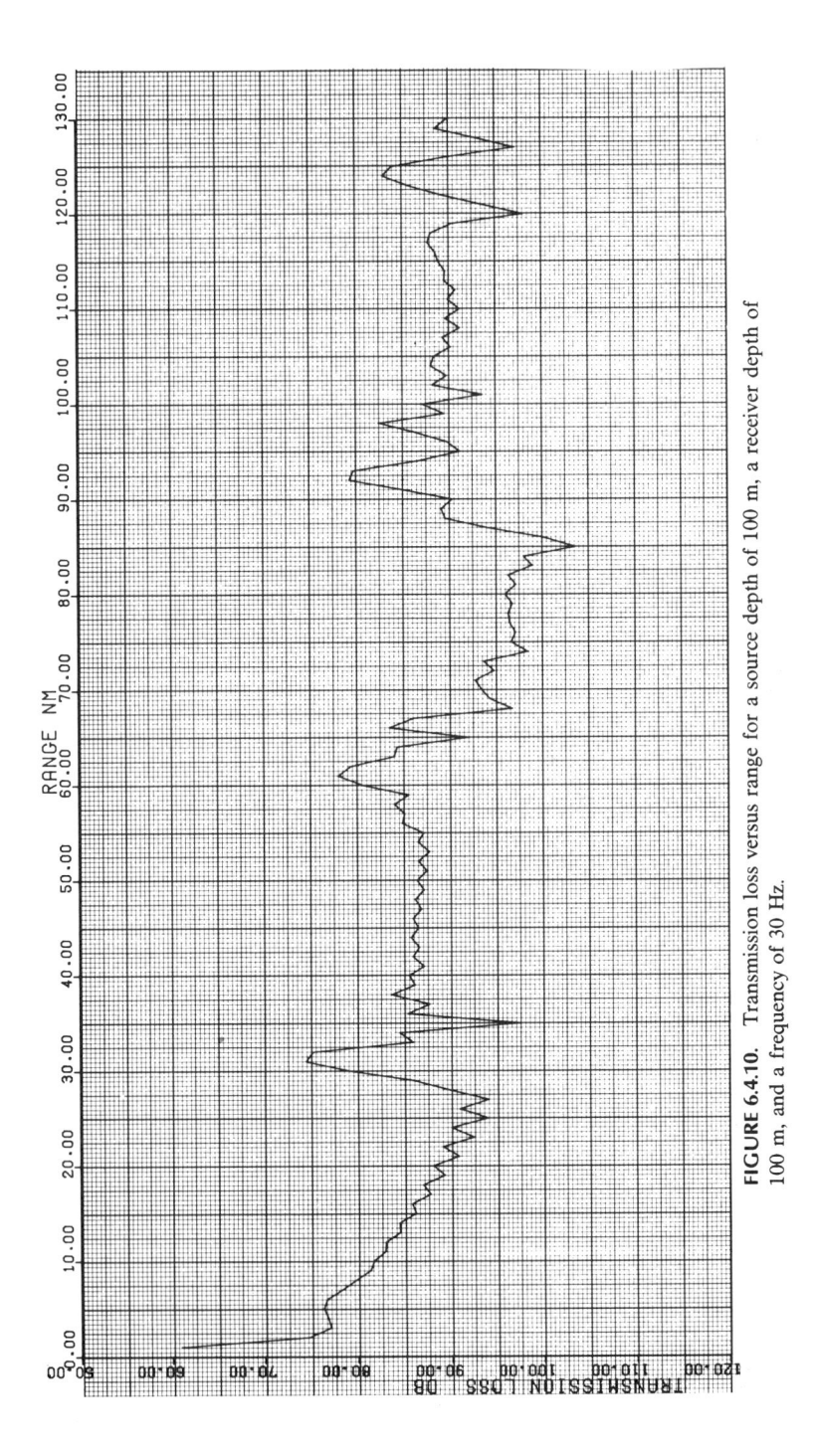

FIGURE 6.4.10. Transmission loss versus range for a source depth of 100 m, a receiver depth of 100 m, and a frequency of 30 Hz.

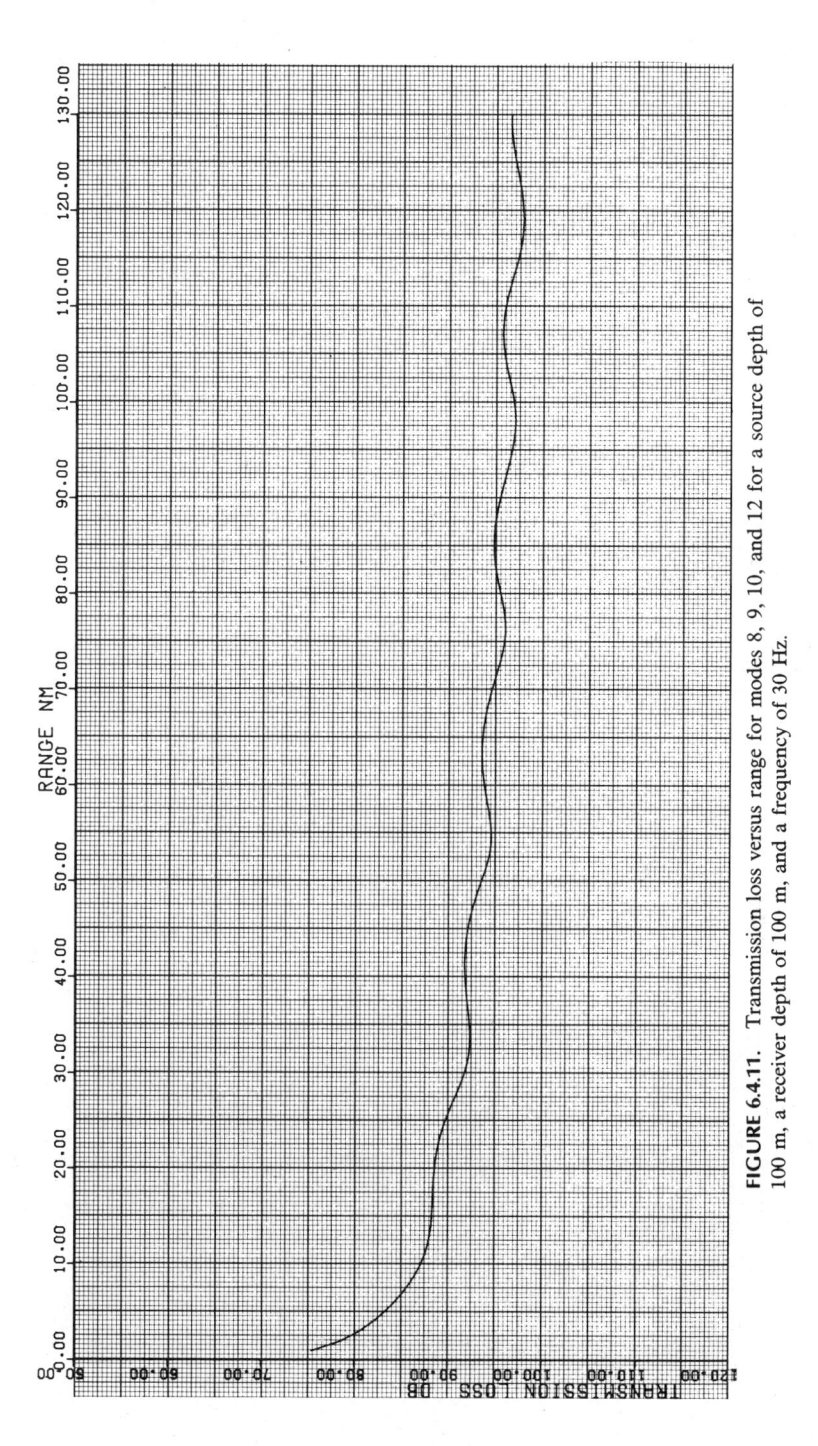

FIGURE 6.4.11. Transmission loss versus range for modes 8, 9, 10, and 12 for a source depth of 100 m, a receiver depth of 100 m, and a frequency of 30 Hz.

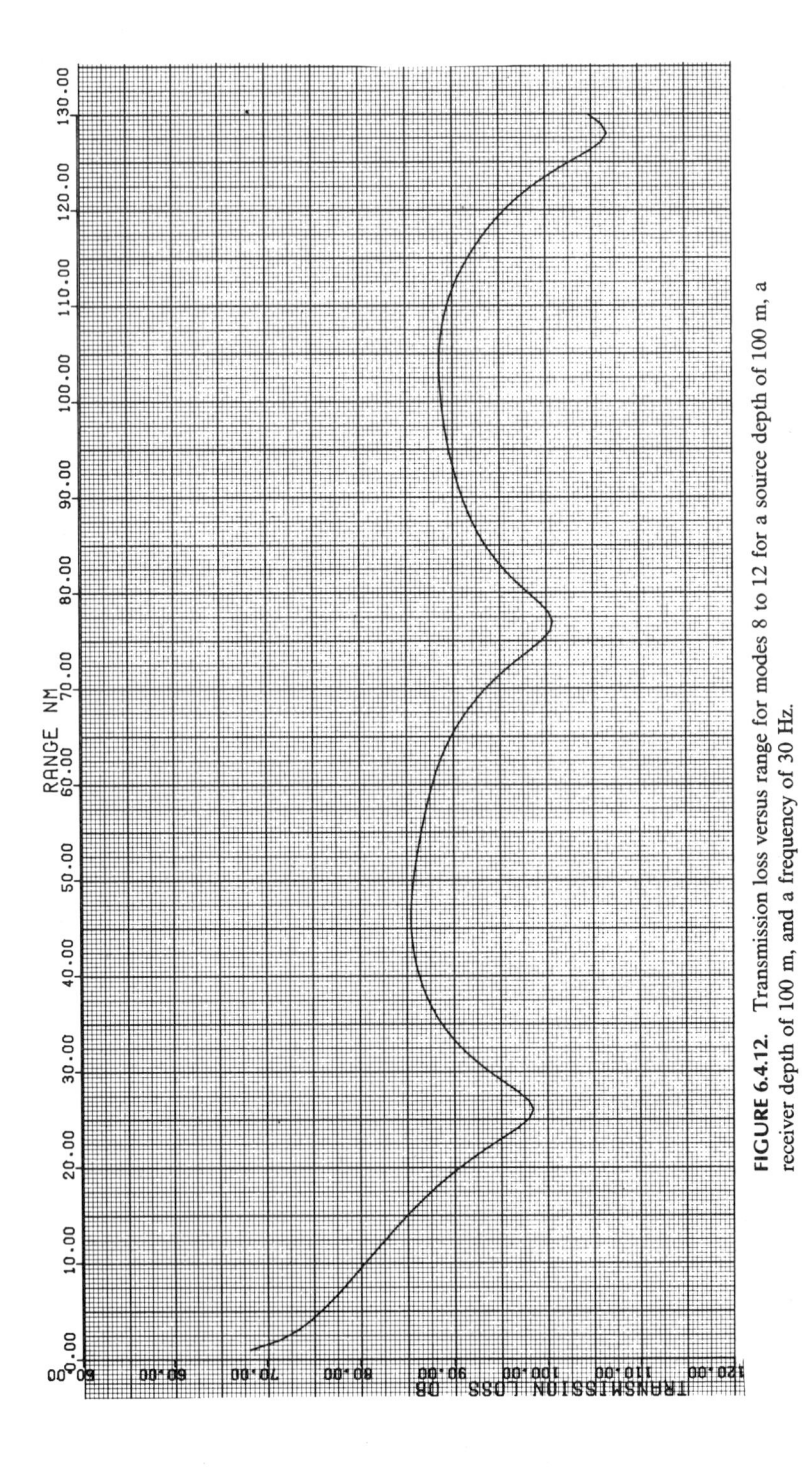

FIGURE 6.4.12. Transmission loss versus range for modes 8 to 12 for a source depth of 100 m, a receiver depth of 100 m, and a frequency of 30 Hz.

288

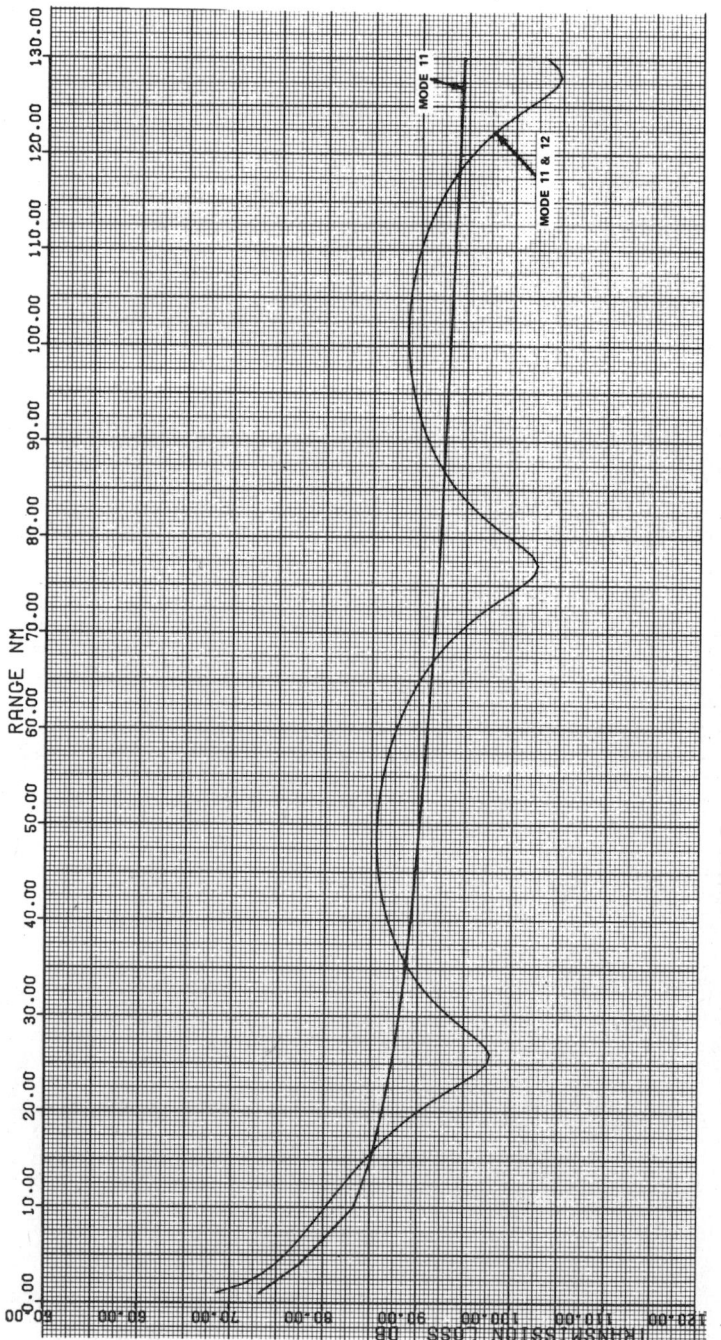

FIGURE 6.4.13. Comparison of transmission loss for mode 11 with that for modes 11 and 12.

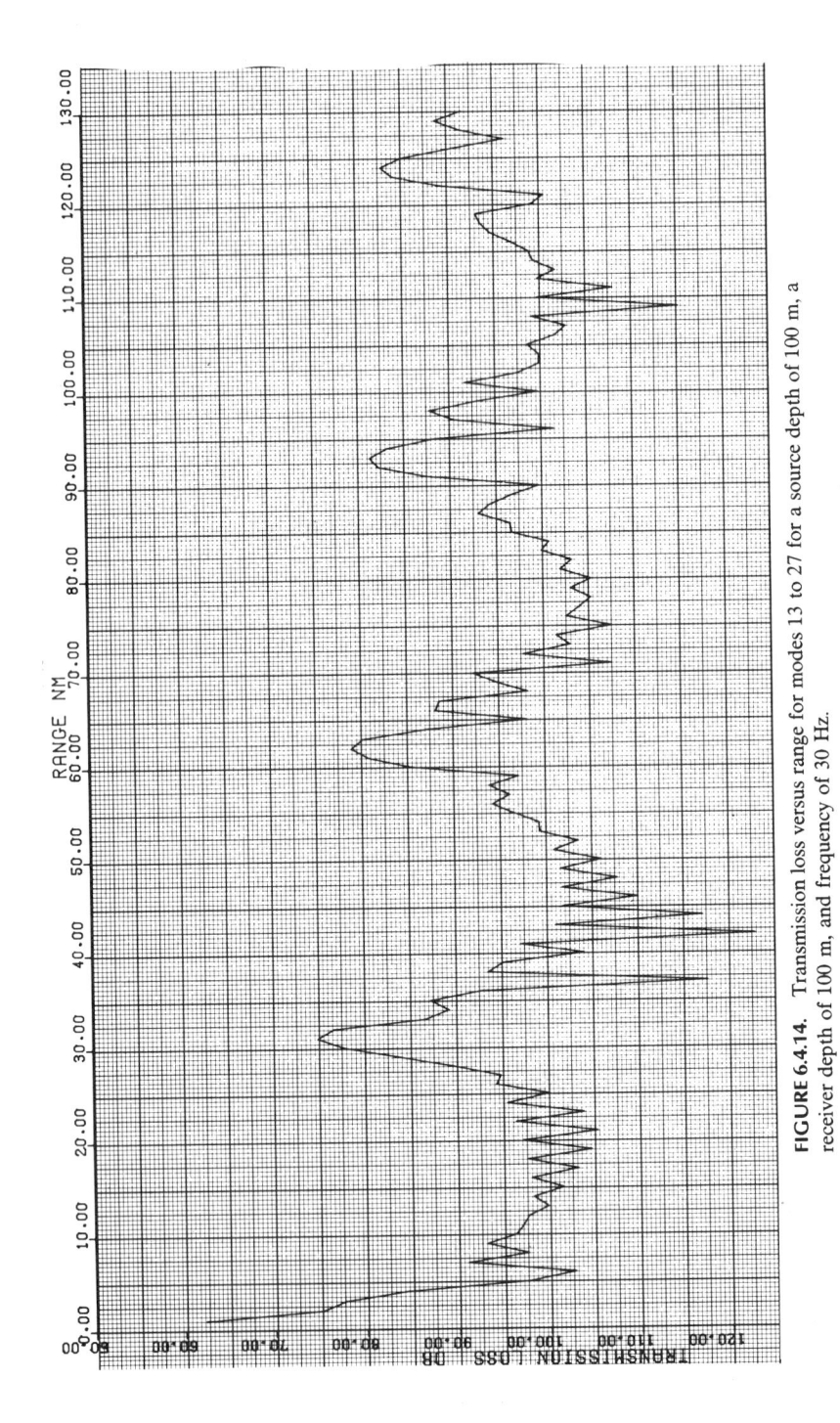

FIGURE 6.4.14. Transmission loss versus range for modes 13 to 27 for a source depth of 100 m, a receiver depth of 100 m, and frequency of 30 Hz.

TABLE 6.4.4. Eigenvalues and Phase Velocities for Modes Trapped in the Water Column at a Frequency of 10 Hz

Mode No.	Eigenvalues (1/meter)	Phase Velocity (meters/second)
1	0.4211326100828338D − 01	0.1491973111734029D + 04
2	0.4196371545484623D − 01	0.1497290037137064D + 04
3	0.4181748909623419D − 01	0.1502525400424521D + 04
4	0.4168419075160556D − 01	0.1507330523608061D + 04
5	0.4156615963618463D − 01	0.1511610733869646D + 04
6	0.4145877679112280D − 01	0.1515525973869290D + 04
7	0.4135473530270835D − 01	0.1519338779752291D + 04
8	0.4125356160459035D − 01	0.1523064933739067D + 04
9	0.4115814761274255D − 01	0.1526595746314472D + 04

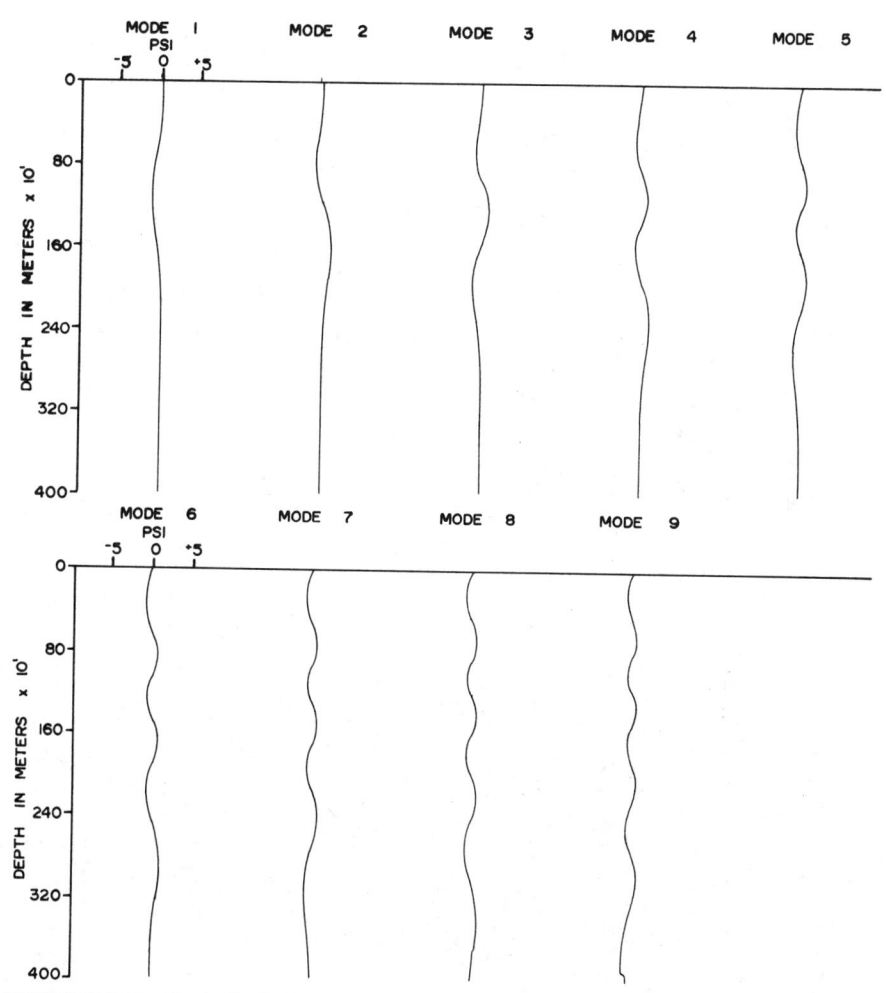

FIGURE 6.4.15. Normalized depth eigenfunctions $\psi_n(z)$ for the surface duct profile for a frequency of 10 Hz.

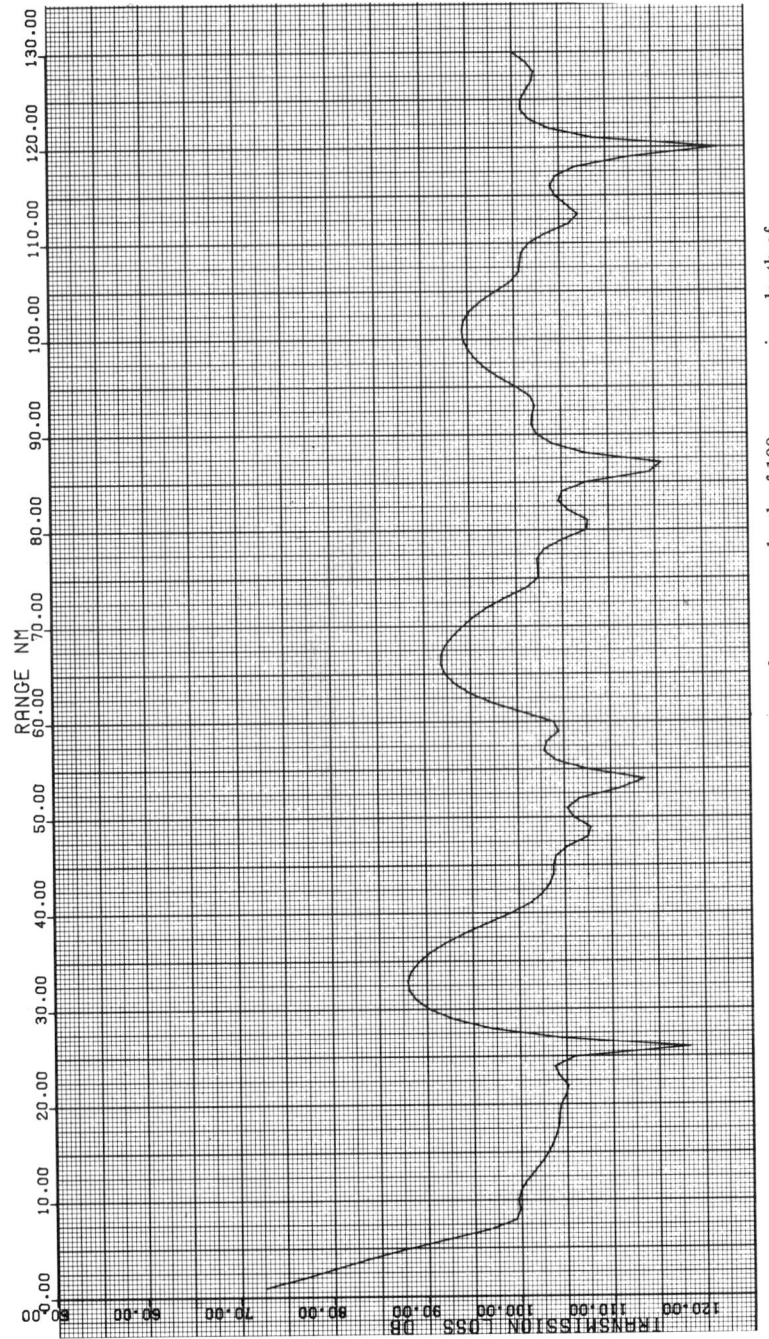

FIGURE 6.4.16. Transmission loss versus range for a source depth of 100 m, a receiver depth of 100 m, and a frequency of 10 Hz.

Modes 3 and 4 have phase velocities that fall between the surface sound speed and the sound speed at the bottom of the duct at 400 meters. We see that these modes are weakly excited for a source in the duct. Moreover, no virtual modes exist for this case. Modes 5 through 9 are the RSR mode. Figure 6.4.16 shows the transmission loss versus range for this case. We see that in the duct between CZs the transmission loss is much greater than for 30 and 60 Hz. We would say that at this low frequency the duct is very "leaky."

REFERENCES

6.1. H. Weinberg, "CONGRATS I: Ray Plotting and Eigenray Generation," NUSL Report No. 1052, 30 October 1969, Navy Underwater Sound Laboratory, New London, Conn.

6.2. K. V. Mackenzie, "Discussion of Sea Water Sound Speed Determinations, *J. Acoust. Soc. Am.* **70**, 801–806 (1981).

6.3. K. V. Mackenzie, "Nine Term Equation for Sound Speed in the Oceans," *J. Acoust. Soc. Am.* **70**, 807–812 (1981).

6.4. F. M. Labianca, "Normal Modes, Virtual Modes and Alternative Representations in the theory of Surface Duct Sound Propagation," *J. Acoust. Soc. Am.* **53**, 1137–1147 (1973).

7 AN OCEANIC WAVEGUIDE WITH A RANGE- AND DEPTH-DEPENDENT REFRACTIVE INDEX AND A TIME VARYING, RANDOMLY ROUGH SEA SURFACE

7.0. INTRODUCTION

In the first six chapters we have dealt exclusively with range-independent problems. By this we mean we have only considered waveguides whose bounding surfaces (planes in our case) coincide with the coordinate surfaces of a separable coordinate system, and whose refractive index had the properties necessary to satisfy Theorem 2.1.1. The bounding surfaces and refractive index of a real oceanic waveguide do not satisfy the conditions of separability stated in Theorem 2.1.1. Therefore, we must find another technique besides separation of variables to solve the wave equation.

One technique to solve nonseparable problems is that of coupled, local, normal modes. The theory of coupled, local, normal modes was introduced to the field of underwater acoustics in 1965 by Pierce [7.1]. Until the work of Boyles [7.2], [7.3] and Dozier [7.4], no one had succeeded in obtaining an exact numerical solution of the coupled differential equations which arise from the Helmholtz equation when it is applied via coupled mode techniques to nonseparable acoustic problems. This is because the set of coupled, second-order differential equations and their associated boundary conditions do not lend themselves to the application of known numerical integration algorithms. This chapter presents the work of Boyles and Dozier and shows how the coupled equations can be reformulated to allow the application of known numerical techniques. A survey of the current work done in this field can also be found in the paper of Boyles [7.2].

7.1. THE COUPLED SECOND-ORDER SYSTEM OF DIFFERENTIAL EQUATIONS IN CYLINDRICAL COORDINATES

We consider the problem of a point source located at $r = 0$ and $z = z_s$ in a cylindrically symmetric oceanic waveguide. The source is emitting a time

THE COUPLED SECOND-ORDER SYSTEM IN CYLINDRICAL COORDINATES 295

harmonic signal of angular frequency ω. The coordinate system in relation to the waveguide is shown in Figure 7.1.1. The bottom is assumed flat and rigid at $z = d$, and the surface is given as a function of range and time by

$$z = s(r, t) \tag{7.1.1}$$

The speed of sound, $c(r, z)$, in the waveguide can be a function of both r and z. The wave equation which governs the acoustic field away from the source is

$$\left[\nabla^2 - \frac{1}{\rho_0} \nabla \rho_0 \cdot \nabla - n^2 \frac{\partial^2}{\partial t^2} \right] \mathcal{P}(\mathbf{r}, t) = 0 \tag{7.1.2}$$

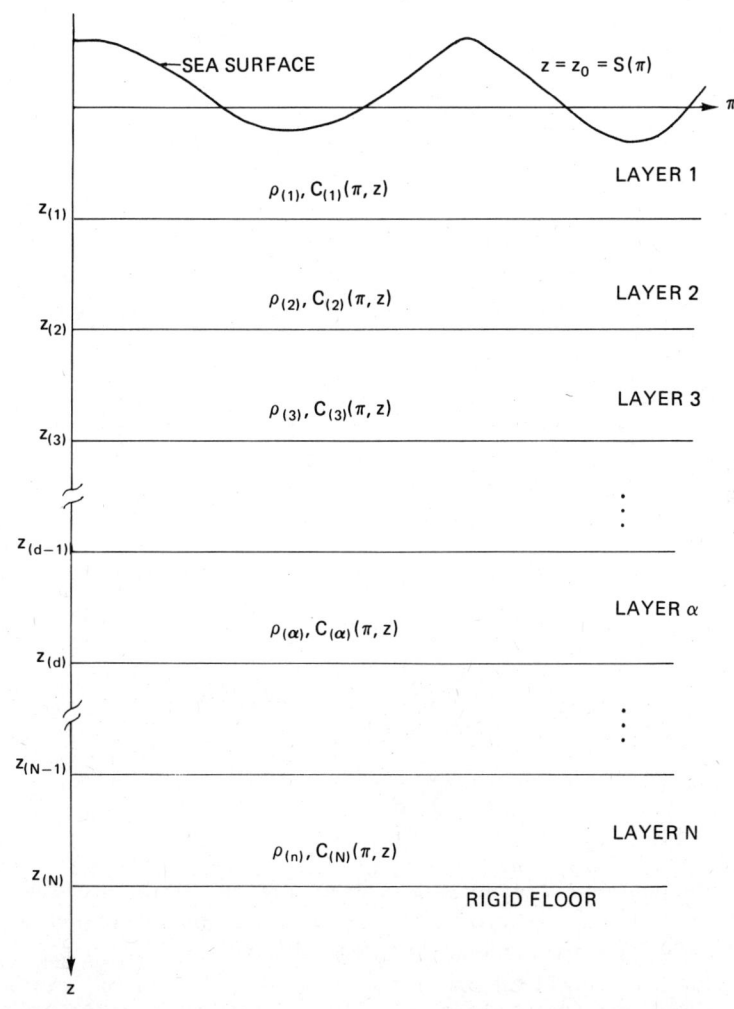

FIGURE 7.1.1. Coordinate geometry for an inhomogeneous, cylindrically symmetric waveguide with a rough sea surface.

In Eq. (7.1.2) $\mathscr{P}$ is the acoustic pressure, ρ_0 is the density of the medium (water and sediment layers) and n is the refractive index defined by

$$n(r, z) = \frac{1}{c(r, z)} \qquad (7.1.3)$$

We now make the narrow-band approximation to the wave equation. This amounts to assuming that

$$\frac{\partial^2 \mathscr{P}}{\partial t^2}(r, t) \approx -\omega^2 \mathscr{P}(\mathbf{r}, t) \qquad (7.1.4)$$

This approximation is discussed in more detail in Appendix 7A. The narrow-band approximation implies that the frequency of the ocean surface waves is smaller than the acoustic frequency. It should be noted that the narrow-band approximation only puts an upper frequency limit on surface waves if we wish to calculate the effects of a time varying surface on the modulation of the acoustic signal. For acoustic frequencies of interest to us, this implies that we probably could not treat the higher frequency capillary waves. If we wish to apply this theory to a single, static surface, then there is no upper limit on the frequency of the surface waves.

If we assume that the density ρ_0 is a function of z only, then in the narrow-band approximation, Eq. (7.1.2) can be written in cylindrical coordinates as

$$\frac{\partial^2 p}{\partial r^2} + \frac{1}{r}\frac{\partial p}{\partial r} + \frac{\partial^2 p}{\partial z^2} - \frac{1}{\rho_0(z)}\frac{d\rho_0}{dz}\frac{\partial p}{\partial z} + k^2(r, z)p = 0 \qquad (7.1.5)$$

where we have put

$$\mathscr{P}(r, z, t) = p(r, z, t)e^{-i\omega t} \qquad (7.1.6)$$

and

$$k(r, z) = \omega n(r, z) \qquad (7.1.7)$$

Following Pierce [7.1], we assume a solution of the form

$$p = \sum_{n=1}^{\infty} \phi_n(r)\psi_n(z, r) \qquad (7.1.8)$$

The functions $\psi_n(z, r)$ are called local, normal modes because of the dependence on the range variable r. Thus, for each range, there is a complete set of orthogonal eigenfunctions $\psi_n(z, r)$ as we shall see.

For simplicity we will suppress the time factor, t, in ϕ_n and ψ_n. Further, we will assume that the local depth mode functions $\psi_n(z, r)$ satisfy the partial differential equation.

$$\frac{\partial}{\partial z}\left[\frac{1}{\rho_0(z)}\frac{\partial \psi_n}{\partial z}\right] + \left(\frac{k^2(r,z)}{\rho_0(z)} - \frac{\kappa_n^2(r)}{\rho_0(z)}\right)\psi_n = 0 \qquad (7.1.9)$$

Note that Eq. (7.1.9) is formally the same as the separated Eq. (4.1.7) except that Eq. (7.1.9) is a partial differential equation whereas Eq. (4.1.7) is an ordinary differential equation.

The following boundary conditions are imposed:

1. On the sea surface:

$$p[r, s(r,t), t] = 0 \qquad (7.1.10)$$

2. On the rigid bottom:

$$\left[\frac{\partial p}{\partial z}(r,z,t)\right]_{z=d} = 0 \qquad (7.1.11)$$

Now it is easy to show that the eigenfunction solutions, ψ_n of Eq. (7.1.9), subject to the boundary conditions given by Eqs. (7.1.10) and (7.1.11), form a complete, orthonormal system relative to the weight function $(1/\rho_0)$ at each range point r. The orthonormality condition is

$$\int_s^d \rho_0^{-1}(z)\psi_n(z,r)\psi_m(z,r)\, dz = \delta_{nm} \qquad (7.1.12)$$

There is still a source condition at $r = 0$, $z = z_s$ which must be satisfied. This will be discussed later.

We now need to derive the system of coupled differential equations for the range-dependent amplitudes $\phi_n(r)$ given in Eq. (7.1.8). To do this, we substitute the expression for the acoustic pressure p, given by Eq. (7.1.8) into the wave equation, Eq. (7.1.5). Then, if we make use of Eq. (7.1.9), we obtain

$$\sum_{n=1}^{\infty}\frac{d^2\phi_n}{dr^2}\psi_n + 2\sum_{n=1}^{\infty}\frac{d\phi_n}{dr}\frac{\partial\psi_n}{\partial r} + \sum_{n=1}^{\infty}\phi_n\frac{\partial^2\psi_n}{\partial r^2} + \frac{1}{r}\sum_{n=1}^{\infty}\frac{\partial\phi_n}{\partial r}\psi_n$$

$$+ \frac{1}{r}\sum_{n=1}^{\infty}\phi_n\frac{\partial\psi_n}{\partial r} + \sum_{n=1}^{\infty}\kappa_n^2\phi_n\psi_n = 0$$

We now multiply this last equation by $\rho_0^{-1}\psi_m$, integrate over z from $z = s(r)$ to $z = d$, and use Eq. (7.1.12) to obtain the desired system of coupled equations for ϕ_m:

$$\frac{d^2\phi_m}{dr^2} + \frac{1}{r}\frac{d\phi_m}{dr} + \kappa_m^2(r)\phi_m = -\sum_{n=1}^{\infty}\left\{A_{mn}(r)\phi_n + B_{mn}(r)\left[\frac{\phi_n}{r} + \frac{2\,d\phi_n}{dr}\right]\right\}$$

$$(7.1.13)$$

In Eq. (7.1.13), we have put

$$B_{mn}(r) = \int_s^d \rho_0^{-1}(z)\psi_m(z,r)\frac{\partial \psi_n}{\partial r}(z,r)\,dz \qquad (7.1.14)$$

$$A_{mn}(r) = \int_s^d \rho_0^{-1}(z)\psi_m(z,r)\frac{\partial^2 \psi_n}{\partial r^2}(z,r)\,dz \qquad (7.1.15)$$

The two coefficients B_{mn} and A_{mn} are called the coupling coefficients. These expressions for the coupling coefficients will subsequently be greatly simplified.

Let us now state the boundary and source conditions for the radial equations. The boundary condition at infinity is the radiation condition which for two dimensions can be written in the form

$$\lim_{r \to \infty} \sqrt{r}\left\{\frac{d\phi_n}{dr} - i\kappa_n \phi_n\right\} = 0 \qquad (7.1.16)$$

The condition imposed by the source at $r = 0$ manifests itself as a condition on the first derivative of ϕ_n. This condition can be stated in the form

$$\lim_{\varepsilon \to 0}\left\{\varepsilon \frac{d\phi_n(\varepsilon)}{dr}\right\} = \frac{\psi_n(z_s, 0)}{2\pi \rho_0(z_s)} \qquad (7.1.17)$$

provided that the coupling coefficients A_{mn} and B_{mn} and their first derivatives are continuous.

Now there are no numerical algorithms for solving the system of coupled equations given by Eq. (7.1.13), subject to the complicated radiation condition Eq. (7.1.16) and source condition Eq. (7.1.17). The more sophisticated and accurate techniques for numerically solving systems of differential equations pertain only to initial value problems. An example of such techniques is the Runge-Kutta methods. There are techniques for solving the two-point boundary value problem—that is, the unknown function is specified at both end points of the interval over which the differential equation is defined—but our system of equations and boundary conditions do not strictly fit these techniques as it is now formulated.

In Section 7.2 we will show how this two-point boundary value problem can be transformed into another two-point boundary value problem which lends itself to known numerical techniques. This will allow an exact numerical integration of the coupled equations. Before doing this let us derive simpler expressions for the coupling coefficients.

The coupling coefficients B_{mn} and A_{mn} are given by Eqs. (7.1.14) and (7.1.15). It is seen that they involve derivatives with respect to r under the integral sign. Since this derivative is difficult to evaluate, an alternate representation is sought.

THE COUPLED SECOND-ORDER SYSTEM IN CYLINDRICAL COORDINATES 299

Differentiate Eq. (7.1.9) with respect to r, multiply the resulting equation by ψ_m, and integrate over z from $z = s(r)$ to $z = d$, and then use Eq. (7.1.12) to obtain

$$\int_s^d \frac{\partial}{\partial z}\left(\frac{1}{\rho_0}\frac{\partial \dot{\psi}_n}{\partial z}\right)\psi_m \, dz + \int_s^d \frac{k^2}{\rho_0}\psi_m \dot{\psi}_n \, dz + 2\int_s^d \frac{k\dot{k}}{\rho_0}\psi_n\psi_m \, dz$$

$$-2\kappa_n \dot{\kappa}_n \delta_{nm} - \kappa_n^2 \int_s^d \frac{1}{\rho_0}\psi_m \dot{\psi}_n \, dz = 0 \qquad (7.1.18)$$

where a "dot" over a function denotes $\partial/\partial r$.

Let us integrate the first term in Eq. (7.1.18) twice by parts:

$$\int_s^d \frac{\partial}{\partial z}\left(\frac{1}{\rho_0}\frac{\partial \dot{\psi}_n}{\partial z}\right)\psi_m \, dz = \left[\frac{1}{\rho_0}\psi_m \frac{\partial \dot{\psi}_n}{\partial z}\right]_s^d$$

$$- \left[\frac{1}{\rho_0}\dot{\psi}_n \frac{\partial \psi_m}{\partial z}\right]_s^d + \int_s^d \dot{\psi}_n \frac{\partial}{\partial z}\left(\frac{1}{\rho_0}\frac{\partial \psi_m}{\partial z}\right) dz$$

$$(7.1.19)$$

If we now use Eq. (7.1.9) to rewrite the last term in Eq. (7.1.19) and then replace the first term in Eq. (7.1.18) by the resulting expression, we obtain

$$(\kappa_m^2 - \kappa_n^2) B_{mn} = 2\kappa_n \dot{\kappa}_n \delta_{nm} + \left[\frac{1}{\rho_0}\dot{\psi}_n \frac{\partial \psi_m}{\partial z}\right]_s^d$$

$$- \left[\frac{1}{\rho_0}\psi_m \frac{\partial \dot{\psi}_n}{\partial z}\right]_s^d - 2\int_s^d \frac{k\dot{k}}{\rho_0}\psi_n\psi_m \, dz \qquad (7.1.20)$$

This expression for B_{mn} is clearly not valid for $m = n$ because of the factor $\kappa_m^2 - \kappa_n^2$ in the denominator. We will have to examine the case $m = n$ separately. For $m \neq n$, the first term on the right in Eq. (7.1.20) vanishes. Hence, for $m \neq n$:

$$(\kappa_m^2 - \kappa_n^2) B_{mn} = \left[\frac{1}{\rho_0}\dot{\psi}_n \frac{\partial \psi_m}{\partial z}\right]_d - \left[\frac{1}{\rho_0}\dot{\psi}_n \frac{\partial \psi_m}{\partial z}\right]_s$$

$$- \left[\frac{1}{\rho_0}\psi_m \frac{\partial \dot{\psi}_n}{\partial z}\right]_d + \left[\frac{1}{\rho_0}\psi_m \frac{\partial \dot{\psi}_n}{\partial z}\right]_s$$

$$- 2\int_s^d \frac{k\dot{k}}{\rho_0}\psi_n\psi_m \, dz \qquad (7.1.21)$$

300 AN INHOMOGENEOUS OCEAN WITH A ROUGH SEA SURFACE

If we invoke the boundary condition at the ocean's surface, Eq. (7.1.10), we see that the fourth term on the right vanishes. If we invoke the boundary condition at the floor, Eq. (7.1.11), we see that the first term on the right vanishes.

Next we want to show that the third term on the right vanishes. The completeness relation for the eigenfunctions ψ_n is given by

$$\sum_{n=1}^{\infty} \frac{1}{\rho_0} \psi_n(z, r) \psi_n(z', r) = \delta(z - z') \qquad (7.1.22)$$

Multiplying Eq. (7.1.22) by $\dot{\psi}_m$ and integrating over z from $z = s$ to $z = d$ results in

$$\dot{\psi}_m(z, r) = \sum_{n=1}^{\infty} B_{nm} \psi_n(z, r) \qquad (7.1.23)$$

Consequently,

$$\frac{\partial \dot{\psi}_m}{\partial z} = \sum_{n=1}^{\infty} B_{nm} \frac{\partial \psi_n}{\partial z} \qquad (7.1.24)$$

since the coupling coefficients B_{nm} do not depend on z. By Eq. (7.1.11), $\partial \psi_n/\partial z$ vanishes on the boundary $z = d$, hence, the third term in Eq. (7.1.21) vanishes.

We now need to evaluate the second term in Eq. (7.1.21). We will need the following version of differentiation by the chain rule. Let $f(r, z)$ be a function of the two variables r and z and in turn suppose $z = z(r)$. Then if we let $g(r) = f[r, z(r)]$, we get for the partial derivative of g with respect to r:

$$\frac{dg}{dr} = \frac{\partial f}{\partial r} + \frac{\partial f}{\partial z} \frac{\partial z}{\partial r} \qquad (7.1.25)$$

Applying the chain rule given by Eq. (7.1.25) to the boundary condition at the surface gives:

$$\psi_n[s(r), r] = 0$$

$$\frac{d\psi_n}{dr} = \frac{\partial \psi_n}{\partial r} + \frac{\partial \psi_n}{\partial z} \frac{ds}{dr} = 0$$

and, finally,

$$\dot{\psi}_n = -\frac{\partial \psi_n}{\partial z} \frac{ds}{dr} \qquad (7.1.26)$$

THE COUPLED SECOND-ORDER SYSTEM IN CYLINDRICAL COORDINATES 301

Combining all of these results, the expression for the coupling coefficient B_{mn} ($m \neq n$) given by Eq. (7.1.21) becomes

$$B_{mn} = \frac{1}{\rho_0(s)(\kappa_m^2 - \kappa_n^2)} \left\{ \frac{\partial \psi_m}{\partial z} \frac{\partial \psi_n}{\partial z} \frac{ds}{dr} \right\}_{z=s}$$

$$- \frac{2\omega^2}{(\kappa_m^2 - \kappa_n^2)} \int_s^d \frac{1}{\rho_0} n \frac{\partial n}{\partial r} \psi_m \psi_n \, dz \qquad (7.1.27)$$

Where we have used Eq. (7.1.7) to obtain the final form for the last term on the right.

We still need an expression for B_{nn}. Differentiating Eq. (7.1.12) with respect to r gives

$$\int_s^d \frac{1}{\rho_0} \dot{\psi}_n \psi_m \, dz + \int_s^d \frac{1}{\rho_0} \psi_n \dot{\psi}_m \, dz - \left[\frac{1}{\rho_0(s)} \psi_n \psi_m \frac{ds}{dr} \right]_{z=s} = 0$$

or

$$B_{mn} + B_{nm} = 0$$

In deriving this result we made use of Eqs. (7.1.10) and (7.1.14). Consequently,

$$B_{nn} = 0 \qquad (7.1.28)$$

We now turn our attention to the second coupling coefficient A_{mn}. We start by differentiating Eq. (7.1.14) with respect to r.

$$\dot{B}_{mn} = \int_s^d \frac{1}{\rho_0} \psi_m \ddot{\psi}_n \, dz + \int_s^d \frac{1}{\rho_0} \dot{\psi}_m \dot{\psi}_n \, dz - \left[\frac{1}{\rho_0} \psi_m \dot{\psi}_n \frac{ds}{dr} \right]_{z=s}$$

The integrated term vanishes because of the boundary condition Eq. (7.1.10). Using Eq. (7.1.15), we can write this last expression in the form

$$A_{mn} = \dot{B}_{mn} - \int_s^d \frac{1}{\rho_0} \dot{\psi}_m \dot{\psi}_n \, dz \qquad (7.1.29)$$

Using Eq. (7.1.23), this last result becomes

$$A_{mn} = \frac{dB_{mn}}{dr} - \sum_{i=1}^{\infty} B_{im} B_{in} \qquad (7.1.30)$$

where use was made of Eq. (7.1.12).

Let us summarize the results for the coupling coefficient B_{mn}. We see that $B_{mn}(r, t)$ is a function of the range r and time t and consists of the sum of two terms, say

$$B_{mn} = S_{mn} + N_{mn} \tag{7.1.31}$$

where for $m \neq n$

$$S_{mn} = \frac{1}{\rho_0(s)(\kappa_m^2 - \kappa_n^2)} \left\{ \frac{\partial \psi_m}{\partial z} \frac{\partial \psi_n}{\partial z} \frac{ds}{dr} \right\}_{z=s} \tag{7.1.32}$$

$$N_{mn} = \frac{-2\omega^2}{(\kappa_m^2 - \kappa_n^2)} \int_s^d \frac{1}{\rho_0} n \frac{\partial n}{\partial r} \psi_n \psi_m \, dz \tag{7.1.33}$$

First consider the coupling coefficient S_{mn}. This term arises because of coupling due to the rough sea surface. We see that it is inversely proportional to the difference of the squares of the modal wave numbers κ_n. Thus, coupling is strongest for two neighboring modes. Since the squares of the modal wave numbers appear, S_{mn} feeds the same power into both forward and backward traveling waves at a given range point. Whether or not this power builds up in either the forward or backscattered wave depends on the phase relationship of the incremental amounts of power fed to the wave at other points. A wave builds up only if the incremental contribution at all points along the waveguide add up in phase. Finally, we note that S_{mn} is proportional to the slope ($\partial \psi_n / \partial z$) of the wave functions at the sea surface. This is a very reasonable result. One way to look at this, as given by E. Murphy [oral communication], is to consider the ray equivalent for a mode. As the order of the mode increases, the slope of the wave function at the surface increases and consequently the strength of the coupling increases. But as the order of the mode increases, the ray equivalent strikes the surface at larger and larger grazing angles and, consequently, we would expect more pronounced scattering.

The second term, N_{mn}, in the expression for the coupling coefficient B_{mn}, is associated with mode coupling due to horizontal gradients ($\partial n / \partial r$) in the index of refraction $n(r, z)$.

7.2. THE COUPLED FIRST-ORDER SYSTEM OF DIFFERENTIAL EQUATIONS IN CYLINDRICAL COORDINATES

We shall make a series of transformations on the coupled second-order system of differential equations [Eq. (7.1.13)] to convert them into a coupled first-order system of differential equations with different, but numerically tractable, boundary conditions. We begin by introducing the radial component of particle velocity, v, given by

$$v = \frac{1}{i\omega\rho_0} \frac{\partial p}{\partial r} \tag{7.2.1}$$

THE COUPLED FIRST-ORDER SYSTEM IN CYLINDRICAL COORDINATES 303

Since the eigenfunctions, ψ_n, form a complete set, we can expand v in terms of them. Thus

$$v = \sum_{n=1}^{\infty} \hat{\phi}_n \rho_0^{-1} \psi_n \qquad (7.2.2)$$

where the coefficients $\hat{\phi}_n$ are given by

$$\hat{\phi}_n = \int_s^d v \psi_n \, dz \qquad (7.2.3)$$

We can evaluate this last expression for $\hat{\phi}_n$ by making use of Eqs. (7.1.8), (7.1.12), (7.1.14), and (7.2.1) to obtain

$$\hat{\phi}_m = \frac{1}{i\omega} \frac{d\phi_m}{dr} + \frac{1}{i\omega} \sum_{n=1}^{\infty} B_{mn} \phi_n \qquad (7.2.4)$$

Using the two variables ϕ_m and $\hat{\phi}_m$, we can write Eq. (7.1.13) as an equivalent system of first-order differential equations:

$$\frac{d\phi_m}{dr} - i\omega \hat{\phi}_m = -\sum_{n=1}^{\infty} B_{mn} \phi_n \qquad (7.2.5)$$

$$i\omega \frac{d\hat{\phi}_m}{dr} + \frac{i\omega}{r} \hat{\phi}_m + \kappa_m^2 \phi_m = -i\omega \sum_{n=1}^{\infty} B_{mn} \hat{\phi}_n \qquad (7.2.6)$$

It is easy to see that Eqs. (7.2.5) and (7.2.6) are equivalent to Eq. (7.1.13) by simply solving Eq. (7.2.5) for $\hat{\phi}_m$ and substituting into Eq. (7.2.6). Notice that the complicated coupling coefficients A_{mn} given by Eq. (7.1.15) no longer appear.

Next, we introduce two new variables, a_m^+ and a_m^-, which we shall call the forward and backscattered waves respectively. We will define these new variables by the following transformation equations:

$$\phi_m(r) = a_m^+(r) + a_m^-(r) \qquad (7.2.7)$$

$$\hat{\phi}_m(r) = \frac{1}{i\omega} \left\{ \left[i\kappa_m(r) - \frac{1}{2r} \right] a_m^+(r) - \left[i\kappa_m(r) + \frac{1}{2r} \right] a_m^-(r) \right\} \qquad (7.2.8)$$

We now substitute the expressions for ϕ_m and $\hat{\phi}_m$ into Eqs. (7.2.5) and (7.2.6). Each of the resulting equations contain both derivatives da_m^+/dr and da_m^-/dr. It is a simple matter to eliminate first da_m^-/dr and then da_m^+/dr

304   AN INHOMOGENEOUS OCEAN WITH A ROUGH SEA SURFACE

between the two equations to obtain

$$\frac{da_m^+}{dr} - \left(i\kappa_m - \frac{1}{2r}\right)a_m^+ + \frac{i}{8\kappa_m r^2}(a_m^+ + a_m^-) = \sum_{n=1}^{\infty} B_{mn}^{++} a_n^+ + \sum_{n=1}^{\infty} B_{mn}^{+-} a_n^-$$

(7.2.9)

$$\frac{da_m^-}{dr} + \left(i\kappa_m + \frac{1}{2r}\right)a_m^- + \frac{i}{8\kappa_m r^2}(a_m^+ + a_m^-) = \sum_{n=1}^{\infty} B_{mn}^{-+} a_n^+ + \sum_{n=1}^{\infty} B_{mn}^{--} a_n^-$$

(7.2.10)

In Eqs. (7.2.9) and (7.2.10) we have put

$$B_{mn}^{++} = \frac{-1}{2\kappa_m}\left\{(\kappa_n + \kappa_m) B_{mn} + \delta_{mn}\frac{d\kappa_n}{dr}\right\}$$

(7.2.11)

$$B_{mn}^{+-} = \frac{1}{2\kappa_m}\left\{(\kappa_n - \kappa_m) B_{mn} + \delta_{mn}\frac{d\kappa_n}{dr}\right\}$$

(7.2.12)

$$B_{mn}^{-+} = B_{mn}^{+-}$$

(7.2.13)

$$B_{mn}^{--} = B_{mn}^{++}$$

(7.2.14)

The only new quantity that has appeared in these equations is $d\kappa_m/dr$. This quantity can be evaluated from Eq. (7.1.20). Recalling that B_{nn} vanishes and $\delta_{nn} = 1$, and using the expression for the other terms in Eq. (7.1.20) that we have already evaluated, we obtain

$$\frac{d\kappa_n}{dr} = -\left\{\frac{1}{2\kappa_n \rho_0(z)}\left(\frac{\partial \psi_n}{\partial z}\right)^2 \frac{ds}{dr}\right\}_{z=s} + \frac{\omega^2}{\kappa_n}\int_s^d \frac{1}{\rho_0} n \frac{\partial n}{\partial r}(\psi_n)^2 dz$$

(7.2.15)

We now need to discuss the transformation of variables given by Eqs. (7.2.7) and (7.2.8) and the new boundary conditions for these variables. First, let us discuss the boundary conditions.

We confine the range-dependent properties of the waveguide to an interval $R_0 \leq r \leq R_1$. Consequently, for $0 \leq r \leq R_0$ and $r > R_1$, the sea surface is smooth and c does not depend on the variable r. The range R_0 can be made as small as we please and R_1 as large as we please.

The boundary conditions on the variable a_n^+ and a_n^- are specified by their values at $a_n^+(R_0)$ and $a_n^-(R_1)$. To satisfy the radiation condition at infinity, we must assume that

$$a_n^-(R_1) = 0$$

(7.2.16)

Since the waveguide has no range-dependent properties from R_1 to infinity, an outgoing wave at infinity is guaranteed. The values for $a_n^+(R_0)$ are calculated from Eq. (4.1.8) where ϕ_n is given by Eq. (4.1.52) since this section of the waveguide is also assumed to be range-independent.

If we work in the region $\kappa_m R_0 \gg 1$, we can simplify the system of coupled equations given by Eqs. (7.2.9) and (7.2.10), by neglecting the third term on the left side, that is, by neglecting the term in $1/\kappa_m r^2$. This results in

$$\frac{da_m^+}{dr} - \left[i\kappa_m - \frac{1}{2r}\right]a_m^+ = \sum_{n=1}^{\infty} B_{mn}^{++} a_n^+ + \sum_{n=1}^{\infty} B_{mn}^{+-} a_n^- \qquad (7.2.17)$$

$$\frac{da_m^-}{dr} + \left[i\kappa_m + \frac{1}{2r}\right]a_m^- = \sum_{n=1}^{\infty} B_{mn}^{-+} a_n^+ + \sum_{n=1}^{\infty} B_{mn}^{--} a_n^- \qquad (7.2.18)$$

Now, let us discuss the reason for the particular form of the transformation given by Eqs. (7.2.7) and (7.2.8). In an inhomogeneous medium there appears, in general, to be no unique splitting of the total field into forward and backward traveling waves. This is discussed more in detail in the works of Schelkunoff [7.5], Sluijter [7.6], and Brekhovskikh [7.7].

In our case the inhomogeneity is expressed by the fact that the sound speed $c(r, z)$ is a function of the variable r, since we are only splitting the radial amplitudes into forward and backward traveling waves. If $c(r, z)$ is a slowly varying function of r or independent of r, then the total radial field can be split. In any case, since a_n^+ and a_n^- are initialized in a range-independent section of the waveguide where the total field can be split uniquely into forward and backscattered waves, the total field in the region $R_0 \leq r \leq R_1$ should be correct. What we cannot do is to identify a_n^+ and a_n^- with the forward and backscattered energy when there is strong range dependence in the sound speed.

The particular rationale for the splitting of the field given by Eqs. (7.2.7) and (7.2.8) is the same as that used in the theory of electromagnetic wave propagation in optical fiber with rough boundaries as discussed by Marcuse [7.8]. Equation (7.2.8) is the solution for $\hat{\phi}_m$ in the absence of coupling when $\kappa_n r \gg 1$ and the asymptotic forms of the Hankel functions may be used; that is, if we assume that the sea surface is flat everywhere and $c(z)$ is not a function of r, then Eq. (7.2.8) is the solution of the wave equation. This suggests that we use the same form in the presence of coupling, but now κ_n is allowed to be a function of r.

7.3. A MODEL FOR A TIME VARYING, RANDOMLY ROUGH SEA SURFACE

It should be borne in mind that the propagation model is independent of any surface model, and any other model simulating a random sea surface could be used instead of the one described in this section.

306 AN INHOMOGENEOUS OCEAN WITH A ROUGH SEA SURFACE

The model that we will use for a random ocean surface is that used by Harper and Labianca [7.9]. We will simply reproduce their model here for the convenience of the reader.

The random ocean surface is simulated by

$$z = s(r, t) = \sum_{j=1}^{M} h_j \cos\left(\Gamma_j r - \Omega'_j t + \gamma_j\right) \tag{7.3.1}$$

where $\Omega'_j > 0$ and the dispersion relation is taken to be

$$\Gamma_j = \frac{\Omega'^2_j}{g} \tag{7.3.2}$$

Here, g is the acceleration of gravity. In Eq. (7.3.1), the phases γ_j are also taken to be statistically independent, random variables, uniformly distributed between 0 and 2π. The frequencies Ω'_j are also taken to be statistically independent, random variables, but the amplitudes h_j are chosen in a deterministic manner. The process, Eq. (7.3.1), is ergodic with respect to its mean and autocorrelation function under a restriction on Ω'_j which we shall describe.

Our aim is to simulate an actual random surface s_{as}, with a given spatially independent, autocorrelation function:

$$R_{as}(\tau) = E\left[s_{as}(r, t) s_{as}(r, t + \tau)\right] \tag{7.3.3}$$

and the corresponding power spectra:

$$S_{as}(\sigma) = \frac{1}{2\pi} \int_{-\infty}^{\infty} R_{as}(\tau) e^{i\sigma\tau} d\tau \tag{7.3.4}$$

In Eqs. (7.3.3), $E[\xi]$ denotes the expected value or mean of the random variable ξ. If the reader is not familiar with the concept of stochastic processes, he should consult Papoulis [7.10].

Under the assumption that the actual power spectrum is band limited, $\sigma_{min} \leq |\sigma| \leq \sigma_{max}$, we define the equal intervals

$$\Delta\sigma = \frac{1}{M}(\sigma_{max} - \sigma_{min}) \tag{7.3.5}$$

and the discrete set of frequencies

$$\Omega_j = \sigma_{min} + (j - \tfrac{1}{2})\Delta\sigma, \quad j = 1,\ldots, M \tag{7.3.6}$$

The set of random frequencies Ω'_j in Eq. (7.3.1) is statistically independent of the γ_j and is defined by

$$\Omega'_j = \Omega_j + \delta\Omega_j, \quad j = 1,\ldots, M \tag{7.3.7}$$

where the $\delta\Omega_j$ are random variables uniformly distributed between $-\Delta\sigma'/2$ and $\Delta\sigma'/2$ with $\Delta\sigma' \ll \Delta\sigma$. This definition ensures the ergodicity, in an approximate sense, of the simulated process $z = s(r, t)$ with respect to its correlation function.

The amplitudes h_j appearing in Eq. (7.3.1) are determined according to the prescription

$$h_j^2 = 4S_{as}(\Omega_j)\Delta\sigma \tag{7.3.8}$$

It has been shown by Shinozuka and Jan [7.11] that the autocorrelation function of the simulated surface

$$R(\tau) = E[s(r, t)s(r, t + \tau)] \tag{7.3.9}$$

where the Ω_j and h_j are defined by Eqs. (7.3.7) and (7.3.8), converges to R_{as} as $1/M^2$ for $M \to \infty$.

The simulated ocean surface $s(r, t)$ is a zero-mean process with autocorrelation function

$$R(\tau) = \frac{1}{2}\sum_{j=1}^{M} h_j^2 \cos\Omega_j \tau \tag{7.3.10}$$

and power spectral density

$$S(\sigma) = \frac{1}{2\pi}\int_{-\infty}^{\infty} R(\tau)e^{i\sigma\tau}d\tau$$

$$= \frac{1}{4}\sum_{j=1}^{M} h_j^2 \delta(|\sigma| - \Omega_j) \tag{7.3.11}$$

The mean square wave height A^2 is given by

$$A^2 = E[s^2] = R(0) = \frac{1}{2}\sum_{j=1}^{M} h_j^2 \tag{7.3.12}$$

It should be noted that had we chosen the Ω_j' to be deterministic, say $\Omega_j' = \Omega_j$, the process $s(r, t)$ would be ergodic exactly with respect to its autocorrelation function. The main reason for requiring Ω_j' to have a small random part $\delta\Omega_j$ is to preclude the possibility of periodicity of $s(r, t)$ in time. Also, by virtue of the central limit therein, the simulation $s(r, t)$ defined above approaches a Gaussian process as $M \to \infty$.

7.4. THE COUPLED SECOND-ORDER SYSTEM OF DIFFERENTIAL EQUATIONS IN CARTESIAN COORDINATES

The main limitation of the work presented in Section 7.1 and 7.2 is the assumption of cylindrical symmetry. A cylindrically symmetric surface will generate no scattering out of the vertical plane. In this section and the next the

308 AN INHOMOGENEOUS OCEAN WITH A ROUGH SEA SURFACE

limitation of cylindrical symmetry will be removed. However, a general random sea surface still cannot be treated. Here we treat the problem of a random sea running in one direction. While this is not the most general state for a sea surface, it does allow for scattering out of the vertical plane. However, it should be noted that the existing models, which treat the ocean as an inhomogeneous waveguide and which use perturbation theory, can also obtain numerical solutions for a sea running in only one direction.

We consider the problem of a point source located in an inhomogeneous, oceanic waveguide bounded above by a random, time-varying sea surface and bounded below by a rigid bottom. The source is emitting a time harmonic signal of angular frequency ω and is located at $x = y = 0$, and $z = z_s$. The orthogonal, Cartesian coordinate system used is shown in Figure 7.4.1 relative to the waveguide. The bottom of the waveguide is assumed flat and rigid at $z = d$. Since this model allows for sediment layers, the rigid bottom is taken as the basement of all the sediment layers.

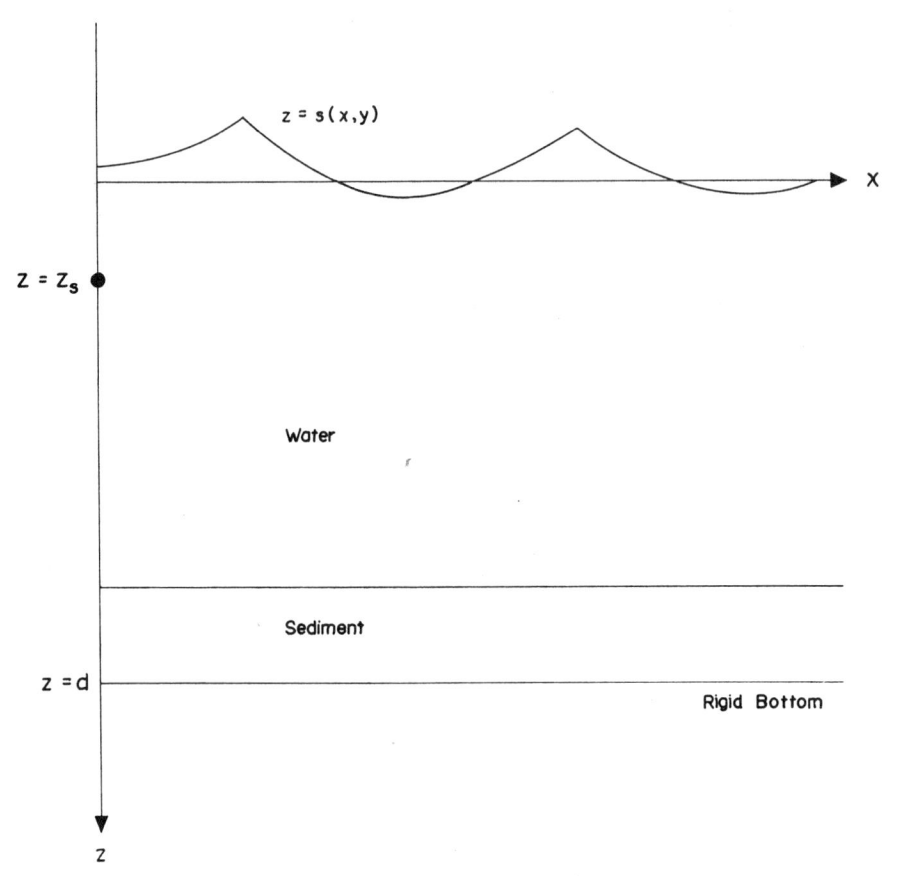

FIGURE 7.4.1. Geometry for an inhomogeneous waveguide with a rough sea surface in Cartesian coordinates.

THE COUPLED SECOND-ORDER SYSTEM IN CARTESIAN COORDINATES   309

We assume the sea surface is characterized as a function of x and t only. Hence, the equation for the sea surface can be written

$$z = s(x, t) \tag{7.4.1}$$

Equation (7.4.1) exemplifies what we shall mean by a sea running in one direction. The line joining the source and the receiver can be at any angle relative to this surface. Here, we only need to say that the xy plane is taken to be coincident with the mean sea surface.

The speed of sound, c, in the waveguide is assumed to be a function of x and z. The index of refraction, n, is defined as

$$n(x, z) = \frac{1}{c(x, z)} \tag{7.4.2}$$

The wave equation which governs the acoustic field away from the source is

$$\left\{ \nabla^2 - \frac{1}{\rho_0} \nabla \rho_0 \cdot \nabla - n^2 \frac{\partial^2}{\partial t^2} \right\} \mathscr{P}(\mathbf{r}, t) = 0 \tag{7.4.3}$$

Here $\mathscr{P}$ is the acoustic pressure and ρ_0 is the density of the water and sediment layers. It is assumed that $\rho_0(z)$ is a function of z only.

As previously, we make the narrow-band approximation to the wave equation. Recall this amounts to assuming that

$$\frac{\partial^2 \mathscr{P}}{\partial t^2}(\mathbf{r}, t) \approx -\omega^2 \mathscr{P}(\mathbf{r}, t) \tag{7.4.4}$$

With this assumption, Eq. (7.4.3) can be written

$$\frac{\partial^2 p}{\partial x^2} + \frac{\partial^2 p}{\partial y^2} + \frac{\partial^2 p}{\partial z^2} - \frac{1}{\rho_0} \frac{d\rho_0}{dz} \frac{\partial p}{\partial z} + k^2(x, z) p = 0 \tag{7.4.5}$$

where we have put

$$\mathscr{P}(x, y, z, t) = p(x, y, z, t) e^{-i\omega t} \tag{7.4.6}$$

and

$$k(x, z) = \omega n(x, z) \tag{7.4.7}$$

Following the usual procedure, we assume a solution of the form

$$p(x, y, z) = \sum_{n=1}^{\infty} \phi_n(x, y) \psi_n(z, x) \tag{7.4.8}$$

310 AN INHOMOGENEOUS OCEAN WITH A ROUGH SEA SURFACE

For simplicity, we shall suppress the time factor, t, in p, ϕ_n and ψ_n. Further, we shall assume that the local depth mode functions $\psi_n(z, x)$ satisfy the partial differential equation

$$\frac{\partial}{\partial z}\left[\frac{1}{\rho_0}\frac{\partial \psi_n}{\partial z}\right] + \left[\frac{k^2(x, z)}{\rho_0} - \frac{\kappa_n^2(x)}{\rho_0}\right]\psi_n = 0 \qquad (7.4.9)$$

The following boundary conditions are imposed:

1. On the sea surface:

$$p[x, y, s(x, t), t] = 0 \qquad (7.4.10)$$

2. On the rigid bottom:

$$\left[\frac{\partial p}{\partial z}(x, y, z, t)\right]_{z=d} = 0 \qquad (7.4.11)$$

Now it can be shown that the eigenfunction solutions, ψ_n, of Eq. (7.4.9), subject to the boundary conditions given by Eqs. (7.4.10) and (7.4.11), form a complete, orthonormal system of functions relative to the weight functions $(1/\rho_0)$ at each point x. The orthonormality condition is

$$\int_s^d \rho_0^{-1}(z)\psi_n(z, x)\psi_m(z, x)\, dz = \delta_{nm} \qquad (7.4.12)$$

There is still a source condition at $x = y = 0$ and $z = z_s$ that must be satisfied. This matter will be discussed later.

We now need to obtain the differential equations governing the amplitudes $\phi_n(x, y)$. We substitute the expression for the acoustic pressure p given by Eq. (7.4.8) into Eq. (7.4.5). After making use of Eq. (7.4.9), we multiply the resulting equation by $\rho_0^{-1}\psi_m$, integrate from $z = s$ to $z = d$, and use the orthonormality condition, Eq. (7.4.12), to obtain the system of coupled, second-order differential equations for ϕ_n, namely,

$$\frac{\partial^2 \phi_m}{\partial x^2} + \frac{\partial^2 \phi_m}{\partial y^2} + \kappa_m^2(x)\phi_m = -\sum_{n=1}^{\infty} A_{mn}\phi_n - 2\sum_{n=1}^{\infty} B_{mn}\frac{\partial \phi_n}{\partial x} \qquad (7.4.13)$$

In Eq. (7.4.13), we have put

$$B_{mn(x)} = \int_s^d \rho_0^{-1}(z)\psi_m(z, x)\frac{\partial \psi_n}{\partial x}(z, x)\, dz \qquad (7.4.14)$$

$$A_{mn}(x) = \int_s^d \rho_0^{-1}(z)\psi_m(z, x)\frac{\partial^2 \psi_n}{\partial x^2}(z, x)\, dz \qquad (7.4.15)$$

THE COUPLED SECOND-ORDER SYSTEM IN CARTESIAN COORDINATES 311

These equations for the coupling coefficients are stated as a function of the parameter x. However, they are formally identical to Eqs. (7.1.14) and (7.1.15), which are a function of the parameter r. So it is simply a matter of replacing r by x in the results obtained in Section 7.1. This gives for B_{mn}:

$$B_{mn} = S_{mn} + N_{mn}, \qquad n \neq m \qquad (7.4.16)$$

$$B_{nn} = 0 \qquad (7.4.17)$$

where for $m \neq n$,

$$S_{mn} = \frac{1}{\rho_0(\kappa_m^2 - \kappa_n^2)} \left\{ \frac{\partial \psi_m}{\partial z} \frac{\partial \psi_n}{\partial z} \frac{ds}{dx} \right\}_{z=s} \qquad (7.4.18)$$

and

$$N_{mn} = \frac{-2\omega^2}{(\kappa_m^2 - \kappa_n^2)} \int_s^d \rho_0^{-1} n \frac{\partial n}{\partial x} \psi_m \psi_n \, dz \qquad (7.4.19)$$

For A_{mn} we get

$$A_{mn} = \frac{dB_{mn}}{dx} - \sum_{n=1}^{\infty} B_{im} B_{in} \qquad (7.4.20)$$

Note that

$$B_{mn} = -B_{nm} \qquad (7.4.21)$$

The term S_{mn} arises because of coupling due to the rough sea surface, and the term N_{mn} arises because of coupling due to the change of the index of refraction with x.

We can eliminate the y-dependence in Eq. (7.4.13) by taking a Fourier transform. If we let $\hat{\phi}_m(z, \gamma)$ be the Fourier transform of $\phi_m(x, y)$, then they are related by

$$\phi_m(x, y) = \frac{1}{2\pi} \int_{-\infty}^{\infty} \hat{\phi}_m(x, \gamma) e^{i\gamma y} \, d\gamma \qquad (7.4.22)$$

$$\hat{\phi}_m(x, \gamma) = \int_{-\infty}^{\infty} \phi_m(x, y) e^{-i\gamma y} \, dy \qquad (7.4.23)$$

If we use Eq. (7.4.20) and take the transform of Eq. (7.4.13), we get

$$\frac{d^2 \hat{\phi}_m}{dx^2} + \left(\kappa_m^2(x) - \gamma^2\right) \hat{\phi}_m = -\sum_{n=1}^{\infty} A_{mn} \hat{\phi}_n - 2 \sum_{n=1}^{\infty} B_{mn} \frac{d\hat{\phi}_n}{dx} \qquad (7.4.24)$$

312 AN INHOMOGENEOUS OCEAN WITH A ROUGH SEA SURFACE

These are the desired second-order, coupled differential equations. We now apply the technique developed in Section 7.2 to reformulate these equations as a system of coupled, first-order differential equations.

7.5. THE COUPLED FIRST-ORDER SYSTEM OF DIFFERENTIAL EQUATIONS IN CARTESIAN COORDINATES

We begin by introducing the x component of particle velocity, v, given by

$$v(x, y, z, t) = -\frac{i}{\rho_0 \omega} \frac{\partial p}{\partial x} \qquad (7.5.1)$$

Next, we take the Fourier transform of Eqs. (7.4.8) and (7.5.1) with respect to y to obtain

$$\hat{p}(x, \gamma, z) = \sum_{n=1}^{\infty} \hat{\phi}_n(x, \gamma) \psi_n(z, x) \qquad (7.5.2)$$

$$\hat{v}(x, \gamma, z) = -\frac{i}{\rho_0 \omega} \frac{\partial \hat{p}}{\partial x} \qquad (7.5.3)$$

Here $\hat{p}(x, \gamma, z)$ and $\hat{v}(x, \gamma, z)$ are the transforms of $p(x, y, z)$ and $v(x, y, z)$ respectively.

Since the eigenfunctions, ψ_n, form a complete set, we can expand $\hat{v}$ in terms of them. Thus,

$$\hat{v} = \sum_{n=1}^{\infty} \hat{\mu}_n(x, \gamma) \rho_0^{-1} \psi_n(z, x) \qquad (7.5.4)$$

where the coefficients $\hat{\mu}_n(x, \gamma)$ are given by

$$\hat{\mu}_n(x, \gamma) = \int_s^d \hat{v} \psi_n \, dz \qquad (7.5.5)$$

We can evaluate this last expression for $\hat{\mu}_n$ by making use of Eqs. (7.4.12), (7.4.14), (7.5.2), and (7.5.3) to obtain

$$\hat{\mu}_m = \frac{1}{i\omega} \frac{d\hat{\phi}_m}{dx} + \frac{1}{i\omega} \sum_{n=1}^{\infty} B_{mn} \hat{\phi}_n \qquad (7.5.6)$$

Using the two variables $\hat{\phi}_m$ and $\hat{\mu}'_m$, we can write Eq. (7.4.24) as an equivalent system of first-order differential equations:

$$\frac{d\hat{\phi}_m}{dx} - i\omega \hat{\mu}_m + \sum_{n=1}^{\infty} B_{mn} \hat{\phi}_n = 0 \qquad (7.5.7)$$

$$i\omega \frac{d\hat{\mu}_m}{dx} + (\kappa_m^2 - \gamma^2) \hat{\phi}_m + i\omega \sum_{n=1}^{\infty} B_{mn} \hat{\mu}_n = 0 \qquad (7.5.8)$$

THE COUPLED FIRST-ORDER SYSTEM IN CARTESIAN COORDINATES 313

That Eqs. (7.5.7) and (7.5.8) are equivalent to Eq. (7.4.24) is easy to see by simply solving Eq. (7.5.7) for $\hat{\mu}_m$ and substituting into Eq. (7.5.8).

Notice that the complicated coupling coefficients A_{mn} given by Eq. (7.4.20) no longer appear.

Consider the case of a flat-sea surface and a refractive index $n(z)$ that is independent of x. Then, Eq. (7.4.24) becomes

$$\frac{d^2\hat{\phi}_m}{dx^2} + \beta_m^2 \hat{\phi}_m = 0 \tag{7.5.9}$$

where, for convenience, we have put

$$\beta_m^2 = \kappa_m^2 - \gamma^2 \tag{7.5.10}$$

The solution of this is

$$\hat{\phi}_m = c^+ e^{i\beta_m x} + c^- e^{-i\beta_m x}$$

$$= a_m^+ + a_m^- \tag{7.5.11}$$

where the c^+ and c^- are constants. The quantities a_m^+ and a_m^- are the forward and backward traveling waves.

Using Eqs. (7.5.6) and (7.5.11), and recalling that for this case B_{mn} vanishes, we see that

$$\hat{\mu}_m = \frac{\beta_m}{\omega}(a_m^+ - a_m^-) \tag{7.5.12}$$

We now assume that for the range-dependent case $\hat{\phi}_m$ and $\hat{\mu}_m$ have the same functional dependence on a_m^+ and a_m^- as they have in the range-independent case. That is, in general, we assume

$$\hat{\phi}_m(x,\gamma) = a_m^+(x,\gamma) + a_m^-(x,\gamma) \tag{7.5.13}$$

$$\hat{\mu}_m(x,\gamma) = \frac{\beta_m(x,\gamma)}{\omega}\{a_m^+(x,\gamma) - a_m^-(x,\gamma)\} \tag{7.5.14}$$

where now $\beta_m(x,\gamma)$ can be a function of x.

We now substitute these expressions for $\hat{\phi}_m$ and $\hat{\mu}_m$, given by Eqs. (7.5.13) and (7.5.14) into Eqs. (7.5.7) and (7.5.8). Each of the resulting equations contains both derivatives da_m^+/dx and da_m^-/dx. It is a simple matter to first eliminate da_m^-/dx and then da_m^+/dx between these equations to obtain

$$\frac{da_m^+}{dx} - i\beta_m a_m^+ = \sum_{n=1}^{\infty} B_{nm}^{++} a_n^+ + \sum_{n=1}^{\infty} B_{nm}^{+-} a_n^- \tag{7.5.15}$$

$$\frac{da_m^-}{dx} + i\beta_m a_m^- = \sum_{n=1}^{\infty} B_{nm}^{-+} a_n^+ + \sum_{n=1}^{\infty} B_{nm}^{--} a_n^- \tag{7.5.16}$$

Here we have put

$$B_{nm}^{++} = \frac{1}{2\beta_m}\left[\beta_m B_{nm} + \beta_n B_{nm} - \delta_{nm}\frac{d\beta_n}{dx}\right] \quad (7.5.17)$$

$$B_{nm}^{+-} = \frac{1}{2\beta_m}\left[\beta_m B_{nm} - \beta_n B_{nm} + \delta_{nm}\frac{d\beta_n}{dx}\right] \quad (7.5.18)$$

$$B_{nm}^{-+} = B_{nm}^{+-} \quad (7.5.19)$$

$$B_{nm}^{--} = B_{nm}^{++} \quad (7.5.20)$$

The only new quantity that has appeared in these equations is $d\beta_m/dx$. From Eq. (7.5.10), we see that $d\beta_m/dx$ is related to $d\kappa_m/dx$ and, by a procedure identical with that in Section 7.1, it can be shown that

$$\frac{d\kappa_m}{dx} = -\left\{\frac{1}{2\kappa_m\rho_0}\left[\frac{\partial\psi_m}{\partial z}\right]^2 \frac{ds}{dx}\right\}_{z=s}$$
$$+ \frac{\omega^2}{\kappa_m}\int_s^d \rho_0^{-1} n(x,z)\frac{\partial n(x,z)}{\partial x}(\psi_n)^2\, dz \quad (7.5.21)$$

For convenience, let $R = x^2 + y^2$. We confine the range-dependent properties of the waveguide to an interval $R_0 < R < R_1$. Consequently, for $R < R_0$ and $R > R_1$, the sea surface is assumed smooth and the speed of sound not dependent on x. The value of R_0 can be made as small as we please, and the value of R_1 can be made as large as we please. For example, if we are studying long-range propagation, R_1 could be several hundred nautical miles.

We now need to specify the boundary conditions for a_m^+ and a_m^-. At this point, we shall make a simplifying assumption. We shall actually assume that the halfspace $x < 0$ is range-independent. The assumption amounts to neglecting the backscattered energy from the halfspace $x < 0$ which propagates into the halfspace $x > 0$. This is not an inherent limitation in the model, since we could remove this assumption by making iterative computer runs. However, the backscattered energy from the halfspace $x < 0$ that reaches a receiver at any appreciable distance from the source is probably negligible.

With this assumption, the boundary conditions on the variables a_n^+ and a_n^- are specified by their values at $a_m^+(x_0)$ and $a_m^-(x_1)$ where, as above, the waveguide is assumed to be range-dependent only in the interval $x_0 < x < x_1$.

To satisfy the radiation condition at infinity, we must assume that

$$a_m^-(x_1) = 0 \quad (7.5.22)$$

Since the initial section of waveguide ($0 < x < x_0$) is range-independent, normal mode theory is used to calculate the Green's function for a source at

$x = y = 0$ and $z = z_s$. The amplitudes of the modes calculated in this manner are then used as the initial conditions for $a_m^+(x_0)$, which are the modes that propagate down the waveguide.

As discussed in Section 7.2 in an inhomogeneous medium, there appears, in general, to be no unique splitting of the total field into forward and backward traveling waves. In this case, the inhomogeneity is expressed by the fact that the sound speed is a function of the variable x, since we are only splitting the amplitudes $a_m^+(x)$ and $a_m^-(x)$ into forward and backward traveling waves. If $c(x, z)$ is a slowly varying function of x or independent of x, then the total field can be split. In any case, since a_n^+ and a_n^- are initialized in a range-independent section of the waveguide where the total field can be split uniquely into forward and backscattered waves, the total field in the region $R_0 \le R \le R_1$ should be correct. What we cannot do is to identify a_n^+ and a_n^- with the forward and backscattered energy when there is strong range dependence of the sound speed.

The model for a random sea surface presented in Section 7.3 can be used for the case of Cartesian coordinates by simply replacing the parameter r by the parameter x.

APPENDIX 7.A. THE NARROW-BAND APPROXIMATION TO THE WAVE EQUATION

We will simply reproduce the discussion of the narrow-band approximation that is given by Labianca and Harper [7.12].

We make the a priori observation that, for the case of ocean surface waves, the scattered acoustic signal will be a narrow-band process about the source (or carrier) frequency. We will assume the process stationary so any realization of the real pressure may be represented as shown by Papoulis [7.13] to be

$$\mathscr{P}(\mathbf{r},\mathbf{r}', t) = U(\mathbf{r},\mathbf{r}', t)\cos \omega t + V(\mathbf{r},\mathbf{r}', t)\sin \omega t \qquad (7A.1)$$

where U and V vary on a time scale $2\pi/\omega$ of the carrier. The derivative of $\mathscr{P}(\mathbf{r},\mathbf{r}', t)$ with respect to t therefore involves these two time scales:

$$\frac{1}{\omega^2}\frac{\partial^2 \mathscr{P}}{\partial t^2} = -\mathscr{P} + \frac{2}{\omega}\left[\frac{\partial V}{\partial t}\cos \omega t - \frac{\partial U}{\partial t}\sin \omega t\right]$$

$$+ \frac{1}{\omega^2}\left[\frac{\partial^2 U}{\partial t^2}\cos \omega t + \frac{\partial^2 V}{\partial t^2}\sin \omega t\right] \qquad (7A.2)$$

where, for any particular sample function of the random process, the second term on the right-hand side of Eq. (7A.2) is $o(\sigma_{max}/\omega)$ while the third term is of $o[(\sigma_{max}/\omega)^2]$. Thus we have

$$\frac{1}{\omega^2}\frac{\partial^2 \mathscr{P}}{\partial t^2} = -\mathscr{P} + o\left(\frac{\sigma_{max}}{\omega}\right) \qquad (7A.3)$$

For the narrow-band approximation, we drop terms of order σ_{max}/ω and higher and write

$$\frac{\partial^2 \mathscr{P}}{\partial t^2} \approx -\omega^2 \mathscr{P} \tag{7A.4}$$

REFERENCES

7.1. A. D. Pierce, "Extension of the Method of Normal Modes to Sound Propagation in an Almost-Stratified Medium," *J. Acoust. Soc. Am*. **37**, 19–27 (1965).

7.2. C. A. Boyles, "Coupled Mode Solution for a Cylindrically Symmetric Oceanic Waveguide with a Range and Depth Dependent Refractive Index and Time Varying Rough Sea Surface," *J. Acoust. Soc. Am*. **73**, 800–805 (1983).

7.3. C. A. Boyles, "Coupled Mode Theory for an Inhomogeneous Oceanic Waveguide with a Time Varying, Randomly Rough Sea Running in One Direction." STD-R-615, The Johns Hopkins Univ., Applied Physics Laboratory, Laurel, Md. (April 1982).

7.4. L. B. Dozier, "Numerical Solution of Coupled Mode Equations," STD-N-076, Science Applications, Inc., McLean, Va. (May 1982).

7.5. S. A. Schelkunoff, "Remarks Concerning Wave Propagation in Stratified Media," *Comm. Pure and Applied Math*, vol. IV, no. 1, 117–128 (1951).

7.6. F. W. Sluijter, "Arbitrariness of Dividing the Total Field in an Optically Inhomogeneous Medium into Direct and Reversed Waves," *J. Optical Soc. Am*. **60**, 8–10 (1970).

7.7. L. M. Brekhovskikh, *Waves in Layered Media* (Academic Press, New York, 1960), pp. 229–230.

7.8. D. Marcuse, *Theory of Dielectric Optical Waveguides* (Academic Press, New York, 1974).

7.9. E. Y. Harper and F. M. Labianca, "Perturbation Theory for Scattering of Sound from a Point Source by a Rough Surface in the Presence of Refraction," *J. Acoust. Soc. Am*. **57**, 1044–1051 (1975).

7.10. A. Papoulis, *Probability, Random Variables and Stochastic Processes* (McGraw-Hill, New York, 1965).

7.11. M. Shinozuka and C. M. Jan, "Digital Simulation of Random Processes and its Applications," *J. Sound Vib*. **25**, 111–128 (1972).

7.12. F. M. Labianca and E. Y. Harper, "Connection Between Various Small-Waveheight Solutions of the Problem of Scattering from the Ocean Surface," *J. Acoust. Soc. Am*. **62**, 1144–1157 (1977).

7.13. See Papoulis, pp. 380–381.

INDEX

Acoustic field equations, 30
Acoustic field variables, 24
Acoustic pressure p, 41, 113
Acoustic pressure ρ, 40, 112
Acoustic source strength, 24
Adiabatic equation of state, 26
Airy functions, 107–110, 168, 205, 229
Airy functions, relationship to Bessel functions, 107, 108, 205
Airy's differential equation, 107, 168, 205
Analytic function, 82
Area of a parallelogram in vector notation, 6
Arrival angle structure, 223, 224
 for North Atlantic convergence zones, 265, 267
 for North Atlantic surface duct, 275–277
 for North Pacific convergence zones, 244, 246
Attenuation, 171–173
Auto correlation function for sea surface, 306

Backscattered wave a_n^- in Cartesian coordinates, 313–315
Backscattered wave a_n^- in cylindrical coordinates, 303–305
Base vectors, unitary, 9
Bessel functions:
 Bessel's differential equation, contour integral solution of, 100–103
 of the first kind, 95–97
 limiting forms, 105, 106
 modified, 106
 recursion formulas, 105
 relationship to Hankel functions, 104, 105
 of the second kind, 97–100
 series solution of, 90–94
 of the third kind, 100–105
Boundary conditions:
 for Helmholtz equation, 35–37
 at infinity, 116, 120–122, 133, 298
 at infinity for backscatter wave, 305, 314
 at pressure release surface, 117, 128, 133, 140
 for random sea surface in Cartesian coordinates, 310
 for random sea surface in cylindrical coordinates, 297
 at rigid bottom, 117, 128, 133, 140, 297, 310
 at source, 298
 at source for forward traveling wave, 305, 314, 315
 for Sturm-Liouville problem, 44
Boundary value problem, two point, 298

Calculus of variations, 49–57
 with constraints, 52–55
 N independent variables, 55–57
 one independent variable, 49–52
Cauchy-Riemann conditions, 82
Caustics, 185, 244
 cusp, 226, 227, 244, 275, 276
 equations for cusp, 231
 equations for smooth, 230, 231

317

Caustics (*Continued*):
  phase shift when ray touches a, 222, 235–238
  ray theory intensity at, 224, 225, 231, 234
  smooth, 220–231, 244, 247, 275, 276
Characteristic equation, 45, 46, 128, 141, 145, 150, 151, 152, 171
Closed thermodynamic system, 20
Completeness, 59–71
  proof of for a restricted set of eigenfunctions, 64
  and uniform convergence, 70, 71
Completeness relation, 79
  for local eigenfunctions, 300
Composite thermodynamic system, 19
Conservation:
  of mass for a perfect fluid, 13, 14
  of momentum for a perfect fluid, 14–16
  of total energy for a perfect fluid, 16–22
Conservation of energy:
  in acoustic approximation, 30–34
  in ray theory, 184, 185
Continuity equation for mass for a perfect fluid, 14
Continuous spectrum, 71–77, 130, 131, 144, 151–154
  and Sturm-Liouville theory, 75–77
Contour:
  C_r^+, 119
  C_z^-, 119
Convected coordinates, 2
Convergence zones, 220, 241–272, 275–278, 282
Coordinate:
  curves, 8
  surfaces, 8
Coupled differential equations:
  first order wave equations in Cartesian coordinates, 313
  first order wave equations in cylindrical coordinates, 304, 305
  Fourier transform of second order wave equations in cylindrical coordinates, 311
  second order wave equations in Cartesian coordinates, 310
  second order wave equations in cylindrical coordinates, 297
Coupling coefficient:
  A_{mn}, 298, 301, 310, 311
  B_{mn}, 298, 301, 302, 310, 311
Coupling coefficients, 298, 310, 311
  generalized in Cartesian coordinates, 314
  generalized in cylindrical coordinates, 304
Cutoff frequency for a waveguide, 131

Cylindrical:
  coordinates, 41, 43, 113
  spreading, 198
  symmetry, 43

Delta function, 76, 77, 112–114
  in cylindrical coordinates, 114
  in spherical coordinates, 176
Density:
  acoustic, 24
  of sea water, 247
  variation in ocean, 240
Depth excess, 243, 244
Determinants, 7, 8
Differential equations, fundamental theorem for solution of, 45
Diffraction, 185, 210, 211, 221, 247, 255, 273, 281, 282, 293
Discrete spectrum, 48, 130, 131, 139, 144, 151, 153, 154
Dispersion relation for gravity waves, 306
Double channel sound speed profile, 263, 267

Eigenfunctions, 46
  depth dependent, 128, 129, 141, 142, 165
  improper, 76
  normalization of, 46, 48
  obtained from a variational principle, 57–59
  orthogonality of, 46–48
Eigenfunction expansion for two dimensional Green's function, 129, 130, 143, 144
Eigenvalues:
  obtained from a variational principle, 57–59
  range derivative in Cartesian coordinates, 314
  range derivative in cylindrical coordinates, 304
  realness of, 48, 49
  for the Sturm problem, 46
Einstein summation convention, 4
Energy:
  conservation of acoustic, 30–34
  conservation of total for a perfect fluid, 16–22
Energy density:
  acoustic, 31
  internal, 18–22
  total fluid, 18
Entropy, 17, 20–22
Epsilon notation, 5
Equation of state:
  acoustic, 26
  for fluid, 17
  linearized, 26

Equations of state, 22
Equilibrium state, 19
Equivalent ray, 199, 200, 201, 203, 204, 253
Euler-Lagrange equations:
  for problem with constraints, 55
  for several independent variables, 55–57
  for single independent variable, 52
Euler's constant, 98
Extensive parameters, 19

First law of thermodynamics, 18
Focal point, 185
Focal surface, 220
Forward wave a_n^+:
  in Cartesian coordinates, 313–315
  in cylindrical coordinates, 303–305
Fourier coefficients, 62
Frequency, angular, 42
Frobenius, method of, 90
Fundamental thermodynamic relation:
  for energy, 21
  for entropy, 20

Gamma function, 94, 95
Gradient operator, 4
Gravitational force, 16, 23–29
Green's function:
  condition on first derivative, 79
  contour integral of, 81
  expansion into eigenfunctions, 79, 80, 129, 130, 143, 144
  expansion of two dimensional as a residue series, 125–127, 137–139
  one dimensional, 77–82
  one dimensional depth, 116, 133, 136
  one dimensional radial, 115, 116, 120, 133
  singularities of one dimensional depth, 123, 137, 148–150
  singularities of one dimensional radial, 124, 125, 137, 147
  in spherical coordinates, 177
  symmetry of, 81, 82, 131
  two dimensional, 117, 123, 136, 142, 144, 145, 153, 160

Hankel functions, 100–105
  asymtotic expression for, 104, 120, 206
  relationship to Bessel function, 104, 105, 207
Harmonic function, 82
Heat flux, 18
Helmholtz equation, 42, 43, 71, 113, 115, 127, 132, 140, 155, 165, 176, 296, 309
  in Cartesian coordinates, 43, 71, 309
  in cylindrical coordinates, 43, 115, 127, 132, 140, 165, 296
  solution by separation of variables, 43, 44
  in spherical coordinates, 176
Homogeneous layer, 111
Hydrodynamic equations, 22, 23

Iconal equation, 180
Implicit function theorem, 2
Improper eigenfunctions, 76
Index of refraction for Cartesian coordinate system, 309
Index of refraction for cylindrically symmetric medium, 42, 165, 296
Inhomogeneous layered waveguide, 163–177
Intensity:
  average acoustic, 34
  average in ray theory, 184–187
  instantaneous acoustic, 31
Intensive thermodynamic parameters, 22
Internal energy, 18–22

Jacobian, 2, 5, 10–12
  in cylindrical coordinate, 113
  in spherical coordinates, 175
  time derivative of, 11, 12

Kinetic energy:
  acoustic, 31
  for perfect fluid, 18
Kronecker delta, 10

Lagrange multipliers, 53–55
Lagrangian coordinate, 2
Leakage out surface duct, 210, 211, 273, 281, 282, 293
Linear independence of solutions, 45
Linearized hydrodynamic equations, 30
Local normal modes:
  in Cartesian coordinates, 309, 310
  in cylindrical coordinates, 296
  differential equation for in Cartesian coordinates, 310
  differential equation for in cylindrical coordinates, 297

Mass:
  conservation of for a perfect fluid, 13, 14
  rate of creation of, 14
Material coordinate of a fluid particle, 2
Material derivative, 4
Material description of fluid motion, 3
Mean square approximation, 62
Mean square wave height, 307

INDEX

Mixed layer, 273
Mole numbers, 19
Momentum, conservation of fluid, 14–16

Narrowband approximation to wave equation, 296, 309, 315, 316
Neumann functions, 97–100
Newton's second law, 14
  for a perfect fluid, 16
Normal mode theory, relationship to ray theory, 198–201

Orthogonality of eigenfunctions, 46–48
Orthonormality of local eigenfunctions:
  in Cartesian coordinates, 310
  in cylindrical coordinates, 297

Particle paths, 3
Particle velocity:
  in Cartesian coordinates, 312
  in cylindrical coordinates, 302
Path length along ray, 192, 194
Perfect fluid, 15
Permutation symbols, 5
Phase velocity, modal, 153
Plane wave, 178
Point source, 112
Pointwise convergence, 67, 69, 70
Position vector for fluid particle, 2
Potential energy, acoustic, 33
Potential energy density, acoustic, 31–34
Power spectra for sea surface, 306
Pressure:
  acoustic, 24
  fluid, 21
  p, acoustic, 41, 113
  n, acoustic, 40, 112
  thermodynamic, 21
Pressure release surface, 112

Ray:
  bundles, 184–187, 220–223
  curvature, 191
  families in the first convergence zone, 247
Ray paths:
  for arbitrary sound speed variation, 188, 189
  for constant velocity gradient, 192–194
  for a North Atlantic sound speed profile, 264, 266, 267
  for a North Atlantic surface duct, 273, 275
  for a North Pacific profile, 244, 245, 247–250
  in the ocean, 194–197
  when sound speed is a function of one coordinate, 189–192
Ray theory:
  approximation to Green's function, 217–220
  average intensity in, 184–187
  conditions for validity of, 183
  conservation of energy in, 184, 185
  correction for smooth caustics, 227–229
  direction of energy flow in, 183–185
  multipath expansion, 217–220
  relationship to modes, 198–201
Reciprocity, 131
Regular operator, 75
Representation of an arbitrary function, 60–62
Reynold's transport theorem, 13
Rigid surface, 112
RR modes, 253, 263, 267, 269
RSR modes, 254, 269, 278, 293

Saddle point, 84–87
Salinity, variation in the ocean, 240, 241
Sea running in one direction, 309
Sea surface:
  equation for in Cartesian coordinates, 309
  equation for in cylindrical coordinates, 295
  equation for random surface in cylindrical coordinates, 306
Separation of variables, 39–44
  used to solve Helmholtz equation in Cartesian coordinates, 71–74
  used to solve Helmholtz equation in cylindrical coordinates, 42–44
  used to solve time dependent wave equation, 41, 42
  theorem on, 39, 40
Shadow zone, 222, 244, 247, 255
Simple thermodynamic system, 19
Single channel sound speed profile, 241, 244
Singular operator, 76
Snell's law, 190, 191
Sommerfeld radiation condition, 116, 120–122, 133, 155–160
Sound speed c_0, reference, 42
Source condition for radial Green's function, 116, 120, 121, 133
Source function for acoustic radiation, 112
Source strength:
  acoustic, 24
  fluid, 14
Spatial description of fluid motion, 3
Spectrum:
  continuous, 130, 131, 144, 151, 153, 154

discrete, 46, 48, 130, 131, 139, 144, 151, 153, 154
of eigenvalues, 46
Speed of sound, 26
  in the ocean, 240, 241
    dependence on depth, 241
    dependence on salinity, 241
    dependence on temperature, 241
Spherical coordinates, 175
Stationary phase, 90
Steepest descent, 82–90
Steepest descents applied to Bessel's equation, 103, 104
Sturm-Liouville:
  equation, 44
  problem, fundamental theorem for, 48
  theory, 44–49
Summation convention, 4
Surface duct, leakage out of, 273, 281, 282
Surface ducts, 273

Taylor series in three dimensions, 4
Temperature, 21
  variation in the ocean, 240, 241, 273
Thermocline, 241
Thermodynamic walls, 19
Time harmonic solution of wave equation, 41, 42
Transmission loss, 174–177
  for a North Atlantic sound speed profile, 269–272
  for a North Atlantic surface duct, 282, 283, 286–290
  for a North Pacific sound speed profile, 254–263
Travel time along a ray, 192, 194
Turning points, 203, 251–254
Two point boundary value problem, 298

Uniform convergence, 67–70
  and completeness, 70, 71
  and continuity, 70
  and differentiability, 70
  and integrability, 70
Uniqueness of splitting total field in forward and backward waves, 305, 315

Unitary base vectors, 9

Vector product, 5
Velocity:
  acoustic particle, 24
  fluid, 1, 3
Virtual modes, 281, 282, 293
Volume element, 9–11
  change of variables, 10
Volume of a parallelepiped in vector notation, 6

Walls, thermodynamic, 19
Wave equation:
  magnitude estimate for coefficients in, 28, 29
  solution for homogeneous layer, 123, 127, 130
  solution for homogeneous layer over halfspace, 153
  solution for inhomogenous ocean, 174
  solution for two homogeneous layers, 138, 139
  time dependent, 28, 29, 40, 112, 139, 295, 309
  time independent, 42, 43, 71, 113, 115, 127, 132, 140, 155, 165, 176, 296, 309
  time independent in Cartesian coordinates, 43, 71, 309
  time independent in cylindrical coordinates, 43, 115, 127, 132, 140, 165, 296
  time independent in spherical coordinates, 176
Wave number:
  in Cartesian coordinates, 309
  in cylindrical coordinates, 165, 296
  k_0, reference, 42
Wave surface, 179, 188
WKB, phase change along totally reflected ray, 209, 210
WKB approximation, 198, 201–204
  to characteristic equation, 214
  correction for turning points, 204–209
  to depth eigenfunctions, 211–213
  to Green's functions, 214–217
Wronskian, 45, 46
  for Bessel's equation, 96–97, 100, 105